D0948179

Biological membranes

a practical approach

TITLES PUBLISHED IN
THE
PRACTICAL APPROACH
SERIES

Biological membranes

a practical approach

Edited by

J B C Findlay

Department of Biochemistry, University of Leeds,
Leeds LS2 9JT, UK

W H Evans

Medical Research Council, National Institute for Medical
Research, Mill Hill, London NW7 1AA, UK

IRL PRESS
Oxford · Washington DC

IRL Press Limited
P.O. Box 1,
Eynsham,
Oxford OX8 1JJ,
England

British Library Cataloguing in Publication Data

Biological membranes : a practical approach.—(Practical approach series)
  1. Membranes (Biology)
  I.   Findlay,J.B.C.   II.   Evans,W.Howard
  III. Series
  574.8′75   QH601

ISBN 0-947946-84-5 (Hardbound)
ISBN 0-947946-83-7 (Softbound)

Printed in England by Information Printing Ltd, Oxford, England

Preface

After decades of relative immobility, the 70's and 80's have seen a dramatic increase in our understanding of membrane phenomena arising from major advances in the selective isolation of membrane fractions and in the discovery and analysis of membrane components. Today's membranologist must possess a wide inter-disciplinary knowledge, ranging from the detailed chemistry and biochemistry of lipids and proteins, through immunology, to complex physical methods and their underlying theory.

Unlike most previous advanced texts on membranes, this volume focuses more on the methodological aspects, allowing a very much wider range to be covered. The inevitable consequence of such a practically-oriented approach is that the theoretical aspects are restricted to the minimum necessary for understanding and explaining the new methodologies. Nevertheless the text, through the examples quoted and the discussion, should supply a bolus of scientific information for the more general reader.

The volume aspires to cover most of the principal approaches for the study of biological membranes. In many cases, it has been possible to provide step by step protocols. In others, general methods are described which can be readily adapted to help achieve a particular goal. Emphasis is placed on extensive referencing so that any small gaps remaining in the text can be plugged. This is particularly important in such a vast area as membrane isolation and characterization where the various methods will require tailoring to the tissues or cells under question. However, methods now exist for isolation of the 10 and 13 biochemically differentiated membrane systems in animal eukaryotic and plant cells respectively.

The rapid advances in the application of mono and polyclonal antibodies to membrane research are detailed, including immunoaffinity purification methods and the use of antibodies in screening cDNA libraries, an approach that is leading to the molecular cloning of an increasing number of membrane proteins. The methods for lipid analysis of membranes are more routine but even here there have been important recent developments, especially with regard to analyses of phosphoinositides and the topology of lipids in membranes. The fractionation and manipulation of proteins have provided fresh scientific challenges and have required the development of new systems, which are given here. Also covered is the art of reconstitution, one of the newer and more important strategies employed to further our understanding of the functional abilities and mechanisms of action of this new class of protein. A comprehensive description of carbohydrate analysis has been omitted since an entire volume in this series deals with the subject.

The wide range of subject matter covered identifies this volume as rather different from most others in the series and this is especially obvious by our inclusion of biophysical approaches. Although we appreciate that it is not possible to detail each of the highly specialized and sophisticated techniques, we felt it important that these approaches, which have contributed so much to our knowledge of membrane structure and function, should not be neglected. We hope that the text in Chapters 7 and 8 will illuminate an otherwise difficult area and spark the experimental imagination into new areas for exploration.

Inevitably, there will be areas which have been omitted or overcondensed. We would appreciate, therefore, receiving any comments on the text and the correction of any errors, subtle or glaring, that might have escaped our notice.

J.B.C.Findlay and W.H.Evans

Contributors

E.M.Bailyes
Department of Clinical Biochemistry, Addenbrooke's Hospital, Cambridge University Medical School, Hills Road, Cambridge CB2 2QR, UK

C.L.Bashford
Department of Biochemistry, St. George's Hospital Medical School, Cranmer Terrace, London SW17 ORE, UK

A.O.Brightman
University of Purdue Cancer Research Center, West Lafayette, IN 47907, USA

J.P.Earnest
Department of Biochemistry and Biophysics, University of California, Davis, CA 95616, USA

W.H.Evans
Medical Research Council, National Institute for Medical Research, The Ridgeway, Mill Hill, London NW7 1AA, UK

J.B.C.Findlay
Department of Biochemistry, University of Leeds, Leeds LS2 9JT, UK

J.A.Higgins
Department of Biochemistry, University of Sheffield, Sheffield S10 2TN, UK

O.T.Jones
Department of Biochemistry and Biophysics, University of California, Davis, CA 95616, USA

J.P.Luzio
Department of Clinical Biochemistry, Addenbrooke's Hospital, Cambridge University Medical School, Hills Road, Cambridge CB2 2QR, UK

M.G.McNamee
Department of Biochemistry and Biophysics, University of California, Davis, CA 95616, USA

D.J.Morré
University of Purdue Cancer Research Center, West Lafayette, IN 47907, USA

P.J.Richardson
Department of Clinical Biochemistry, Addenbrooke's Hospital, Cambridge University Medical School, Hills Road, Cambridge CB2 2QR, UK

A.S.Sandelius
Botanical Institute, University of Göteborg, 41319 Göteborg, Sweden

D.Schubert
*Max-Planck-Institut für Biophysik, D-6000 Frankfurt am Main 70, Kennedy-
Allee 70, FRG*

A.Sivaprasadarao
Department of Biochemistry, University of Leeds, Leeds LS2 9JT, UK

Contents

Abbreviations

AChR	acetylcholine receptor
ANS	1-anilino-8-naphthalene sulphonate
BSA	bovine serum albumin
CFM	continuous fluorescence microphotolysis
CHAPS	3-[(3-cholamidopropyl)-dimethylammonio] 1-propane-sulphonate
CMC	critical micelle concentration
CTAB	cetyltrimethylammonium bromide
DDISA	3,5-diido-4-diazobenzene sulphonate
DEPC	dielaidoylphosphatidylcholine
DMEM	Dulbecco's modified Eagle's medium
DMPC	dimyristoylphosphatidylcholine
DMSO	dimethyl sulphoxide
DOC	deoxycholate
DOPC	dioleoylphosphatidylcholine
DSC	Differential Scanning Calorimetry
DTAB	dodecyltrimethylammonium bromide
DTNB	dithionitrobenzoate
DTT	dithiothreitol
EDTA	ethylenediaminetetraacetic acid
ELISA	enzyme-linked immunosorbent assay
EM	electron microscopy
e.s.r.	electron spin resonance
FACE	formic acid: acetic acid: chloroform: ethanol
FCCP	carbonyl cyanide p-trifluoromethoxyphenylhydrazone
FP	flavoprotein
f.p.l.c.	fast protein liquid chromatography
FRAP	fluoresence recovery after photobleaching
HAT	hypoxanthine, aminopterin, thymidine
h.p.l.c.	high performance (pressure) liquid chromatography
IMP	intra-membranous particle
INT	2-p-iodophenyl-3-p-nitrophenyl-5-phenyl-2H-tetrazolium chloride
NEPHGE	non-equilibrium pH gradient electrophoresis
NPA	N-1-naphthylphthalamic acid
PAGE	polyacrylamide gel electrophoresis
PBS	phosphate-buffered saline
PC	phosphatidylcholine
PEG	polyethylene glycol
PMSF	phenylmethylsulphonyl fluoride
PN	pyridine nucleotide
POPC	palmitoyloleoylphosphatidylcholine
PTA	phosphotungstic acid
PVP	polyvinylpyrrolidone
SDS	sodium dodecyl sulphate
TFA	trifluoroacetic acid
TID	3-trifluoromethyl-3-(m-iodophenyl)diazirine
t.l.c.	thin layer chromatography

CHAPTER 1

Organelles and membranes of animal cells

W.HOWARD EVANS

1. INTRODUCTION

The membranes of animal cells are organized to encompass morphologically distinctive organelles with well-defined functions, or appear to extend through the cytoplasm as reticular, or vesicular structures. Approximately 10 categories of membranes have been isolated and biochemically characterized from nucleated animal cells. By the late 1960s the major organelles and membranes were prepared and characterized especially from liver tissue, and their relative areas or volumes derived by morphometric techniques (1). However, analyses of nuclei-free liver homogenates by centrifugation in sucrose gradients using zonal rotors indicated that an appreciable amount of membrane protein remained unassigned. Studies have now shown the complexity of the Golgi apparatus (2), especially the extensive trans-tubular axis (3), the functional mosaicism of the plasma membrane of tissue (4,5) and cultured cells (6,7). Also the identification of 'coated' membranes and studies on the labyrinthian complexity of the endocytic compartment (8) have resulted in most, if not all, of the cellular membranes being functionally characterised.

This chapter addresses in general terms the strategies underlying the isolation and characterization of animal cell membranes. Amplified information is available in specialized monographs (9) or in the original papers referenced. The isolation and characterization of plant membranes and organelles, where similar general principles apply, are addressed in Chapter 2.

2. ISOLATION METHODS

2.1 Non-disruptive methods

2.1.1 Cell surface shedding

In these methods, patches of the cell surface become detached from cells mainly as vesicular structures. Cells, especially tumour cells (10), release spontaneously into media microvesicles originating from the plasma membrane. For example, leukaemic cells release vesicles enriched in tumour antigens (11), mouse mastocytoma cells release H-2 antigens (12), reticulocytes release vesicles containing transferrin receptors (13) and A-431 cells release vesicles containing epidermal growth factor receptor−kinase complexes (14). The shedding of microvesicles enriched in plasma membrane marker components can also be induced by exposing cells to a variety of chemical reagents. Thus, fibroblast monolayers and myoblasts release plasma membrane vesicles after exposure for 15 min to 2 h to protein-free media containing 25 mM formaldehyde − 2 mM dithiothreitol (DTT), or 10 mM N-ethylmaleimide (15). Lymphocytes incubated in the presence of colchicine or cytochalasin release vesicles containing IgM and IgD (16). Erythrocytes incubated in EDTA − CaCl$_2$ media release, without haemolysis, vesicles impoverished in actin and spectrin (17), and platelets in the presence of low amounts of detergents, for example, dilaurylglycerol phosphocholine (Sigma) also release vesicles (18). Isolated hepatocytes release vesicles, originating mainly from the plasma membrane into saline media when subjected to centrifugation at high speed (19). In addition to cells, whole organisms, for example schistosomes, trematode worms, release membrane fragments from their tegumental surface, a process accelerated by mechanical treatment (e.g. vortexing the worms) (20).

At the practical level, released vesicles are recovered simply by allowing cells, etc., to settle at unit gravity or at low centrifugal forces, for example 500 *g* for 2 min, and collecting the supernatants.

2.1.2 *Attachment to solid supports*

This is a rapid method for preparation of plasma membranes from cells and one that avoids the need for subcellular fractionation. The generalized procedure is described in *Table 1* and is illustrated in *Figure 1*.

This method, used to prepare membranes from red cells, hepatoma and HeLa cells (21−24), has a wide potential. Although the initial experiments with erythrocytes indicated that the plasma membranes attached to the negatively-charged beads after washing off the cells and cellular debris have their cytoplasmic aspect exposed to the medium, the proclivity of plasma membranes of nucleated cells to form vesicles with a right-side-out configuration should be borne in mind. Membranes attached to beads can be analysed directly enzymically and chemically; however, the release of membrane fragments from the beads is difficult to achieve and usually requires detergents or extremes of pH that can destroy biological activities.

Other solid supports used besides polylysine-coated beads include nitrocellulose-treated Sephadex beads to prepare mouse L-cell plasma membranes (25), nylon wool fibres

Table 1. Membrane isolation on polylysine-coated beads[a].

1.	Derivatize Bio-Gel P-2 beads with polylysine (mol. wt. 84−91 K, Miles Laboratories) or polyethyleneimine (mol. wt 50−100 K, Polyscience) as described in ref 24. Alternatively polylysine-coated beads can be purchased from Biorad. Cytodex 1 beads (Pharmacia) may be used as a substitute.
2.	Attach the cells to the beads as follows. Wash the beads (~0.5 ml) five times in 10 vol of 0.15 Tris pH 7.4 (bench centrifuge) and four times in 0.14 M sorbitol, 20 mM acetate pH 5.0 (buffer 1). Add the beads suspended in 5 ml of buffer 1 to an equal volume of cells also suspended in buffer 1. Leave for 10 min at 4°C in a centrifuge tube, gently inverting the tube to ensure constant mixing.
3.	Dilute the tube to about 50 ml with buffer 1 and allow the beads to settle. Remove the supernatant and repeat this washing procedure twice to remove unbound cells.
4.	Vortex tube vigorously for 5−10 s, add ice-cold 10 mM Tris-HCl, pH 7.4 (buffer 2), and mix by inversion. Wash the beads a further four times in buffer 2 to release cells. The number of cells still bound can be assessed by scanning electron microscopy; further washings are necessary if cells are still bound to the beads.
5.	After a final wash, suspend the membrane-coated beads in buffer 2 and sonicate at 20 W (or lowest setting) for 5 s. Wash the beads a further three times in buffer 2. Beads may now be used for biochemical analysis of attached membranes.

[a]Taken from ref. 21; see *Figure 1*.

2

CELL ATTACHMENT-BARE SITE NEUTRALIZATION

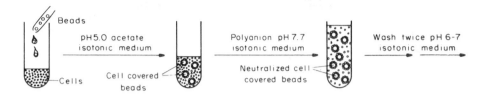

CELL DISRUPTION-PLASMA MEMBRANE ISOLATION

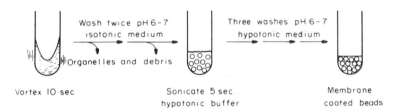

Figure 1. Isolation of plasma membranes on polycation-coated beads. For full details, see *Table 1*. Figure reproduced from ref. 24.

(Baxter-Travenol Labs, Deerfield, IL 60015, USA) to prepare peritoneal leucocyte plasma membranes (26), colloidal silica to isolate *Dictyostelium* plasma membranes (27) and various lectins attached to Sepharose beads to isolate plasma membranes from erythrocytes (28) and mouse lymphoma cells (6). The biological specificity of this approach is open to further development by attachment of antibodies to solid supports, and the immunoabsorption technique for isolation of cell membranes is described in Chapter 3.

2.2 Methods involving cell disruption

2.2.1 Tissue and cell breakage

A wide range of methods have been described for the breakage of tissues (see *Table 2* for their relative merits).

(i)   Isolated cells, especially cultured cells are, in general, more difficult than tissues to disrupt efficiently and to prepare subcellular fractions of acceptable quality in good yield. The simplest and most common method involves the application of tight-fitting Dounce homogenizers of small capacity (5 – 12 ml) to disrupt cells first subjected to hypotonic shock; this involves suspending cells for approximately 10 min at 4°C in 5 mM Tris-HCl, pH 7.6. The extent of cell breakage after homogenization is monitored by phase-contrast microscopy and should exceed 80%. Cells grown in spinner cultures are easier to disrupt, for geometrical reasons, than monolayers.

(ii)   Of greater utility for disruption of cultured cell lines has been gas cavitation. Cells are suspended in media in a nitrogen-filled metal cylinder for 5 – 30 min at high pressures (7 – 65 atmospheres). When the cells are rapidly returned to atmospheric pressure, nitrogen gas dissolved in the cytoplasm is released and

3

Table 2. Techniques for tissue cell breakage.

Apparatus	Comments on use
Dounce homogenizers	Widely available; various sizes (3−30 ml) with 'tight' (0.07 mm clearance) or 'loose' pestles. Hand operated, thus low shear forces. 'Tight' useful for homogenizing cultured cell pre-swollen in hypotonic media.
Potter−Elvehjem homogenizers	Widely available; various sizes (5−50 ml). Require high torque motor. Water-cooled models available (B.Braun, Melsinger, FRG). Clearance usually 0.10−0.23 mm, but machining Teflon pestle can increase. Useful for tissues, but first mince or cut into small portions with scissors. Apparatus allowing continuous homogenization available (Yamamoto Scientific Co., Minato-ku, Tokyo 105, Japan, and Northbrook, IL 60062, USA).
Polytron, Ultraturrax	Widely available. Shafts of various dimensions available; can be difficult to standardize. Utilizes circular saw at tip of shaft to disperse tissue — useful for fibrous and muscle tissue. With large amounts of tissue, a Waring Blender can be used or a food processor. For volumes below 0.5 ml, a Tissue-Tearor is available from Biospec Inc., P.O. Box 722, Bartesville, OK 74005, USA.
Nitrogen bombs	Allow single cycle dispersion of cells. No heat damage. Require calibrating to minimize organellar damage. Pressures used range from 20 to 100 atmospheres and times 10 to 30 min. Available from Parr, Kontes, Braun and Yeda. 'Mini'-sized vessels (capacity 5−15 ml) also available.
Pumps	Not widely available. Electrically or hydraulically operated with spring-operated orifices can produce good plasma membrane preparations. Stansted and Locke pumps in current use.
Presses	A number of variants are available in addition to the French Press for homogenizing tissues by forcing them through orifices or mesh wire. Biox X-press (AB-Biox, Box 143, Jarfalla, Sweden S17523, Life Science Labs Ltd, Sarum Road, Luton, Beds., UK) uses frozen material. A tissue press is also available from Edco Scientific Inc., PO Box 64c, Sherborn, MA 01770, USA.

this, combined with the rapid expulsion from the cylinder via a narrow exit pipe, results in extensive disruption. Under optimal conditions that require careful evaluation for each cell type, the plasma membrane and endoplasmic reticulum are converted into small vesicles whereas organelles are released intact (30).

(iii) Sonication is a further method used to break cells recalcitrant to conventional approaches. Mast cells were disrupted by sonic energy and the parameters (discontinuous versus continuous and the number of pulses) for optimal results have been explored (31). It was noted, for example, that some populations of mast cells were disrupted by few pulses, whereas others required increased amounts of sonication, and that the Ca^{2+} concentration of the medium was important. These generalizations probably apply to most cell types.

(iv) Many organelles and membrane fractions after isolation are subjected to further disruption to yield subfractions separated by density gradient centrifugation. Thus, homogenization of liver plasma membranes using tight-fitting Dounce homogenizers releases bile canalicular vesicles and junction-containing components that are then resolved on the basis of different densities by centrifugation in sucrose gradients (32,33). The French Press, used at pressures as high as 600 atmospheres releases from skeletal muscle microsomal fractions the component parts of the T-tubules that, by virtue of their low sedimentation rate and density, are purified in sucrose gradients (34).

2.2.2 Media

The type of media in which to carry out cell breakage is usually chosen in conjunction with the homogenization method used. With tissues, iso-osmotic sucrose at a ratio of 10 g wet weight to 100 ml medium is used; in general it is better to err on the side of a lower tissue to medium ratio especially when large volume centrifuge rotors are available. Hypotonic media aid in the disruption of cells, but can cause osmotic damage to organelles if used for prolonged periods. Addition of $1-5$ mM $MgCl_2$ helps to minimize damage to nuclei, but the use of media of high salt concentrations is to be avoided owing to the production of membrane aggregates. In contrast, brush border vesicles can be prepared from post-nuclear homogenates of rat intestine or kidney using 10 mM $MgCl_2$, 5 mM EGTA to selectively remove other membranes (35,38).

In general, Ca^{2+} salts should be avoided owing to their activation of proteases (see below). The use of media of pH $8.0-9.6$ helps in some instances to stabilize membranes, reducing, for example, the degree of vesicle formation by plasma membrane (36). The stability of clathrin-coated vesicles is enhanced at pH $6.5-6.8$. The above examples indicate that generalizations are difficult to make about the choice of media and each membrane preparation has to be assessed individually.

2.2.3 Proteolysis of membranes

The problems generated during subcellular fractionation by the action of proteolytic enzymes have been underestimated. This has become clear when purified membrane preparations have been used as starting points for the isolation of membrane proteins, for example, epidermal growth factor receptor (39) and gap junctions (40). After cell disruption, many cytosolic, membrane-bound, and lysosomal enzymes previously compartmentalized, are released. Membrane-bound proteases are activated during prolonged preparative procedures especially if detergents feature. The meagre knowledge of the precise nature of many of these proteolytic enzymes complicates the selection of specific inhibitors.

The various classes of proteases, the properties of corresponding inhibitors, and advice on their use are given in *Table 3*. Phenylmethylsulphonyl fluoride (PMSF) is probably the most widely used proteolytic inhibitor. Its half-life in aqueous media at pH 7 is 100 min, with hydrolysis occurring faster at higher pH and in phosphate media. PMSF, kept as an ethanolic stock solution, is added to media shortly before use and further additions are necessary at regular intervals.

3. SEPARATION OF SUBCELLULAR COMPONENTS

3.1 Centrifugation

Methods for the separation of subcellular components on the basis of differences in sedimentation rate (determined by weight, size and shape) and density have been described extensively (9,41) and are also discussed in Chapter 2. The ideal gradient

Table 3. Protease inhibitors.

Inhibitor	Specificity	Normal concentration range	Comments
Amastatin (also epiamastatin)	Amino exopeptidases	$1-10$ μg/ml	
Antipain	Cathepsin B, papain trypsin	$1-10$ μg/ml	
Benzamidine	Serine proteases	Up to 10 mM	
Benzylmalic acid	Carboxypeptidases	$1-10$ μg/ml	
Bestatin (also epibestatin)	Amino exopeptidases	Up to 1 μg/ml	
Chymostatin	Cathepsin B, chymotrypsin, papain	$1-10$ μg/ml	
Diisopropylphosphorofluoridate (DFP)	Serine proteases	Up to 0.1 mM	Very toxic
Diprotin A and B	Dipeptidylamino peptidases	$10-50$ μg/ml	
Elastatinal	Elastase	10 μg/ml	
Ethylenediaminetetraacetic acid (EDTA)	Metalloproteases	$0.1-5$ mM	Useful general inhibitor
Leupeptin	Cathepsin B, papain plasmin, trypsin	$1-100$ μg/ml	
Pepstatin A	Carboxyl proteases (e.g. pepsin, renin)	$1-10$ μg/ml	Soluble in ethanol or methanol rather than water
Phenylmethylsulphonyl fluoride (PMSF)	Serine proteases	Up to 0.1 mM	Limited half-life in aqueous buffers. Useful general inhibitor
Phosphoramidon	Collagenase, thermolysin	$1-10$ μg/ml	Not a general Zn-protease inhibitor
Sodium tetrathionate	Thiol proteases	Up to 5 mM	
Tosyl-methyl ketone	Papain, trypsin		
Tosyl-phenylalanine chloromethyl ketone	Chymotrypsin		
Trypsin inhibitors Types I–IV	Chymotrypsin, trypsin		From various sources (e.g. chicken egg white, soybean) and with various activities

In addition to the above, other inhibitors to note include parachloromercuribenzoate (PCMB), iodoacetamide and heavy metal ions (for inhibition of -SH groups), α_2-macroglobulin (for inhibition of collagenases), aprotinin (from lung tissue) and other chelating agents besides EDTA (e.g. 2-phenanthroline, α,α^1-bipyridyl Na fluoride). Many of the inhibitors are available from Sigma and Boehringer. Aprotonin is available as Trasylol® from Bayer, FRG with subsidiaries in various countries. Nupercaine or tetracaine (0.15 mM) prevent lipid hydrolysis.

Proteolytic enzymes are classified according to type (37). Thus, serine proteases include trypsin, chymotrypsin, elastase, cathepsin G and plasminogen activators active at pH $7-9$ and are inhibited by DFP. Thiol proteases include various cathepsins active at pH $3-8$ and inhibited by PCMB and iodoacetamide. Carboxyl proteases include cathepsin D and pepsin active at $2-7$ and inhibited by pepstatin. Metallo-proteases include microvillus proteases and collagenases active at pH $7-9$ and inhibited by EDTA and dithiothreitol.

A commonly used proteinase inhibitor cocktail includes A: pepstatin, 0.5 mg; leupeptin, 5.0 mg; chymostatin, 5.0 mg; antipain, 5.0 mg; aprotinin, 5.0 mg in 10 ml of water as a suspension and used at 1:100 dilution. All these reagents are available from Sigma. B: PMSF, 4.3 mg dissolved in 1 ml of ethanol and used at 200 μl/100 ml. C: 1 mM EDTA, DFP, and PCMB, p-chloromercuribenzoate.

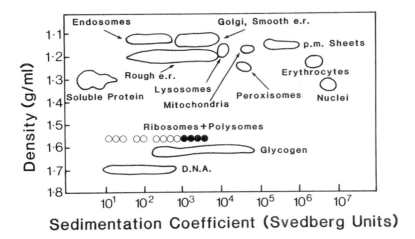

Figure 2. Sedimentation coefficient and banding densities of subcellular components. These can be separated by various combinations of differential and density gradient centrifugation. Because many subcellular components differ only slightly in weight and density, other techniques, e.g. phase separations (Section 3.2), electrophoresis (Section 3.4) or affinity (Section 3.5) are applied. Density perturbation techniques are described in Section 3.1.2. See also Chapter 2.

material should be freely soluble in water, physiologically innocuous and chemically inert, non-viscous, transparent in visible and u.v. light and possess a low osmotic pressure. Most subcellular separations have been carried out in sucrose. Although sucrose fulfills many of the desired criteria, its high viscosity at high concentrations coupled with the osmotic effects created are clear disadvantages (see Appendix). Thus, subcellular vesicles dehydrate on contact with hyperosmotic sucrose solutions, and secretory granules are liable to destructive lysis limiting their usefulness for functional studies. High concentrations of sucrose cause uncoating of clathrin-coated vesicles, but substitution with $^1H_2O/^2H_2O-8\%$ sucrose gradients limits this process (42). Dextran-H_2O gradients have also proved useful for distinguishing between sealed and leaky membrane vesicles (43).

Most density gradient separations of membrane fractions are carried out over $3-4$ h or overnight at 80 000 *g* or higher in swing-out rotors (e.g. Beckman SW28 or equivalent). Vertical rotors may provide, in certain instances, the advantage of lower hydrostatic pressure and facilitate higher *g*-forces owing to the shorter path length, for example, in the rate-zonal separation of peroxisomes (44). However, streaking of membranes along the walls of the centrifuge tubes can be a disadvantage.

The complexity of the subcellular components of similar sedimentation coefficient and density in the 'window' comprising endoplasmic reticulum, Golgi and endosome membranes (*Figure 2*) has led to the adoption of other separation methods (multidimensional cell fractionation) that exploit different parameters, for example counter current distribution (45,46).

3.1.1 *New density gradient materials*

(i) *Percoll.* Silica sols coated with polyvinylpyrolidone (Percoll®) offer the important advantages of speed and iso-osmotic conditions for the preparation of cells and subcellular

Table 4. List of the use of Percoll gradients for organelle and membrane separations.

Fraction	Source	Density (g/ml)	Reference
Chromaffin granules	Bovine adrenal	1.067−1.081	52
Endocytic vesicles	K562 cells	1.03	53
	Rabbit alveolar		
	macrophages	1.05	49, 50
Glycosomes	*Trypanosoma brucei*	1.087	54
Golgi	Rat liver	1.042, 1.057	55, 56
		1.028, 1.051	57
Lysosomes	Rat liver	1.087	55, 58
	Rat kidney cortex	1.15	51
	Human lymphoblasts	1.085	59
	Pig thyroid	1.14	60
Mitochondria	Rat liver	1.085−1.100	61, 62
	Bovine skeletal muscle	1.035−1.070	63
Peroxisomes	Rat liver	1.075	64
Plasma membranes	Human platelets	n.s.	65
	Rat liver sinus.	1.02−1.04	66
	Rat aorta	n.s.	67
	Krebs ascites cells	n.s.	48
	Rabbit kidney cortex b.l.	1.037	68
	Rabbit kidney cortex b.b.	1.042	68
	Neuroblastoma	n.s.	69
Synaptosomes	Rat brain	1.04−1.05	70

Note that density applies to Percoll in 0.25 M sucrose; in ref. 52, 0.27 M sucrose was used; in refs. 48 and 69, pH 8.8−9.2 was used as opposed to pH 7.6. n.s., density not stated; b.l., basolateral plasma membranes; b.b., brush borders; sinus., sinusoidal domain plasma membranes.

particles. These are desirable requirements for the preparation of fragile organelles and vesicles for transport studies. Most separations use Percoll suspended in iso-osmotic sucrose at neutral pH, but in certain instances (see *Table 4*) improved resolution of subcellular components is obtained at pH 9 (36). The separations obtained in Percoll are, in many instances, only marginally better than those achieved in gradients of sucrose.

Percoll interferes with protein determination by the Lowry procedure, but the Coomassie Blue (Bradford) method can be used with modifications (see Chapter 6). Percoll has minimal effects on most enzyme assays. Removal of the colloidal silica particles is usually effected by differential centrifugation or by gel filtration using Sephacryl S-1000 superfine beads as illustrated by the purification of secretory granules from bovine adrenal medulla (47).

The density of Percoll fractions can be ascertained by running simultaneously gradients containing coloured density marker breads obtained from Pharmacia. Also, there is a linear relationship between density and refractive index (see Appendix). Examples of the use of Percoll gradients to isolate subcellular membranes are given in *Table 5*. *Table 4* lists the application of Percoll gradients for isolation of subcellular components from a range of tissues and cells.

(ii) *Metrizamide® and Nycodenz®*. The application of these iodinated density gradient materials available from Nycomed, Oslo 4, Norway with distributors in various countries

Table 5. Examples of procedures using Percoll for rapid separation of membranes and organelle fractions.

A. *Separation of fat cell plasma membranes and mitochondria*[a]

1. Homogenize isolated fat cells (0.5−0.7 g wet weight) by agitating for 30 sec in a Vortex mixer in 2−3 ml of 0.25 M sucrose, 10 mM Tris−HCl, 2 mM EGTA, pH 7.4 (medium A).
2. Centrifuge at 1000 g for 30 sec and collect, by aspiration, the infranatant and small pellet located below the fat plug. Centrifuge this crude extract at 30 000 g for 30 min.
3. Remove the supernatant and resuspend the pellet in 400−500 µl of medium A (small Dounce homogenizer) and disperse in 8 ml of iso-osmotic Percoll.
4. Prepare iso-osmotic Percoll by mixing with 2 M sucrose, 80 mM Tris-HCl, 8 mM EGTA, pH 7.4 (medium B) and medium A in the ratio 7.1.32 (by vol) to yield density 1.05 g/ml. Suspend 400 µl of crude membranes in 8 ml of this solution.
5. Centrifuge in a fixed-angle rotor at 10 000 g for 15 min. Plasma membranes are collected at the top of the gradient well separated from mitochondria close to the bottom. Collect, using a syringe, and dilute 5-fold in 10 mM Tris−HCl, 1 mM EGTA pH 7.4 containing 0.15 M NaCl (for plasma membranes) or 0.25 M sucrose (for mitochondria).
6. Centrifuge at 10 000 g for 25 min to remove Percoll beads.

B. *Preparation of endosomes and lysosomes from macrophages*[b]

A similar method applies for isolation of endosomes from cultured cells[c].

1. Prepare alveolar macrophages from rabbit pulmonary lavage and suspend in 0.25 M sucrose, 3 mM imidazole pH 7.4 (medium A).
2. Homogenize by nitrogen cavitation (Section 2.2.1) by suspending at 250 p.s.i. (1.73 MPa) for 10 min.
3. Centrifuge at 1000 g for 5 min, collect the supernatant and centrifuge this at 17 000 r.p.m. for 10 min (Beckman JA-17 rotor) (2.8×10^5 g/min).
4. Resuspend the pellet in medium A (∼2.5 ml).
5. Layer over 30 ml of Percoll (density 1.07 g/ml) and centrifuge at 17 000 r.p.m. for 1 h (JA-17 rotor). Unload the gradient as 1 ml fractions; within the density range 1.02−1.12 g/ml, endosomes are recovered at 1.05 g/ml and lysosomes at 1.10 g/ml.
6. Remove Percoll beads by differential centrifugation.

C. *Isolation of kidney lysosomes*[d]

1. Mince rat kidney cortex and homogenize using four strokes of a Potter−Elvehjem homogenizer rotating at 1000 r.p.m. (Section 2.2.1) in 0.3 M sucrose, 1 mM EDTA, adjusted to pH 7.4 with Tris (Buffer A). Ratio of medium to tissue is 10 ml/g.
2. Centrifuge at 270 g for 5 min, collect the supernatant and centrifuge this at 3000 g for 10 min (Sorvall SS.34 rotor). Remove the supernatant and the loose upper portion of the pellet, and resuspend the pellet in a large volume (× 20) of buffer A. Centrifuge this at 500 g for 5 min twice to remove red blood cells, as a pellet, and then centrifuge the supernatant at 3000 g for 10 min to yield a crude lysosomal fraction.
3. Resuspend the pellet in the minimum volume of buffer A (300 µl/g tissue). Prepare a Percoll gradient by centrifuging 14 ml of 75% Percoll, 15% 2 M sucrose and 10% 0.1 M 3 (N-morpholino) propanesulphonic acid, pH 7.4 at 48 000 g for 50 min (glass tubes, SS.34 rotor).
4. Layer the resuspended crude lysosomes into a Percoll gradient and centrifuge at 48 000 g for 30 min. Separation is shown in *Figure 3*. Collect Zone 4, dilute with buffer A and centrifuge at 3000 g for 10 min to yield lysosomes leaving the beads in the supernatant.

[a]Taken from ref. 71.
[b]Taken from refs 49 and 50.
[c]See refs 72 and 73.
[d]Taken from ref 51.

(American distributors 'Accurate Chemical and Scientific Corp.', 300 Shames Drive, Westbury, NY 11590, USA) has been the subject of another monograph in this series (74). Both substances yield solutions of lower osmotic activity and viscosity than sucrose

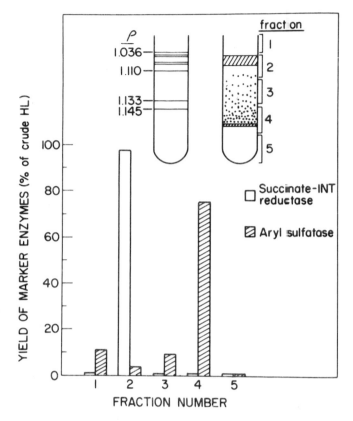

Figure 3. Separation of kidney mitochondria and lysosomes on a Percoll gradient. A 'mitochondrial–lysosomal' crude fraction was layered onto a Percoll gradient as described in the text. The density was determined by using colour-coded beads. Fraction 4, corresponding to a dense zone of turbidity was collected using a syringe, diluted in sucrose/EDTA and centrifuged at 3000 *g* for 10 min to sediment lysosomes (a dark pellet). Percoll beads do not sediment at these low centrifugation speeds. Reproduced from (51).

solutions of comparable density (see Appendix I) and have found their widest applications thus far in the separation of mitochondria, lysosomes and peroxisomes (75) and in the isolation of secretory granules that maintain transport functions (76). Nycodenz possesses some advantages over Metrizamide; it is easier to dissolve in sucrose or salt solutions and to construct density gradients; it contains no glucosamine and thus is less likely to interfere with glycosyl transferase assays and in general, it interferes less with enzyme analyses. The high E_{280} absorption of these reagents necessitates the estimation of protein by dye-binding assays, and they also interfere with nucleic acid measurements. Nycodenz does not precipitate at low pH, and can be removed from samples by dialysis, ultrafiltration or gel filtration. It is more stable and less expensive than Metrizamide.

Although it is possible to construct Nycodenz gradients *in situ* by centrifugation in angle or vertical rotors, it is generally advantageous to use pre-formed gradients containing 0.25 M sucrose buffered to pH 7.0–8.0. Continuous iso-osmotic gradients are prepared using a two-chamber gradient mixer; gradients are unloaded by passing a thin tube to the bottom of the gradient and then pumping the fractions out directly

Table 6. Banding densities of various subcellular fractions in gradients of Nycodenz or sucrose.

Subcellular fraction	Density (g/ml)	
	Sucrose	Nycodenz[a]
Guinea pig enterocyte		
Mitochondria	1.19−1.21	1.16−1.19
Lysosomes	1.20−1.21	1.14−1.17
Plasma membranes	1.14−1.16	1.11−1.15
Endoplasmic reticulum	1.10−1.23	1.11−1.18
Rodent liver		
Nuclei	>1.30	1.23
Endosomes	1.11−1.13	1.09−1.11
Peroxisomes	1.23	1.22

[a]Nycodenz in 0.25 M sucrose

or by displacement upwards by using heavy sucrose or dense inert non-viscous Maxidens, density 1.9 g/ml immiscible with aqueous media and available from Nycomed. Metrizamide or Nycodenz solutions are made up as 40% stock in 0.25 M sucrose, and diluted with 0.25 M sucrose as required. For densities below 1.10 g/ml, water may be substituted. Examples of fractionation in Nycodenz gradients are given below.

(i)   Subfractionation of rat liver endocytic vesicles. Layer a 'low density' microsomal fraction (prepare this as described in *Table 12*) onto a continuous Nycodenz gradient constructed by mixing in a gradient maker 6 ml each of 27.6% and 13.8% (w/v) Nycodenz dissolved in 0.25 M sucrose and water respectively. Centrifuge at 98 000 *g* for 2 h in a swing-out rotor (Beckman, SW28) to separate two ligand-containing endosome fractions (see Section 5.7) of density 1.090 and 1.115 g/ml (77,78).

(ii)   Isolation of lysosomes or peroxisomes. Construct a gradient extending from 19% (density 1.105 g/ml) to 30% (density 1.16 g/ml). Apply a rat liver mitochondrial/lysosomal fraction (75) (see Sections 5.3 and 5.4 for fuller experimental details) and collect lysosomes at the 1.110−1.135 g/ml interface and peroxisomes at the 1.21−1.26 g/ml interface after centrifuging at 97 000 *g* for 2 h.

Densities of iodinated density gradient samples are calculated as follows: η = refractive index. The sucrose is iso-osmotic (0.25 M) and the NaCl concentration is 0.15 M.

density (g/ml) at 20°C = 3.453η − 3.601 for Metrizamide
3.287η − 3.383 for Nycodenz (with NaCl)
3.410η − 3.555 for Nycodenz (with sucrose).

A comparison of the banding densities of various subcellular fractions in gradients of Nycodenz and sucrose is given in *Table 6*.

3.1.2 *Density perturbation*

Because of similarities in the sedimentation rates and densities of a number of subcellular components (*Figure 2*), a variety of approaches have been developed for modifying selectively their density. These fall into two major classes. First there are reagents that combine selectively with components present in some membranes but absent in others.

11

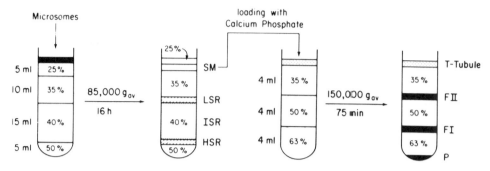

Figure 4. Isolation of transverse tubule membranes from rabbit skeletal muscle by a density perturbation technique. Rabbit microsomes were prepared from skeletal muscle and the smooth membrane (SM) fraction recovered from the sucrose gradient was incubated in a loading solution (see text for composition). Fractionation on sucrose gradients yielded a T-tubule fraction. Reproduced from (83).

Second, a wide range of methods exist that introduce into the vesicle interior substances or modifications that grossly change their sedimentation/density characteristics.

(i) *Membrane perturbants*. Digitonin binds preferentially to cholesterol, causing cholesterol-rich cell membranes to increase their banding density without appreciably solubilizing intrinsic membrane proteins. Optimum displacement of plasma membranes to higher densities occurs at equimolar ratios of digitonin to cholesterol; in the absence of precise information on cholesterol content of fractions (e.g. in muscle membranes), digitonin is used at 0.1 mg/mg protein (79). Treatment with digitonin of a post-nuclear lymphocyte fraction showed that 5'-nucleotidase, alkaline phosphodiesterase and leucine aminopeptidase were shifted markedly to a higher density, whereas galactosyl transferase was only moderately shifted. (See Appendix II for location of subcellular marker enzymes.) In contrast, the distributions of mannosyl transferase (a smooth endoplasmic reticulum marker), cytochrome oxidase and N-acetyl-β-glucosaminidase were not changed (80).

The procedure used is as follows. Determine cholesterol content of membranes (e.g. by gas chromatography of lipid extracts of fractions, see Chapter 4). Add digitonin solution (0.3%) to the stirred subcellular fraction to give a 1:1 molar ratio of digitonin to cholesterol. The final digitonin concentration should not exceed 0.01%. At higher concentrations, digitonin solubilises membrane components and can cause sealed vesicles to become leaky.

(ii) *Content perturbants*

1. *Mitochondria*. Incubation of brain synaptosome fractions after hypotonic shock in 1 mM p-iodonitrotetrazolium dissolved in 80 mM sodium succinate, pH 7.4 for 20 min at 3°C generates formazan crystals in mitochondria rendering them heavier than the synaptosome plasma membranes. In addition, the reaction, as a useful preparative by-product, cross-links components of the post-synaptic density, stabilizing them and aiding in their recovery (81). Treatment of rat liver homogenates with 10 mM K_2PO_4 accelerates the sedimentation rate of mitochondria without affecting lysosomes, thereby allowing the latter to be purified on Percoll gradients (82).

2. *T-tubules*. The use of the calcium oxalate or phosphate loading technique to separate T-tubule-derived vesicles from other reticular and sarcolemmal membranes in muscle

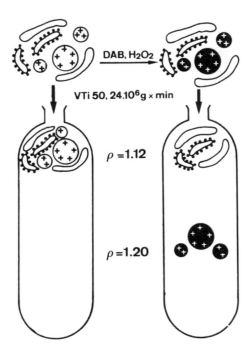

Figure 5. Diagrammatic description of the diaminobenzidine-induced density shift. Diaminobenzidine (DAB) diffuses into rat liver endosome vesicles containing horseradish peroxidase conjugated to galactose−bovine serum albumin. In the presence of H_2O_2, the peroxidase−ligand−DAB is polymerized and thus entrapped. The high equilibrium density of the polymerized DAB results in a large increase in the buoyant density of the endosome vesicles, resulting in separation from other vesicles, e.g. Golgi not containing internalized ligands. Reproduced from (84).

microsomal fractions is a well-substantiated perturbation protocol (*Figure 4*). The presence of higher Ca^{2+}-ATPase activities in the non-T-tubule derived membranes results in greater Ca^{2+} uptake and the calcium phosphate then renders the vesicles denser in sucrose gradients.

In this procedure (83), rabbit muscle, cut into small pieces, is homogenized in a Waring Blendor for 50 sec in 0.3 M sucrose, 0.02 M Tris-maleate pH 7.0, centrifuged at 3000 *g* for 20 min and the resultant supernatant centrifuged at 10 000 *g* for 20 min. The final supernatant is filtered through muslin cloth and centrifuged at 100 000 *g* for 1 h to yield a microsome fraction.

Incubate the microsomal fraction (0.1 mg protein/ml) for 20 min at 22°C in 5 mM $MgCl_2$ 0.15 M KCl, 0.3 mM $CaCl_2$ and 2 mM ATP, buffered to pH 7.4. Add the fraction to the sucrose gradient shown in *Figure 4* (Beckman SW40 Ti rotor or equivalent). The lightest fraction is enriched in T-tubules, and vesicles heavily loaded with calcium phosphate are recovered at higher densities and especially in the pellet.

3. *Endocytic and secretory vesicles.* The similarity in the equilibrium densities of endocytic vesicles, Golgi membranes and other smooth membranes has led to the development of methods for specifically modifying the density of endocytic vesicles based on their ligand content.

One method (84) is based on the introduction of horseradish peroxidase into selected

Table 7. The DAB-induced density shift procedure for preparation of endosomes[a].

1. For uptake by receptor-mediated endocytosis, conjugate the appropriate ligand to horseradish peroxidase (HRP, RZ=3), and separate this from unconjugated ligand and HRP.
2. Define the conditions of endocytosis in order to concentrate ligand−HRP into the endocytic compartment of interest [i.e. ligand species, competitors, time of uptake (usually 3−10 min) etc.].
3. Whenever possible, remove external and membrane-adsorbed HRP by washing the cells or perfusing the tissue with cold saline containing 1 mM EGTA.
4. Isolate the subcellular fraction most enriched in ligand by differential centrifugation. (For a liver light mitochondrial fraction, see Section 5.3). Resuspend the sample gently to avoid release of HRP from damaged organelles.
5. Further purify the ligand-containing organelles by a first equilibration in a density gradient (e.g. 10−60% w/v sucrose buffered with 3 mM imidazole/HCl, metrizamide, Percoll, etc.). This also removes free HRP, which should remain in the loading zone.
6. Incubate for 15−30 min, at 25°C, in the dark, in a freshly prepared, Milipore-filtered, solution containing 5.5 mM 3,3′-diaminobenzidine (e.g. Sigma, Grade II) and 11 mM H_2O_2, dissolved in the medium corresponding to the density where organelles were initially recovered. Measure activities to be tested before and after incubation.
7. Equilibrate the sample again in a density gradient starting at the density where organelles were initially recovered (step 5).
8. To subtract contamination by the centrifugation vessel, incubate controls in the absence of H_2O_2.
9. To test for organelle aggregation, incubate in DAB and H_2O_2 a mixture of the fraction enriched in ligand−HRP, and a fraction isolated similarly from cells not exposed to ligand−HRP and labelled with a different isotope. The latter material should not be displaced.

[a]P.Courtoy, personal communication. See also ref. 84.

endocytic populations followed by incubation in a peroxidase cytochemical medium that selectively increases the equilibrium density of the structures containing the enzyme. Briefly, allow the liver to internalize into endocytic vesicles (density in sucrose 1.11−1.13 g/ml) galactosylated bovine serum albumin conjugated to horseradish peroxidase. Prepare galactosylated serum albumin as described in ref (85). Perfuse rat livers with this ligand, remove 10 min later and fractionate as described in Section 5.3 to yield the supernatant of the 'light-mitochondrial' pellet. Pellet the membrane vesicles in this supernatant (100 000 g for 1 h), re-suspend in 0.25 M sucrose and centrifuge in a 10−60% (w/v) sucrose gradient (Beckman SW-28 or equivalent) for 3 h at 100 000 g. Collect fractions in the density range 1.11−1.13 g/ml and incubate these for 30 min at 25°C in 5.5 mM 3,3′-diaminobenzidine, 11 mM H_2O_2. Re-centrifuge in a 23−60% w/v (1.1−1.3 g/ml) sucrose gradient; collect endocytic vesicles enclosing modified ligands in the 1.19−1.23 g/ml density range. For a generalized procedure, see *Table 7*.

An important caveat in this procedure is the requirement that the fractions treated are free of soluble or adsorbed horseradish peroxidase, since this leads to aggregation. Another approach exploits the high density of colloidal gold and uses gold-labelled antibodies bound to transferrin receptors to isolate endosomes from A431 cells (86). Karnovsky-Roots reagents that combine selectively with acetylcholinesterase thereby depositing iron−copper when added to crude subcellular fractions also allows separation on density gradients of enzyme-containing coated vesicles from endocytic vesicles and this approach has been extended further to separate endocytic and exocytic coated vesicles in perfused rat liver (87, 223).

3.1.3 *Computer simulation of centrifugal separations*

In carrying out subcellular fractionations by centrifugation, the operator has to decide, in advance, such operating parameters as rotor type (angle, swing-out, vertical), run time, sample size, gradient volume, and material, density range and profile, etc. Computer programmes are now available from centrifuge manufacturers that allow virtually any centrifugal separation to be simulated, thereby reducing and possibly eliminating trial and error runs (88). The Centrisim programme (software packages available from E.I. du Pont de Nemours, CT, USA; Sorvall Equipment) operates on Apple or IBM computers and such approaches can be used to exploit fully the centrifuge's potential for effecting sophisticated preparative and analytical separations. A similar programme is also available from Beckman. The application of microcomputer programmes to subcellular separations by centrifugation has been discussed in detail in another volume of this series.

3.2 Two-polymer phase partitioning

Partitioning in dextran−polyethyleneglycol (PEG) aqueous−aqueous phase systems is a useful rapid method for preparation of membranes, especially when ultracentrifuges and the associated rotors are unavailable (89). This method in which capacity is easily scaled up exploits a combination of membrane properties, including electrical charge, density, weight and hydrophobicity. Purification is based upon differential distributions of the constituents in a sample between the three compartments of the two phases, that is the upper and lower phases and the interface. The order of affinity of animal cell membranes for the upper phase is: endoplasmic reticulum < mitochondria < lysosomes < Golgi < plasma membranes.

The method is described in *Table 8* and illustrated in *Figure 3*, Chapter 2. A number of operational points require attention for optimal yields and purity. The salt concentration, osmotic and pH conditions, and temperature govern critically the purity of plasma membranes recovered at the interface. Also, differences occurring between various batch numbers of dextran require the standardization of stock solutions by optical rotation using a polarimeter.

Examples of the application of this method to kidney (92) and liver Golgi (93), bile canalicular and blood sinusoidal liver plasma membranes (94) have been described. Mitochondria and synaptosomes from adult rat brain can be separated, with the mitochondria recovered in the dextran-rich lower phase and the synaptosomes in the upper phase (95). Inclusion of detergents, for example N-lauryl sarcosinate, in the two phases provides a rapid method for the purification of post-synaptic densities from cerebral cortex synaptosomes of animals (96).

To speed up the batch-type separations and to improve the resolution a number of apparatuses have been devised that allow multiple extractions and transfers. These fall into two categories; those based on countercurrent distribution using the Craig discrete mixing−settling transfer approach and those based on continuous flow chromatography. In both categories low-speed centrifugation can be used to speed up the separation of the phase systems, especially when Ficoll/dextran phases close in density are used. Rat liver homogenates have been fractionated using a counter-current distribution apparatus (97).

Table 8. Rapid isolation of plasma membrane by the two-phase procedure[a]

1. Prepare stock solutions as follows:

    20% (w/w) dextran. Layer 220 g of dextran T-500 on 780 g of water and heat with gentle stirring until dissolved. Adjust to 20% in a polarimeter, by addition of small amounts of water.

    30% (w/w) polyethyleneglycol (PEG) Carbowax 8000. Make 300 g up to 1 l in water.

    Prepare stock salt and buffer solutions 10 times stronger than the concentration used in phase systems.

2. Wash the cultured cells ($10^6 - 10^8$ cells) and homogenize them in various ways (see *Table 2*). Usually cells are suspended in a hypotonic medium (e.g. 30 mM $NaHCO_3$, pH 7.0), and are homogenized in a tight-fitting Dounce homogenizer; up to 60 strokes may be required to break >80% cells as assessed using a phase-contrast microscope. Plasma membranes may be stabilized by inclusion of 1 mM $ZnCl_2$ in medium.

3. Centrifuge at $500-1000$ g for 15 min. Resuspend the pellet in the top phase prepared as described below.

4. Mix by repeated inversions ($10-40$) 200 g of 20% dextran, and 103 g of PEG stock solutions with 33 ml of 0.2 M sodium phosphate, pH 6.5 and 179 ml of distilled water. Leave at 4°C for 24 h and collect the two phases.

5. Add the pellet or crude membrane fraction to 30 ml of top phase and aliquot into three polycarbonate 50 ml test tubes. Add 10 ml of bottom phase to each tube, mix, and centrifuge at 2000 g for 15 min in a swing-out rotor (Sorvall HB-4). This may be scaled down to allow separation in 5 ml centrifuge tubes.

6. Collect membranes using a syringe from the interface. Any plasma membranes still present in the supernatant in step 3 can be sedimented at higher speeds and the resuspended pellet used in the two-phase protocol[b].

7. The method of cell breakage and the speeds for centrifugation may require modification for various cells and tissues. See discussion in text.

[a]Adapted from ref. 90.
[b]See ref. 91.
[c]Dextran, predominantly poly(α-1,6-glucose) is available in a range of molecular weight fractions from 10 000 to 2 000 000; the T500 fraction from Pharmacia is the most widely used, with a molecular weight of approximately $450-500$ 000. When dissolving powder, allow for about 10% water content. It is prudent to purchase sufficient amounts to allow completion of a set of experiments, since batches may differ. PEG is usually supplied in powder or flake form by Union Carbide Corp. Note that PEG 8000 was formerly designated 6000. A booklet from Union Carbide lists physical properties.

Success with phase systems depends on the ability to manipulate phase composition. In addition to changing the polymers (confined mainly to dextran and PEG presently), varying the polymer molecular weight, the salt concentration and pH, as well as chemical modification of polymers are being explored. For example, the PEG can be modified by esterification with fatty acids (98) or by attachment of ligands to facilitate affinity-based separations. Separation of *Torpedo* particles containing nicotinic cholinergic receptors (99) and calf brain synaptosomes on the basis of opiate receptor density (100) have been successful following attachment of the relevant ligands to the PEG.

3.3 Permeation chromatography

The availability of porous glass beads has allowed populations of small membrane vesicles of different diameters to be separated. Synaptic vesicles (101) and intestinal brush border vesicles (102) serve as examples of particles separated by passage through columns of carbowax-pre-treated glass beads with mean pore diameter 300 nm. These are available from Electronucleonics Inc., Fairfield, NJ, USA.

Similar size separation of membrane and phospholipid vesicles can be effected in columns of Sephacryl S-1000 (103). Further application of the method to the size determination and separation of coated vesicles using high performance liquid chromatography has also been described (104).

3.4 Electrophoretic separations

High voltage free-flow electrophoresis can produce subcellular membranes and organelles of high purity (105). The latest models, for example Elphor-VaP.22 available from Bender and Hobein (Lindwurmstr. 71, D-8000 Munich 2, FRG) (American distributor Protein Technologies Inc., 1700E, 18th St., Tucson, AZ 85719, USA) have featured in the fractionation of subcellular components obtained by centrifugation methods. A mixture of components to be separated is pumped as a fine jet into a separation buffer moving through an electrical field; membranes carrying different electrical charges migrate different distances across the separation chamber and are thus resolved with 100% recovery (see Figure 4, Chapter 2). The technique is gentle and has a high preparative throughput.

The free-flow electrophoretic technique was first applied to the isolation of lysosomes from rat liver and has been used to prepare plasma membrane subfractions from lymphocytes (106), and leucocytes (107), lysosomes and mitochondria from rat liver (108), lysosomes from cultured cells (109) and to separate inside-out and right-side-out erythrocyte vesicles (110). The technique is being applied to resolve the component parts of the Golgi apparatus (111) and the endocytic compartment (78) of rat liver that are difficult to separate by other methods.

An example of the application of the technique to separate surface and intracellular membranes of platelets is shown in *Figure 6*. Platelets, dispersed by sonication, were fractionated in a linear $(1-3.5 \text{ M})$ sorbitol gradient and the mixed membrane populations then resolved by free-flow electrophoresis (112). The advantages of pre-treatment of cells with neuraminidase are evident from the improved separation obtained, allowing the enzymic characterization of the purified membranes to be carried out.

The free-flow electrophoresis technique is highly specialized and cannot be described adequately here. The following practical hints are worth noting.

(i)   Careful pre-treatment of subcellular fractions applied to the apparatus (e.g. removal of cations) help to ensure reproducible separations and it is essential to prevent aggregation.

(ii)   As indicated in *Figure 6* pre-treatment of fractions with EDTA enables further subfractionation of the platelet surface membranes.

(iii)   Pre-treatment of fractions with low amounts of trypsin (0.5 μg/ml) can also aid in, for example, separating endosomes and lysosomes.

(iv)   Application of higher potential differences across the chamber while maintaining adequate temperature control and electrical stability, possible with the latest models, results in increased resolution.

The high cost of the free-flow electrophoresis apparatus has led to the development of alternative electrophoretic methods. For example, electrophoresis in agarose gels (concentration $0.15-0.3\%$), although used mainly as an analytical tool, has found preparative applications, especially in separating small coated vesicles from smooth-surfaced vesicles (113). A major limitation of this preparation method was the difficulty in eluting particles from the gel, but a device for continuous elution from a downstream collection well permits a more complete recovery (114,115) (*Figure 7*).

17

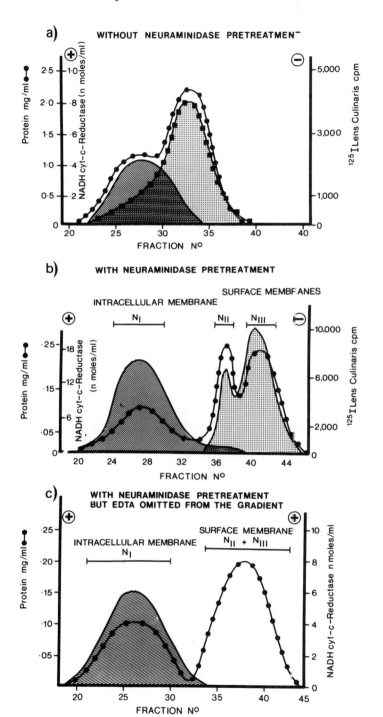

Figure 6. Separation of platelet surface and intracellular membranes by free-flow electrophoresis. The figure illustrates the increased resolution obtained with neuraminidase treatment or when EDTA is included in the sorbitol gradient used to prepare the membranes. For further details, see text. Reproduced from (112).

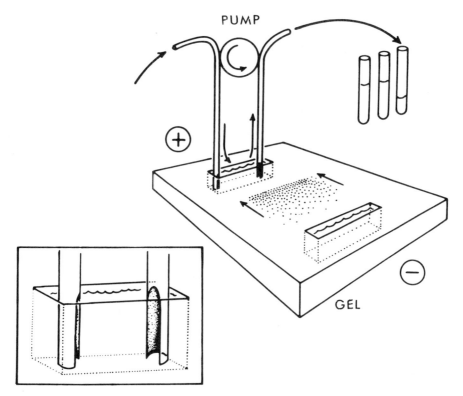

Figure 7. Agarose gel electrophoresis apparatus allowing a more quantitative recovery of small organelles and particles. The apparatus is a modified standard horizontal electrophoresis apparatus. A pump arrangement allows recovery of components as they collect in an elution well. Insert shows details of elution well; the inner side of the tubing is cut away to help direct the flow of buffer across the well, perpendicular to the electrophoretic movement of particles. Coated vesicles (see text) are resolved and collected by this method. The buffer used was 0.02 M 4-morpholinoethane sulphonic acid (MES), 0.03 M phosphate, 0.25 M sucrose adjusted to a pH range of 6.5−7.8. Agarose concentrations are varied from 0.15 to 0.2%. Applied samples should in general contain small-diameter vesicles suspended in a small volume of MES buffer containing 6% sucrose and 6% Ficoll-70 (Sigma). The gel is run at 1.0 v/cm (1 cm deep agarose) allowing a flow-rate of 12 ml/h. See references 114 and 115 for further details..

3.5 Immunoaffinity methods

These methods contrast with the major methods described above, for they rely not on physical differences but on biological differences in the particles to be separated. Immunoaffinity methods have a number of potential advantages, as they can be applied to small amounts of material, yielding fractions rapidly and in high yield. The orientation of the vesicles is of prime importance, and the selection of plasma membranes of the same orientation as in the cell is possible. Full practical details and the current status of this methodology are reported in Chapter 3.

4. IDENTIFICATION AND ASSESSMENT OF THE PURITY OF SUBCELLULAR FRACTIONS

4.1 Definition of subcellular markers

The practice of cell fractionation is based on the premise that each organelle and membrane network has constituents or a combination of features that confer uniqueness

(116). Thus, enzyme markers were predicted by de Duve to be located predominantly on specific organelles and were uniformly distributed in the membranes with which they were associated. It is a requirement, especially in analytical studies of tissue homogenates, that a range of markers are measured in all the fractions prepared, permitting the production of a balance sheet that allows the distribution and recovery of subcellular markers, their enrichment (relative to the homogenate), and the extent of contamination by other cell components to be computed (see also Chapter 2). Markers employed are predominantly enzymic, supported in particular instances by chemical markers and applied radiolabelled markers (primarily of the plasma membrane). Morphological monitoring of fraction purity is carried out extensively in qualitative terms. An assessment of the methods for quantifying these morphological features in a subcellular fraction (morphometry) is presented in Chapter 2.

The original premises (116) have been instrumental in allowing the preparation in most animal cells of the major organelles and membranes (see *Table 13*), but in retrospect they were too restrictive. Thus, it is now appreciated that most markers have a primary location in one cell component, but they may also have one or more secondary locations where the amounts are low but nevertheless significant. This is a consequence of membrane trafficking, a process whereby proteins and lipids are transferred from their sites of synthesis to their functional locations and onwards to sites of degradation. Also, the compartmentation of complex organelles, for example, Golgi apparatus, plasma membrane, into multiple membrane domains can lead to heterogeneity shown by analytical centrifugation studies of tissue homogenates (117).

4.2 Choice of markers and criteria for use

The primary subcellular locations of standard marker enzymes are shown in Appendix II. The suitability of these markers will vary between cell types; for example, glucose-6-phosphatase is confined to cells which carry out gluconeogenesis. The organization of the plasma membrane of tissue cells into functional domains with varying biochemical properties can result in a bimodal distribution in density gradients of marker enzymes owing to their differential location in the plane of the membrane and the segregation maintained by the tight junctions of epithelia. Thus, membranes derived from microvillar plasma membrane regions are of density $1.12 - 1.14$ g/ml in sucrose, whereas those derived from the junction-containing lateral regions have a density $1.16 - 1.18$ g/ml (4). Loss of outer mitochondrial membranes during subcellular fractionation can lead to the appearance of membrane vesicles of density $1.12 - 1.14$ g/ml in addition to the intact mitochondria located at density $1.19 - 1.21$ g/ml.

4.2.1 Enzymic markers

These are in general relatively stable, allowing measurement at convenience, but activities should, as a rule, be carried out as soon as possible using samples kept at 4°C and not subjected to freezing. Galactosyl transferase activity of Golgi membranes should be measured in freshly prepared subcellular fractions, but even glucose-6--phosphatase and NADPH-cytochrome *c* reductase can be labile, losing their activity in frozen liver membranes (118). The topographical position of the substrate-binding sites of enzymes in closed (sealed) vesicles is crucial for obtaining accurate estimates of the distribution of marker enzymes, and whether an enzyme activity is latent can be established by measuring activity in the presence and absence of 0.1 % deoxycholate or Triton X-100. Methods for studying the orientation of membrane vesicles and leakiness of vesicles are described in Chapter 4.

4.2.2 Chemical markers

These are usually supportive of enzymic markers, but many can be tedious to measure. Sialic acid and cholesterol are located predominantly at the plasma membrane, but are also found in endocytic membranes (78,119). Cardiolipin is a phospholipid confined to inner mitochondrial membranes and clathrin is a 180-kd protein located exclusively on coated membranes (coated vesicles). The content of secretory granules often functions as a convenient marker when subcellular functions are analysed by polyacrylamide gel electrophoresis, for example albumin in liver Golgi secretory vesicles.

5. PREPARATION OF VARIOUS ORGANELLES AND MEMBRANE SYSTEMS

The general guidelines for the isolation of subcellular membranes described above are now amplified in the context of specific organelles and membrane systems. Although fractionation schemes exist in which various membranes and organelles are derived from the same tissue homogenate, for example rat liver homogenate yielding plasma membranes, endoplasmic reticulum and Golgi membranes (120); rabbit enterocytes yielding brush borders, baso-lateral plasma membranes, smooth and rough endoplasmic reticulum and Golgi membranes (121), they involve some compromise of fraction purity. In general, specific media, homogenization and separation routines have been developed to obtain the desired fractions.

5.1 Nuclei and nuclear membranes

Nuclei are large fragile organelles consisting of the nuclear envelope, chromatin, nucleolus and various granules. A number of measures help minimize breakage and the consequent release of chromatin that attaches, during subcellular separations, to other cell constituents, especially the plasma membrane. Homogenizers with a wide clearance (0.12 – 0.16 mm) are used to disperse tissues, and 0.32 M sucrose, 1 mM $MgCl_2$ is a medium often employed. Glycerol-containing media (50 % w/w) often yield superior functionally intact nuclei but the application of Percoll and iodinated gradient materials has not found wide application owing to their ability to penetrate into the interiors of isolated nuclei.

The nuclear envelope consists of two 7.5 – 10 nm diameter membranes, with the outer membrane studded with ribosomes. It is a structure that is extremely susceptible to proteolysis during its isolation. A high-ionic strength extraction procedure (122) described in *Table 9* provides nuclear envelopes with well preserved morphology that show low contamination with histones and DNA. Factors affecting the reproducibility of the method, and measures used to control the extent of proteolytic degradation during isolation have been discussed (123).

Nuclear envelopes lack distinctive markers, but nucleoside triphosphatases, some of which are stimulated by poly(A)$^+$ mRNA (124) as well as specific polypeptides (122) are present. The nuclear pore complex, used as a morphological marker, remains to be characterized biochemically. The apparent continuity of the outer nuclear membrane

Table 9. Preparation of nuclear envelopes from rat liver[a].

1.	Wash rat livers in 0.25 M sucrose, 50 mM Tris-HCl, pH 7.4 (4%), 5 mM MgSO$_4$ (buffer 1), mince, and homogenize in a Dounce homogenizer (loose-fitting pestle). Filter through muslin cloth and centrifuge at 800 g for 10 min.
2.	Wash the pellet twice by resuspension and centrifugation as described in step 1.
3.	Resuspend the pellet in 2.1 M sucrose, 50 mM Tris-HCl, pH 7.4, 5 mM MgSO$_4$ (buffer 2) and layer over cushions of the same buffered sucrose. Centrifuge at 70 000 g for 60 min (Beckman SW27 or equivalent) and collect the pellet. Repeat this step once to obtain nuclei using a Dounce homogenizer to resuspend the pellet.
4.	Resuspend nuclei at a concentration of 5 × 10^8/ml in buffer 1 supplemented with 1 mM PMSF, and incubate with DNase I (250 μg/ml) and boiled RNase A for 60 min. Centrifuge at 800 g for 10 min and collect the pellet.
5.	Resuspend the nuclei at the same concentration in 0.2 mM MgSO$_4$, 10 mM Tris-HCl, pH 7.4, 1 mM PMSF (buffer 3) and immediately add 2 M NaCl dissolved in buffer 3 to a concentration of 1.6 M NaCl. Add β-mercaptoethanol with gentle agitation to give a final concentration of 1%. Leave for 15 min and centrifuge at 1600 g for 30 min.
6.	Extract the pellet again with buffer 3 and NaCl as described in step 5 but omit the use of reducing agent. Centrifuge at 1600 g for 30 min to yield a nuclear envelope fraction consisting mainly of uniformly empty, mainly spherical structures.

[a]Taken from ref. 122.

with the endoplasmic reticulum results in a commonality of enzyme constituents, for example glucose-6-phosphatase.

5.2 Mitochondria

Mitochondria are prepared from a variety of sources by standard methods (125). Subfractionation to yield inner and outer membranes and matrix components is achieved by freezing and thawing and/or sonication (126). Sonication of a standard rat liver mitochondrial pellet dispersed in 1.2 M sucrose−2 mM ATP (10 mg membrane protein/ml) disintegrates mitochondria allowing, using a discontinuous sucrose gradient, outer mitochondrial membranes to be collected at a 0.45−1.12 M sucrose interface, inner membrane vesicles and matrix components to be collected as a pellet in 1.2 M sucrose, and an intermediate vesicular fraction to be recovered at the intervening 1.12−1.20 M sucrose interface (126). Electron-transport enzymes serve as markers for inner mitochondrial membranes (Appendix II). The outer mitochondrial membrane, that fragments to form vesicles of similar density and morphology to plasma membranes, contains a monoamine oxidase. A 'micro' method for the isolation of outer mitochondrial membranes from rat liver has also been described (127).

5.3 Lysosomes

These are discrete organelles involved in proteolysis and contain up to 50 hydrolytic enzymes active at low pH, including various phosphatases, nucleases, glycosidases, proteases, peptidases, sulphatases and lipases (116).

All methods for the preparation of lysosomes aim to minimize cross-contamination by mitochondria and peroxisomes. Examples of the separation of lysosomes from other organelles have already been described, for example in Percoll gradients (Section 3.1.1) by density pertubation (Section 3.1.2) and by free-flow electrophoresis (Section 3.4). Other methods developed include the use of iron−sorbitol−citric acid complex (AB

Astra, Sweden) and colloidal gold (25 nm mean diameter) injected intramuscularly that accumulate inside lysosomes of liver increasing their buoyant density (128). A more direct method for increasing the density of lysosomes involves addition of 10 mM Na^+ phosphate to the homogenizing medium (129).

However, iso-osmotic gradient materials, especially Metrizamide and Nycodenz have found wider application in the purification of lysosomes and peroxisomes (see below). Briefly, a 'light mitochondrial' fraction is prepared as follows.

(i)   Homogenize rat livers using a Potter – Elvehjem homogenizer (1000 r.p.m., ~5 strokes) at a ratio of 1 g of tissue to 3 volumes of 0.25 M sucrose, and centrifuge the homogenate at 1000 g for 10 min using a Sorvall SS-34 rotor or equivalent.

(ii) Collect the supernatant and resuspend the pellet in 3 volumes of 0.25 M sucrose. Centrifuge at 1000 g for 10 min and combine this supernatant with the previous supernatant.

(iii) Centrifuge the combined supernatants at 3000 g for 10 min and then centrifuge this supernatant at 10 000 g for 20 min to yield a 'light-mitochondrial' fraction. This is a crude fraction enriched in lysosomes and mitochondria, but note that it also contains Golgi membranes, peroxisomes and fragments of endoplasmic reticulum.

(iv) Resuspend this pellet (Dounce homogenizer) in 45% (1.25 g/ml) iso-osmotic Metrizamide to yield a density of 1.18 g/ml (see Section 3.1 for details of density calculations) and a volume of approximately 10 ml.

(v)   Add overlays (6 ml each) of 30, 26, 24 and 19% (1.16, 1.145, 1.135 and 1.105 g/ml) Metrizamide and centrifuge for 2 h at 97 000 g (Beckman S.W.28 or equivalent). Lysosomes are collected at the 1.105 – 1.135 g/ml interface (75).

A yield of 0.5 – 1.0 mg protein/g liver wet weight is regarded as satisfactory. As with the preparation of kidney lysosomes on Percoll gradients (Section 3.1), lysosomal marker enzymes are increased 60- to 80-fold in specific activity relative to the homogenate. Lysosomal enzymes show structure-linked latency and full enzymic activity is obtained only after rupture of the membranes, usually by suspension in 0.1% Triton X-100.

Lysosomes have also been prepared from atheromatous aorta of rabbits (130) and from cultured cells (109). Autophagic vacuoles — large vacuoles enclosing mitochondria, secretory granules, etc. are prepared from animals pre-treated with leupeptin, chloroquine or vinblastin. Methods for isolation follow closely those used for lysosomes, involving fractionation of the 'light' mitochondrial fraction in Metrizamide (131) or Percoll (132) gradients. The fractions possess characteristic morphology and are enriched in lysosomal enzymic markers but to a lesser extent than purified lysosomal fractions.

5.4 Peroxisomes

These have been defined as organelles containing catalase together with one or more H_2O_2-producing enzymes (133). They have been studied mostly in rodent liver, where their numbers can be increased up to 9-fold by pre-feeding animals with hypolipidaemic drugs such as clofibrate. Methods for their isolation from kidney and small intestine (134), lung (135) and heart (136) have been reported.

The general strategy for the preparation of liver peroxisomes involves the subfractionation of a 'light' mitochondrial fraction (see Section 5.3 for method of preparation) in Metrizamide or Nycodenz gradients in which the iso-osmotic conditions and the lower hydrostatic pressure applied are extremely advantageous for the isolation of fragile organelles. Peroxisomes are adequately separated from lysosomes and mitochondria when the 'light' mitochondrial fraction is centrifuged in 20−50% (w/v) Metrizamide or self-generating 35% (w/v) Nycodenz gradients; peroxisomes sediment to a buoyant density of 1.220−1.245 g/ml; lysosomes are collected in this procedure at 1.120 g/ml. The major contaminant is endoplasmic reticular vesicles (marker: glucose-6-phosphatase). The increase in specific activity, relative to the tissue homogenate, was 36- to 40-fold for catalase, α-hydroxyacid oxidase and β-oxidation enzymes (137). The polypeptide and phospholipid compositions of rat liver peroxisomes have been investigated (138). Percoll gradients have also featured in the isolation of peroxisomes (64), but contamination by mitochondria is greater The stability of peroxisomes during isolation was enhanced by treatment of the post-nuclear supernatants with charcoal-filtered 1 mM glutaraldehyde in 0.5 mM cacodylate, pH 7.4 (139).

5.5 Golgi apparatus

The Golgi apparatus, first identified as a morphological entity consisting of stacks of cisternae, has been prepared mainly from liver tissue, although it is prominent in all non-dividing cells of secretory tissues. Its involvement in terminal glycosylation and sulphation of proteins and lipids, the processing of sugar moieties of glycoproteins and the generation of phosphomannosyl recognition markers attest to the morphological complexity and have led to attempts to subfractionate the Golgi apparatus.

Various methods exist for the rapid preparation of Golgi membranes from rat liver (140,141) where they are recovered from post-nuclear supernatants at low density on sucrose or Percoll (55) gradients (*Table 10*).

Attempts to subfractionate the Golgi apparatus, especially along its functional polarity to determine the sequential distribution of enzymes, are in their infancy. One extensively used method carried out without ethanol pre-feeding of animals (141) separates a Golgi 'light' fraction enriched in secretory droplets and a 'heavy' fraction containing cisternal elements that includes components derived from the *cis*-face of the organelle. Similarly, two fractions are separated on Percoll gradients (57) that contain similar amounts of galactosyl and sialyl transferases, but showing unequal distribution of 1, 2, mannosidase, mannosidase 11 and β-N-acetylglucosaminylphosphotransferase activities. Immunolocalization studies show galactosyl and sialyl transferases to be located mainly at the *trans* face of the Golgi. No markers are yet identified for the *cis*-face. The enzymes are located at the luminal side of the membrane. The free-flow electrophoresis technique (Section 3.4) shows promise in separating rat liver Golgi apparatus fractions unstacked after treatment with amylases followed by mild physical disruption before the sample is injected into the apparatus (111).

The Golgi apparatus fractions obtained by these procedures from liver are heterogeneous. They are contaminated by components destined for secretion and occluded inside the cisternae and vesicles. A method for the release of soluble secretory products involving washing the fractions suspended in 0.1 M Na_2CO_3, pH 11.0 by centrifugation and is described in *Table 10* (142,143). The 'heavy' Golgi fractions are

Table 10. Rapid preparation of rat liver Golgi membranes[a] and separation of membranes and soluble cisternal contents.

1.	Homogenize each rat liver in 26 ml of 0.5 M sucrose, 1% dextran (Sigma, average mol. wt 252 000) 38 mM Tris-maleate, pH 5.4 for 60 sec at 8000 r.p.m. in a Polytron homogenizer (see *Table 2*, Potter−Elvehjem/Dounce homogenizers may substitute).
2.	Centrifuge the homogenate at 6000 *g* for 15 min (Sorvall SS-34 rotor or equivalent), collect the upper yellow−brown portion of the pellet and resuspend it in 5−6 ml of the collected supernatant.
3.	Layer 1 ml portions of the above on 7.5 ml of 1.2 M sucrose (dissolved in 38 mM maleate−Tris, pH 6.4, in Beckman SW28 13 ml (thin bucket) tubes; overlay with distilled water. Centrifuge at 97 000 *g* for 30 min and collect a crude Golgi fraction at the 1.2 M sucrose/applied fraction interface. Collect membranes using a syringe and pellet by centrifuging at 6000 *g* for 20 min.
4.	To separate Golgi membranes from content, resuspend the fractions in approximately 5 ml of 100 mM Na_2CO_3, pH 11.0 and leave at 4°C for 30 min. Protein concentration should be 0.02−1 mg/ml. A small tight-fitting Dounce homogenizer is suitable for this extraction. Centrifuge at 100 000 *g* for 1 h (Beckman 50 Ti rotor or equivalent) and resuspend the pellet in 10 mM Tris-HCl, pH 7.4.
5.	The method for removal of cisternal components is also applicable to microsomal fractions, peroxisomes, etc. Ribosomes are released from rough endoplasmic reticulum by this method[b]. Sealed vesicles are converted into flat sheets by this procedure, and integral membrane proteins are retained. Inclusion of 1 mM PMSF (*Table 3*) in alkaline media helps suppress proteolysis that may occur under these conditions. Note that many enzyme activities may be destroyed by the alkaline extraction procedure. An alternative procedure for separating integral membrane from soluble and peripheral proteins involves two-phase partitioning in Triton X-114 (Sigma) — see Chapter 6 for further details.

[a]Taken from ref. 140.
[b]See ref. 142.

contaminated by plasma membranes, whereas the 'light' fractions are grossly contaminated by endosome membranes (56). A procedure that provides 'endosome'-depleted Golgi fractions is described in Section 5.7 (*Table 12, Figure 8*).

5.6 Coated vesicles

The origin of coated vesicles recovered from tissue homogenates is unclear, but they probably arise mainly from coated membrane regions, in a manner akin to the production of microsomes from endoplasmic reticulum. They are usually 70−200 nm diameter vesicles with a characteristic 'basket-like' coating; the coats have a tendency to dislodge from underlying membrane (see below). Their size can vary according to their origin from the plasma membrane (diameter in excess of 100 nm) or the Golgi apparatus (diameter <75 nm) (144). Biochemically they are characterized as cholesterol-enriched (145) and in containing clathrin, a protein of molecular weight 180 000 (146).

Methods for isolation take advantage of their high density when the coats are attached, and these can be elaborate involving a sequence lasting up to 36 h of rate-zonal, isopycnic−rate-zonal centrifugation steps in sucrose gradients (42). A method for the preparation of human placental coated vesicles is shown in *Table 11*. A similar procedure applies to the brain (146). The coats show a tendency to disassemble during isolation to produce clathrin baskets, a procedure that is accelerated during prolonged centrifugation in 20−60% w/v sucrose gradients. The effects of various agents on coated vesicle disassembly and reassembly have been investigated (148). An isolation procedure involving centrifugation in H_2O/D_2O−8% sucrose gradients that minimize disassembly of coats has been described for bovine brain (42) and rat liver (149). Procedures for the isolation of coated vesicles require large amounts of tissues; with liver,

Table 11. Preparation of human placental coated vesicles[a].

1.	Collect full term placentae after delivery and dissect 250 *g* of villus tissue free of membranes, cord and blood vessels.
2.	Homogenize the tissue in an equal volume of buffer 1 [1 mM EGTA, 5 mM MgCl$_2$, NaN$_3$ (0.2 mg/ml), 0.1 M 4, MES adjusted to pH 7.4] in a tissue blender in three 10-sec bursts at full speed.
3.	Centrifuge the homogenate at 15 000 *g* for 40 min. Collect the supernatant and centrifuge this at 85 000 *g* for 1 h.
4.	Resuspend the pellet (Dounce homogenizer) in 40 ml of buffer 1 and layer it onto discontinuous sucrose gradients constructed of 60% (w/v), 4.5 ml, 50%, 4.5 ml, 40%, 9 ml, 10%, 9 ml, 5%, 4.5 ml; dissolve sucrose in buffer 1. Centrifuge in a Beckman SW28 rotor (or equivalent) at 100 000 *g* for 70 min.
5.	Collect material at the 10−40% sucrose interface together with material in 10% and 40% sucrose. Dilute 3-fold with buffer 1 and centrifuge at 100 000 *g* for 1 h.
6.	Resuspend the pellet in 9 ml of buffer 1, estimate the protein content, and add wheat germ agglutinin (1 mg/10 mg membrane protein). Leave for 2 h at room temperature or overnight at 4°C, and sediment agglutinated material by centrifuging at 20 000 *g* for 15 min. collect the supernatant and centrifuge at 165 000 *g* for 1 h.
7.	Resuspend the final pellet in 1.5 ml of buffer 1. A yield of approximately 3 mg of protein is obtained.

[a]Taken from ref. 147.

Table 12. Preparation of hepatic endosomes and endosome-depleted Golgi membranes (*Figure 8*).

1.	Remove two livers from rats (~15−20 g wet weight), cut into small pieces with scissors and homogenize in a Dounce homogenizer (volume 35 ml clearance 0.12 mm; Blaessig, Rochester, NY, USA) with 10 strokes in 0.2 M sucrose (3 vol/g wet weight).
2.	Filter the homogenate through nylon gauze 50−100 mesh (Swiss Silk Bolting Cloth Manufacturing Co. Ltd, Zurich, Switzerland) and re-homogenize using six strokes with the tight-fitting plunger (clearance 0.07 mm).
3.	Centrifuge the homogenate in polycarbonate tubes in a 8 × 50 ml angle rotor (Sorvall SS-34 or equivalent) at 1000 *g* for 10 min; wash this pellet twice by resuspension (1.5 vol/g liver wet weight) and centrifugation in 0.25 M sucrose, and combine the supernatants (the volume is ~100 ml).
4.	Centrifuge the combined supernatants at 33 000 *g* for 8 min in a Beckman Type 30 rotor. If the centrifuge is programmable, set to 2.61 × 10^9 rad^2/sec. Collect supernatants taking care not to collect the loose material resting on the multilaminate pellet.
5.	Layer the supernatant on six sucrose gradients made in 38 ml tubes (Beckman SW28 rotor). Construct gradients as follows: 1 ml of 70% w/v sucrose, 5 ml of 43% sucrose, a continuous gradient made by mixing 7.5 ml of 40% and 10% sucrose.
6.	Centrifuge at 100 000 *g* for 4 h. Unload gradients from the bottom of the tubes by using a hollow needle attached to a peristaltic pump, collecting 40 × 0.7 ml fractions.
7.	Pool fractions corresponding to the ranges 1.095−1.117 g/ml and 1.117−1.135 g/ml to yield endosome fractions. When livers have endocytosed radio-iodinated ligands (e.g. asialoglycoproteins, polypeptide hormones) injected 2−10 min previously into the portal vein, these are recovered in undegraded form in these density ranges. For relation between sucrose concentration and density, see *Appendix I*.
8.	Subfractionate the endosomes as described in Section 3.1.1 using Nycodenz gradients. Golgi fractions (endosome depleted) are prepared by collecting the loose region of the pellet (step 4), and resuspending, using a loose-fitting Dounce homogenizer, in 50 ml of 39% (w/v) sucrose. Overlay with equal volumes of 29.5, 20.5 and 8% sucrose using six 38 ml centrifuge tubes (Beckman SW28). Centrifuge at 100 000 *g* for 3 h and collect Golgi heavy fraction at the 29.5−39% interface and Golgi intermediate fraction at the 20.5−29.5% interface.

for example, a yield of 1−2 mg protein/40 g liver is obtained (150). Final purification is obtained as described in *Table 11* or by a permeation chromatography step using a Sephacryl S-1000 column that separates smooth-surfaced membrane vesicles from

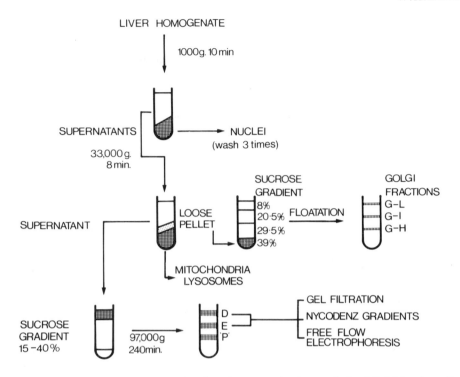

Figure 8. Scheme for the preparation of hepatic endosomes and endosome-depleted Golgi fractions. For explanation see *Table 12*.

coated vesicles (149) (Section 3.3) or by agarose gel electrophoresis (Section 3.4) (113,115) or by immunoabsorption (Chapter 3).

5.7 Endosomes (endocytic vesicles)

Endosomes (receptosomes, pinosomes) are membrane vesicles originating from the endocytic networks. These consist of a vesico-tubular membrane system extending into the cell interior from the plasma membrane where receptor − ligand complexes are transferred, and sorted after endocytic uptake (8).

A density perturbation method for the preparation of a single population of liver endosomes is described in Section 3.1.1. A method that avoids using a peroxidase-linked ligand to perturb the density of endosomes and that yields up to three hepatic endosome subfractions is described in *Table 12* and *Figure 8*. The details of the use of Nycodenz gradients to separate ligand containing receptor-negative endosomes from receptor-containing vesicles are described in Section 3.1.1.

The biochemical characterization of endosome membranes is incomplete. The various radio-iodinated ligands perfused into liver tissue and used to trace the subcellular location of endosomes (asialoglycoproteins, polypeptide hormones, etc.) indicate that these ligands remain undegraded, suggesting an absence of highly active proteolytic enzymes in these vesicles of low density. Endosomes are deficient in those enzymic markers characteristic of intracellular organelles and membranes (77,78,84,150). They are difficult to separate completely from Golgi components owing to the similarities in density. A monensin-activated Mg^{2+}-ATPase, that is equated with the proton-pump that ac-

Table 13. Source list of methods for preparation of subcellular organelles and membranes.

Source (organ, species, etc.)		References
A. Plasma membranes		
Adipose tissue	fat pads, brown fat	71,160
Adrenal medulla	bovine	161
Brain	rat axolemma	162
	rat cortex	163
	bovine caudate nucleus	164
Colonocytes	rat	172
Corpus luteum	bovine, rat	165,166
Eggs	sea urchin	167
Embryos	mouse mesenchymal	168
Epidermis	pig Malpighian cells	169
Epithelial cells	renal LLC-PK, pig	170
Erythrocytes	human	171
Exorbital gland	rat, baso-lat, brush-border	46
Fibroblasts	hamster lung, mouse	173,174
Heart	rat, canine, myocytes	175,176
HeLa		177
Intestine	rat brush-border	178,180
	rat baso-lateral	179,121
Kidney	mouse brush-border	181
	rat baso-lateral	182,197
Lens	various species	183
Lettre cells		184
Liver	sinusoidal, rat	185
	canalicular, baso-lateral	32, 33, 94,185,186
Lung	rat	187,188
Lymphocytes	calf thymus, mouse	189,190
Mast cells	rat	31
Neuroblastoma	murine	191
Neutrophils	human, canine, porcine	192–194
Parotid	rat	195,196
Placenta	human	198,199
	syncytiotrophoblast	200
Platelets		201
Skeletal muscle	rabbit	202
Smooth muscle	various	203
Stomach	pig	204
Sperm	mouse	205
Tumour cells	HSV, hamster, ascites	206,207
B. Junctional complexes		
Desmosomes	cow nose	157,158,208
Gap junctions	rat liver, heart, uterus	29,159,209
Tight junctions	mouse liver	210
Post-synaptic densities	dog, mice, various	96,211,212
Skeletal triad structure	rabbit	213
C. Lysosomes and autophagic vacuoles		
Fibroblasts	human	109
Thyroid	pig	214
Liver	rat	75,108
Liver	rat autophagic vacuoles	131,132,215
Kidney	rat	51
Lung	rabbit	226

D. Peroxisomes

Liver	mouse, various, carp	44,136,139
Heart	rat	136

E. Mitochondria

Liver	rat	127,216
Heart	cow	217
Myometrium	human	218

F. Nuclear membranes

Liver	rat	122,123

G. Golgi apparatus

Liver	rat	55,111,140,141
Brain	rat	227
Leukaemic cells		219

H. Vacuoles, granules, bodies, etc.

Pancreas islet cells	Secretory zymogen granules, rat	220
Parotid	rat	221
Adrenal medulla	Secretory granules	47,222
Pituitary	secretory vesicles	76
Vas deferens	nor-adrenaline granules	224
Growth cones	rat forebrain	225
Lamellar bodies	rabbit lung	226

Further details and sources are to be found in *Table 4* and in Section 5.

counts for acidification in the endosomes (77), when taken with the presence of undegraded ligands, can serve as linked markers for hepatic endosome fractions. Analysis of hepatic endosome fractions indicates that they resemble plasma membranes more closely than other subcellular membranes, for they have a high cholesterol and sphingomyelin content; endosome membranes contain fewer proteins than plasma membranes (119).

Although mainly characterized in liver, methods for their isolation from cultured cells (151), macrophages (49,50) (see *Table 5*) and other cells have been described.

5.8 Plasma membranes

The plasma membrane is probably the most complex and differentiated of the cell's membrane systems for it consists of a continuum of functionally distinctive zones. This is especially the case with cells in tissues and organs where major apical and baso-lateral zones separated by tight junctions and possessing differing biochemical properties exist (4,5). Further micro-domains such as microvilli and coated membranes (Section 5.6) as well as the various categories of intercellular junctions are also present. A domain-structure also exists in isolated cultured cells, and a separation and biochemical analysis has been effected in HeLa cells (7), lymphocytes (152,153) and Chinese hamster ovary cells (154).

Preparation methods for plasma membranes are extremely varied and only some general guidelines can be provided here. *Table 13* lists the source references for the isolation of plasma membranes and other fractions from a variety of tissues, organs and cell lines.

(i) Conditions for the homogenization of tissues and cells for the preparation of plasma membranes require careful choice and monitoring. The plasma membrane can be disassembled during homogenization to form a permutation of polydisperse particles with varying physical and biochemical properties, creating problems in recovery when differential centrifugation steps are used in their isolation. The general strategy to follow with tissues and organs is to homogenize these using a procedure that favours the production of large membrane fragments that sediment at low speed (with nuclei). Plasma membrane vesicles of small diameter present in microsomal fractions are more difficult to purify.

(ii)   Rapid methods for isolation of plasma membranes have been developed that avoid the necessity for subcellular fractionation. These involve release into media (Section 2.1.1) or attachment to solid phases of portions of the cell surface (Section 2.1.2). These methods are restricted to isolated cells. They suffer from the disadvantage of a probable lack of compositional representatives of released vesicles and the difficulties associated with the further release of attached plasma membrane vesicles attached to solid supports, for example polylysine-coated beads.

(iii)   The use of iso-osmotic gradients of Percoll, Nycodenz, etc. in the rapid isolation of plasma membranes, especially from post-nuclear fractions is a major advance (see Section 3.1.1). Although the speed of separation provided by these new gradient materials is clearly advantageous, it should be borne in mind that fractions collected willy-nilly from the light region (e.g. Percoll of density $1.02 - 1.04$ g/ml) of such gradients are likely to be contaminated by Golgi and endosome membranes.

(iv)   Vesicles prepared rapidly and under isomotic conditions (e.g. in Percoll gradients) are useful for further studies on the transport of ions and molecules across the plasma membrane. The fact that the plasma membrane of nucleated cells, owing to its highly asymmetric organisation, forms vesicles that are predominantly of a right-side-out configuration has greatly facilitated such studies.

(v)   In the isolation of plasma membranes from tissues, the different functional and enzymic properties of the various regions, that is baso-lateral and apical of epithelial cells (4,5), makes it increasingly desirable to identify the domain origin of recovered fractions (see *Table 13*).

(vi)   In addition to the enzymic markers described in Appendix II, it may also be advantageous to monitor recovery of plasma membranes from cells covalently labelled with radioactive tracers. These techniques involve reagents that label either cell surface proteins (155,156) or carbohydrates (157). To control and restrict the permeation of free and combined reagents into cells, cell surface labelling is carried out at 4°C to minimise the extent of endocytic uptake of plasma membrane constituents. Covalent labelling of membrane proteins and glycoproteins is described in Chapter 6.

5.8.1 *Plasma membrane junctions*

The isolation of intercellular junctional complexes from various tissues usually involves initial preparation of a plasma membrane fraction containing membrane fragments from the lateral domain, followed by extraction with detergents under conditions that allow selective solubilization of non-junctional membranes leaving the junctions as insoluble pellets. Alternative methods that avoid the use of detergents also feature. For example, desmosomes are isolated from cow nose epidermis by extraction with a citrate buffer pH 2.6 (158) and gap junctions are isolated from liver plasma membranes extracted with 20 mM NaOH (159).

The extraction of homogenates and subcellular fractions with salt solutions of high concentration (as with muscle tissue or in nuclear envelope isolation), with detergents or with aqueous solutions of alkaline pH (9 – 11), accentuates proteolysis of membrane constituents despite the addition of inhibitors of proteolysis (39,40). This is a consequence of subcellular fractionation that is becoming more evident as isolated membranes are used increasingly as a source of specific membrane proteins for amino acid sequence determination and topographical analysis (Chapter 6).

6. CONCLUSION AND PROSPECTS

A wide range of techniques are described for the isolation and characterization of the major membrane and organellar systems of animal cells. These techniques fall into two general categories. First, there are the relatively simple and rapid methods that involve release of membrane fragments that are free or attached to charged, ligand or antibody-covered beads and the utilization of new gradient materials for the fast (30 min – 1 h) fractionation of homogenates under iso-osmotic conditions in one or two aqueous phase systems. Although speed of preparation can compromise fraction purity, these methods undoubtedly provide a useful source of material, for example for transport and receptor studies, etc., especially when the biological processes and the molecules under study are unstable. Second, there are methods that have been developed primarily to allow further dissection (subfractionation) of primary fractions, for example free-flow electrophoresis, density perturbation, etc. Although these involve prolonged handling and can have more damaging effects on membrane constituents they have proved invaluable in the isolation of endosomes, coated vesicles and Golgi fractions.

Finally, although a diverse range of biochemical markers can be applied to assess the quality of the isolated fractions, it is anticipated that antibodies produced against major constituents in the isolated fraction (organelle-specific antibodies) feature in future as immunodiagnostic kits that may allow the investigator to assess and control the quality of the material prepared. The successful application of magnetized beads or bio-engineered polymers coated with organelle membrane-specific antibodies to isolate highly purified membrane domains is a further development to be anticipated (Chapter 3). Indeed, advances in hybrid technologies — the combination of immunological methods that exploit functional as opposed to physical properties with flow cytometrical methods (fluorescent-activated particle sorters) offer boundless opportunities for the separation of discrete membrane domains of parent organelles.

7. REFERENCES

1. Weibel,E.R., Staubl,W., Gnagl,H.R. and Hess,F.A. (1969) *J. Cell Biol.*, **42**, 68.
2. Farquhar,M. and Palade,G.E. (1981) *J. Cell Biol.*, **91**, 77s.
3. Griffiths,G. and Simons,K. (1986) *Science*, **234**, 438.
4. Evans,W.H. (1980) *Biochim. Biophys. Acta*, **604**, 27.
5. Simons,K. and Fuller,S.D. (1985) *Annu. Rev. Cell Biol.*, **1**, 243.
6. Szamel,M., Goppelt,M. and Resch,K. (1985) *Biochim. Biophys. Acta*, **821**, 479.
7. Mason,P.W. and Jacobson,B.S. (1985) *Biochim. Biophys. Acta*, **821**, 264.
8. Wileman,T., Harding,C. and Stahl,P. (1985) *Biochem. J*, **232**, 1.
9. Evans,W.H. (1978) *Preparation and Characterisation of Mammalian Plasma Membranes*. Elsevier, Amsterdam.

10. Coleman,P.S. and Lavietes,B.B. (1981) *CRC Crit. Rev. Biochem.*, **11** 341.
11. Van Blitterswijk,W.J., Emmelot,P., Hilkmann,H.A.M., Oomen-Meulemans,E.P.M. and Inbar,M. (1977) *Biochim. Biophys. Acta*, **467**, 309.
12. Koch,G.L.E. and Smith,M.J. (1978) *Nature*, **273**, 274.
13. Pan,B.T. and Johnstone,R.M .(1983) *Cell*, **33**, 967.
14. Cohen,S., Ushiro,H., Stoscheck,C. and Chinkers,M. (1982) *J. Biol. Chem.*, **257**, 1523.
15. Scott,R.E., Perkins,R.G., Zschunke,M.A., Hoerl,B.J. and Maercklein,P.B. (1979) *J. Cell Sci.*, **35**, 229.
16. Stephen,S.G. and Cone,R.E. (1979) *Proc. Natl. Acad. Sci. USA*, **76**, 5582.
17. Leonards,K.S. and Ohki,S. (1983) *Biochim. Biophys. Acta*, **728**, 383.
18. Kobayashi,T., Okamoto,H., Yamada,J.I., Setaka,M. and Kwan,T. (1984) *Biochim. Biophys. Acta*, **778**, 210.
19. Grivell,A.R., Berry,M.N., Henly,D.C., Phillips,J.W., Wallace,P.G., Gannon,P.J., Henderson,D.W., Mukhergee,T.M. and Swift,J.G. (1986) *Exp. Cell Res.*, **165**, 11.
20. Simpson,A.J.G., Schreyer,M.D., Cesari,I.M., Evans,W.H. and Smithers,S.R. (1981) *Parasitology*, **83**, 163.
21. Kalish,D.I., Cohen,C.M., Jacobson,B.S. and Branton,D. (1978) *Biochim. Biophys. Acta*, **506**, 97.
22. Ohnishi,T., Suzuki,T., Suzuki,Y. and Ozawa,K. (1982) *Biochim. Biophys. Acta*, **684**, 67.
23. Gotlib,L.J. (1982) *Biochim. Biophys. Acta*, **685**, 21.
24. Jacobson,B.S. (1980) *Biochim. Biophys. Acta*, **600**, 769.
25. Gotlib,L.J. and Searls,D.B. (1980) *Biochim. Biophys. Acta*, **602**, 207.
26. Stewart,D.I.H. and Crawford,N. (1981) *FEBS Lett.*, **126**, 175.
27. Chaney,L.K. and Jacobson,B.S. (1983) *J. Biol. Chem.*, **258**, 10062.
28. Lindsay,J.G., Reid,G.P. and D'souza,M.P. (1981) *Biochim. Biophys. Acta*, **640**, 791.
29. Zervos,A.S., Hope,J. and Evans,W.H. (1985) *J. Cell Biol.*, **101**, 1363.
30. Culvenor,J.G., Mandel,T.E., Whitelow,A. and Ferber,E. (1982) *J. Cell. Biochem.*, **20**, 127.
31. Amende,L.M. and Donlon,M.A. (1985) *Biochim. Biophys. Acta*, **812**, 713.
32. Evans,W.H. (1970) *Biochem. J.*, **116**, 833.
33. Meier,P.J., Sztul,E.S., Reuben,A. and Boyer,J.L. (1984) *J. Cell Biol*, **98**, 991.
34. Brandt,N. and Basset,A. (1986) *Arch. Biochem. Biophys.*, **244**, 872.
35. Stieger,B. and Murer,H. (1983) *Eur. J. Biochem.*, **135**, 95.
36. Record,M., Bes,J.C., Chap,H. and Douste-Blazy,L. (1982) *Biochim. Biophys. Acta*, **688**, 57.
37. Barret A.J. (1980) In *Protein Degradation in Health and Disease. CIBA Foundation Symposium 75*, p. 1.
38. Aubry,H., Merrill,A.R. and Proulx,P. (1986) *Biochim. Biophys. Acta*, **856**, 610.
39. Yeaton,R.W., Lipari,M.T. and Fox,C.F. (1983) *J. Biol. Chem.*, **258**, 9254.
40. Manjunath,C.K., Goings,G.E. and Page,E. (1985) *J. Membr. Biol.*, **85**, 159.
41. Sheeler,P. (1981) *Centrifugation in Biology and Medicine*, John Wiley New York.
42. Nandi,P.K., Irace,G., Van Jaarsveld,P.P., Lippoldt,R.E. and Edelhoch,H. (1982) *Proc. Natl. Acad. Sci. USA*, **79**, 5881.
43. Johnstone,A.P. and Crumpton,M.J. (1980) *Biochem. J.*, **190**, 45.
44. Hajra,A.K. and Wu,D. (1985) *Anal. Biochem.*, **148**, 233.
45. Conteas,C.N., McDonough,A.A., Kozlowski,T.R., Hensley,C.B., Wood,R.L. and Mircheff,A.K. (1986) *Am. J. Physiol.*, **250**, C430.
46. Mircheff,A.K., Lu,C.C. and Conteas,C.N. (1983) *Am. J. Physiol.*, **245**, G661.
47. Gratzl,M., Krieger-Brauer,H. and Ekerdt,R. (1981) *Biochim. Biophys. Acta*, **649**, 355.
48. Loten,E.G. and Redshaw-Loten,J.C. (1986) *Anal. Biochem.*, **154**, 183.
49. Wileman,T., Boshans,R.L., Schlesinger,P. and Stahl,P. (1984) *Biochem. J.*, **220**, 665.
50. Diment,S. and Stahl,P. (1985) *J. Biol. Chem.*, **260**, 15311.
51. Harikumar,P. and Reeves,J.P. (1983) *J. Biol. Chem.*, **258**, 10403.
52. Carty,S.E., Johnson,R.G. and Scarpa,A. (1980) *Anal. Biochem.*, **106**, 438.
53. Renswoude,J.V., Bridges,K.V., Harford,J.B. and Klausner,R.D. (1982) *Proc. Natl. Acad. Sci. USA*, **79**, 6186.
54. Opperdoes,F. (1981) *Mol. Biochem. Parasitol.*, **3**, 181.
55. Khan,M.N., Posner,B.I., Khan,R.J. and Bergeron,J.J.M. (1982) *J. Biol. Chem.*, **257**, 5969.
56. Kay,D.G., Khan,M.N., Posner,B.I. and Bergeron,J.J.M. (1984) *Biochem. Biophys. Res. Commun.*, **123**, 1144.
57. Deutscher,S.L., Criek,K.E., Meirion,M. and Hirschberg,C.B. (1983) *Proc. Natl. Acad. Sci. USA*, **80**, 3938.
58. Yamada,H., Hayashi,H. and Natori,Y. (1984) *J. Biochem. (Tokyo)*, **95**, 1155.
59. Harms,E., Kartenbeck,J., Darai,G. and Schneider,J. (1981) *Exp. Cell Res.*, **131**, 251.
60. Alquier,C., Guenin,P., Munari-Silem,Y., Audebet,C. and Rousset,B. (1985) *Biochem. J.*, **232**, 529.
61. Stocco,D.M. (1983) *Anal. Biochem.*, **131**, 453.

62. Reinhart,P.H., Taylor,W.M. and Bygrave,F.L. (1982) *Biochem. J.*, **204**, 731.
63. Mickelson,J.R., Greaser,M.L. and Marsh,B.B. (1980) *Anal. Biochem.*, **109**, 255.
64. Neat,C.E., Thomassen,M.S. and Osmundsen,H. (1980) *Biochem. J.*, **186**, 369.
65. Perret,B., Hugnes,J.C. and Douste-Blazy,L. (1979) *Biochim. Biophys. Acta*, **556**, 434.
66. Epping,R.J. and Bygrave,F.L. (1984) *Biochem. J.*, **223**, 733.
67. Brockbank,K.J. and England,P.J. (1980) *FEBS Lett.*, **122**, 67.
68. Boumendil-Podevin,E.F. and Podevin,R.A. (1983) *Biochim. Biophys. Acta*, **735**, 86.
69. Chakravarthy,B.R., Spence,M.W., Clarke,J.T.R. and Cook,H.W. (1985) *Biochim. Biophys. Acta*, **812**, 223.
70. Nagy,A. and Delgado-Escueta,A.V. (1984) *J. Neurochem.*, **43**, 1114.
71. Belsham,G.J., Denton,R.M. and Tanner,M.J.A. (1980) *Biochem. J.*, **192**, 457.
72. Storrie,B., Pool,R.R., Sachdeva,M., Maurey,K.M. and Oliver,C. (1984) *J. Cell Biol.*, **98**, 108.
73. Magun,B.E., Planck,S.R. and Wagner,H.N. (1982) *J. Cell. Biochem.*, **20**, 259.
74. Rickwood,D., ed. (1983) *Iodinated Density Gradient Media*. IRL Press, Oxford.
75. Wattiaux,R., Wattiaux-De Coninck,S., Ronveaux-Dupal,M.F. and Dubois,F. (1978) *J. Cell. Biol.*, **78**, 349.
76. Loh,Y.P., Tam,W.W.H. and Russell,J.T. (1984) *J. Biol. Chem.*, **259**, 8238.
77. Saermark,T., Flint,N. and Evans,W.H. (1985) *Biochem. J.*, **225**, 51.
78. Evans,W.H. and Flint,N. (1985) *Biochem. J.*, **232**, 25.
79. Lange,Y. and Steck,T.L. (1985) *J. Biol. Chem.*, **260**, 15592.
80. Harrison,E.H. and Bowers,W.E. (1983) *J. Biol. Chem.*, **258**, 7134.
81. Cotman,C. and Kelly,P.T. (1980) In *The Cell Surface and Neuronal Function*. Cotman,C.W., Poste,G. and Nicholson,G.L. (eds), Elsevier, North Holland, Amsterdam, p. 506.
82. Lardeux,B., Gouhot,B. and Forestier,M. (1983) *Anal. Biochem.*, **131**, 160.
83. Rosemblatt,M., Hidalgo,C., Vergara,C. and Ikemoto,N. (1981) *J. Biol. Chem.*, **256**, 8140.
84. Courtoy,P.J., Quintart,J. and Baudhuin,P. (1984) *J. Cell Biol.*, **98**, 870.
85. Wilson,G. (1978) *J. Biol. Chem.*, **253**, 2070.
86. Beardmore,J.M., Howell,K.E., Miller,K. and Hopkins,C.R. (1986) *EMBO J.*, **5**, 3091.
87. Helmy,S., Porter-Jordan,K., Dawidowicz,E.A., Pilch,P., Schwartz,A.L. and Fine,R.E. (1986) *Cell*, **44**, 497.
88. Sheeler,P. and Schultz,S. (1984) *Am. Lab.*, Oct. issue.
89. Albertsson,P.A., Andersson,B., Larsson,C. and Akerlund,H.E. (1982) *Methods Biochem. Anal.*, **28**, 115.
90. Brunette,D.M. and Till,J.E. (1971) *J. Membr. Biol.*, **5**, 215.
91. Gruber,M.Y., Cheng,K.H., Lepock,J.R. and Thompson,J.E. (1984) *Anal. Biochem.*, **138**, 112.
92. Glossmann,H. and Gips,H. (1974) *Naunyn-Schmiedeberg's Arch. Pharmacol.*, **282**, 439.
93. Hino,Y., Asano,A. and Sato,R. (1978) *J. Biochem. (Tokyo)*, **83**, 935.
94. Gierow,P., Sommarin,M., Larsson,C. and Jergil,B. (1986) *Biochem. J.*, **235**, 685.
95. Lopez-Perez,M.J., Paris,G. and Larsson,C. (1981) *Biochim. Biophys. Acta*, **635**, 359.
96. Gurd,J.W., Gordon-Weeks,P. and Evans,W.H. (1982) *J. Neurochem.*, **22**, 281.
97. Heywood-Waddington,D., Sutherland,I.A., Morris,I.B. and Peters,T.J. (1984) *Biochem. J.*, **217**, 751.
98. Johansson,G. (1976) *Biochim. Biophys. Acta*, **451**, 517.
99. Flanagan,S.D., Johansson,G., Yost,B., Ito,Y. and Sutherland,I.A. (1984) *J. Liquid Chromatogr.*, **7**, 385.
100. Olde,B. and Johansson,G. (1985) *Neuroscience*, **15**, 1247.
101. Carlson,S.S., Wagner,J.A. and Kelly,R.B. (1978) *Biochemistry*, **17**, 1188.
102. Ohsawa,K., Kano,A. and Hoshi,T. (1979) *Life Sci.*, **24**, 669.
103. Reynolds,J., Nozaki,Y. and Tanford,C. (1983) *Anal. Biochem.*, **130**, 471.
104. Steer,C.J., Bisher,M., Blumenthal,R. and Steven,A. (1984) *J. Cell Biol.*, **99**, 315.
105. Hannig,K. (1982) *Electrophoresis*, **3**, 235.
106. Brunner,G., Heidrich,H.G., Golecki,J.R., Bauer,H.C., Suter,D., Pluckhahn,P. and Ferber,E. (1977) *Biochim. Biophys. Acta*, **471**, 195.
107. Stewart,D.I.H. and Crawford,N. (1983) *Biochim. Biophys. Acta*, **733**, 154.
108. Hostetler,K.Y., Reasor,M. and Yazaki,P.J. (1985) *J. Biol. Chem.*, **260**, 215.
109. Harms,E., Kern,H. and Schneider,J.A. (1980) *Proc. Natl. Acad. Sci. USA*, **77**, 6139.
110. Heidrich,H.G. and Leutner,G. (1974) *Eur. J. Biochem.*, **41**, 37.
111. Morré,D.J., Morré,D.M. and Heidrich,H.G. (1983) *Eur. J. Cell. Biol.*, **31**, 263.
112. Crawford,N. (1985) In *Cell Electrophoresis*. Schutt,W. and Klinkmann,H. (eds), Walter de Gruyter, Berlin, p. 225.
113. Rubenstein,J.L.R., Fine,R.E., Luskey,B.D. and Rothman,J.E. (1981) *J. Cell Biol.*, **89**, 357.
114. Kedersha,N.L. and Rome,L.H. (1986) *Anal. Biochem.*, **156**, 161.
115. Kedersha,N.L., Hill,D.F., Kronquist,K.E. and Rome,L.H. (1986) *J. Cell Biol.*, **103**, 287.
116. de Duve,C. (1983) *Eur. J. Biochem.*, **137**, 391.

117. Smith,G.D. and Peters,T.J. (1980) *Eur. J. Biochem.*, **104**, 305.
118. Howell,K., Ito,A. and Palade,G.E. (1978) *J. Cell Biol.*, **79**, 581.
119. Evans,W.H. and Hardison,W.G.M. (1985) *Biochem. J*, **232**, 33.
120. Croze,E.M. and Morré,D.J. (1984) *J. Cell Physiol.*, **119**, 46.
121. Moktari,S., Feracci,H., Gorvel,J.P., Mishal,Z., Rigal,A. and Maroux,S. (1986) *J. Membr. Biol.*, **89**, 53.
122. Kaufmann,S.H., Gibson,W. and Shaper,J.H. (1983) *J. Biol. Chem.*, **258**, 2710.
123. Richardson,J.C.W. and Agutter,P.S. (1980) *Biochem. Soc. Trans.*, **8**, 459.
124. Bachmann,M., Bernd,A., Schroder,H.C., Zahn,R.K. and Muller,W.E.G. (1984) *Biochim. Biophys. Acta*, **773**, 308.
125. Johnson,D. and Lardy,H.A. (1967) *Methods Enzymol.*, **10**, 94.
126. Sandri,G., Panfili,E. and Sottocassa,G.L. (1978) *Bull. Mol. Biol. Med*, **3**, 179.
127. Gellerfors,P. and Linden,M. (1981) *FEBS Lett.*, **127**, 91.
128. Henning,R. and Plattner,H. (1974) *Biochim. Biophys. Acta*, **354**, 114.
129. Lardeux,B., Gouhot,B. and Forestier,M. (1983) *Anal. Biochem.*, **131**, 160.
130. Takano,T., Muto,K. Imanaka,T. and Ohkuma,S. (1982) *Biochem. Int.*, **4**, 485.
131. Marzella,L., Ahlberg,J. and Glaumann,H. (1982) *J. Cell Biol.*, **93**, 144.
132. Furuno,K., Ishikawa,T. and Kato,K. (1982) *J. Biochem. (Tokyo)*, **91**, 1943.
133. Lazarow,P.B. and Fujiki,Y. (1985) *Annu. Rev. Cell Biol.*, **1**, 489.
134. Small,G.M., Hocking,T.J., Sturdee,A.P., Burdett,K. and Connock,M.J (1981) *Life Sci.*, **28**, 1875.
135. Goldenberg,H., Kollner,U., Krammer,R. and Pavelka,M. (1978) *Histochemistry*, **56**, 253.
136. Connock,M.J. and Perry,S.R. (1983) *Biochem. Int.*, **6**, 545.
137. Volkl,A. and Fahimi,H.D. (1985) *Eur. J. Biochem.*, **149**, 257.
138. Fujiki,Y., Fowler,S., Shio,L., Hubbard,A. and Lazarow,P. (1982) *J. Cell Biol.*, **93**, 103.
139. Crane,D.I., Hemsley,A.C. and Masters,C.J. (1985) *Anal. Biochem.*, **148**, 436.
140. Morré,D.J. (1971) *Methods Enzymol.*, **22**, 130.
141. Bergeron,J.J.M. (1980) *Biochim. Biophys. Acta*, **555**, 493.
142. Fujiki,Y., Hubbard,A.L., Fowler,S. and Lazarow,P.B. (1982) *J. Cell Biol.*, **93**, 97.
143. Howell,K.E. and Palade,G.E. (1982) *J. Cell Biol.*, **92**, 822.
144. Steer,C.J., Wall,D.A. and Ashwell,G. (1983) *Hepatology*, **3**, 667.
145. Alfsen,A., de Paillerets,C., Prasad,K., Nandi,P.K., Lippoldt,R.E. and Edelhoch,H. (1984) *Eur. Biophys. J.*, **11**, 129.
146. Pearse,B.M.F. (1978) *Proc. Natl. Acad. Sci. USA*, **79**, 451.
147. Booth,A.G. and Wilson,M.J. (1981) *Biochem. J.*, **196**, 355.
148. Woodward,M.P. and Roth,T.F. (1979) *J. Supramol. Struct.*, **11**, 237.
149. Steer,C.J., Bisher,M., Blumenthal,R. and Steven,A.C. (1984) *J. Cell Biol.*, **99**, 315.
150. Hadjiivanova,N., Flint,N., Evans,W.H., Dix,C. and Cooke,B.A. (1984) *Biochem. J.*, **222**, 749.
151. Sun,I., Morré,D.J., Crane,F.L., Safranski,K. and Croze,E.M. (1984) *Biochim. Biophys. Acta*, **797**, 266.
151. Dickson,R.B., Hanover,J.A., Pastan,I. and Willingham,M. (1985) *Methods Enzymol.*, **109**, 257.
152. Resch,K., Loracher,A., Mahler,B., Stoeck,M. and Rode,H.N. (1978) *Biochim. Biophys. Acta*, **511**, 176.
153. Hoessli,D.C. and Rungger-Brandle,E. (1983) *Proc. Natl. Acad. Sci. USA*, **80**, 439.
154. Horst,M.N., Braumbach,G.A., Olympio,M.A. and Roberts,R.M. (1980) *Biochim. Biophys. Acta*, **600**, 48.
155. Morrison,M. (1974) *Methods Enzymol.*, **32B**, 103.
156. Hubbard,A.L. and Cohn,Z.A. (1976) in *Biochemical Analysis of Membranes*. Maddy,A.H. (ed), Chapman and Hall, London, p. 427.
157. Carraway,K.L. (1975) *Biochim. Biophys. Acta*, **415**, 379.
158. Skerrow,C.J.and Matoltsy,A.G. (1974) *J. Cell Biol.*, **63**, 524.
159. Hertzberg,E.L. (1984) *J. Biol. Chem.*, **259**, 9936.
160. Giacobino,J.P. and Perrelet,A. (1971) *Experimentia*, **27**, 259.
161. Zinder,O., Hoffman,P.G., Bonner,W.M. and Pollard,H.B. (1978) *Cell Tissue Res.*, **188**, 153.
162. de Vries,G.H., Matthieu,J.M., Beny,M., Chickeportiche,R., Lazdunski,M. and Dalivo,M. (1978) *Brain Res.*, **147**, 339.
163. Dodd,P.R., Hardy,J.A., Oakley,A.E., Edwardson,J.A., Perry,E.K. and Delaunoy,J.P. (1981) *Brain Res.*, **226**, 107.
164. Gregg,M.R., Spanner,S. and Ansell,G.B. (1982) *Neurochem. Res.*, **7**, 1045.
165. Gospodarowicz,D. (1973) *J. Biol. Chem.*, **248**, 5042.
166. Bramley,T.A. and Ryan,R.J. (1980) *Mol. Cell Endocrinol.*, **19**, 21.
167. Ribot,H., Decker,S.J. and Kinsey,W.H. (1983) *Dev. Biol.*, **97**, 494.
168. Wright,J.T., Elmer,W.A. and Dunlop,A.T. (1982) *Anal. Biochem.*, **125**, 100.
169. Gray,G.M., King,I.A. and Yardley,H.J. (1980) *Br. J. Dermatol.*, **103**, 505.

170. Brown,C.D.A., Bodmer,M., Biber,J. and Murer,H. (1984) *Biochim. Biophys. Acta,* **769**, 471.
171. Hanahan,D.J. and Ekholm,J.E. (1978) *Arch. Biochem. Biophys.,* **187**, 170.
172. Brasitus,T.A. and Keresztes,R.S. (1984) *Biochim. Biophys. Acta,* **773**, 290.
173. Gruber,M.Y., Cheng,K.H., Lepock,J.R. and Thompson,J.E. (1984) *Anal. Biochem.,* **138**, 112.
174. Marggraf,W.D., Anderer,F.A. and Kanfer,J.N. (1981) *Biochim. Biophys. Acta,* **664**, 61.
175. Mansier,P., Charlemagne,D., Rossi,B., Preteseille,M., Swynghedauw,B. and Lelievre,L. (1983) *J. Biol. Chem.,* **258**, 6628.
176. Weglicki,W.B., Owens,K., Kennett,F.F., Kessner,A., Harris,L., Wise,R.M. and Vahouny,G.V. (1980) *J. Biol. Chem.,* **255**, 3605.
177. Constantino-Ceccarini,E., Novikoff,P.M., Atkinson,P.H. and Novikoff,A. (1978) *J. Cell Biol.,* **77**, 448.
178. Stieger,B. and Murer,H. (1983) *Eur. J. Biochem.,* **135**, 95.
179. Scalera,V., Storelli,C., Storelli-Joss,C., Haase,W. and Murer,H. (1980) *Biochem. J.,* **186**, 177.
180. Fujita,M., Ohta,H. and Uezato,T. (1981) *Biochem. J.,* **196**, 669.
181. Fujita,M., Ohta,H., Kawai,H. and Nakao,M. (1972) *Biochim. Biophys. Acta,* **274**, 336.
182. Scalera,V, Huang,Y.K., Hildmann,B. and Murer,H. (1981) *Membr. Biochem.,* **4**, 49.
183. Roy,D., Rosenfeld,L. and Spector,A. (1982) *Exp. Eye Res.,* **35**, 113.
184. Graham,J.M. and Coffey,K.H.M. (1979) *Biochem. J.,* **182**, 165.
185. Wisher,M.H. and Evans,W.H. (1975) *Biochem. J.,* **146**, 375.
186. Inoue,M., Kinne,R., Tran,T., Biempica,L. and Arias,I.M. (1983) *J. Biol. Chem.,* **258**, 5183.
187. Nijjar,M.S. and Ho,J.C.H. (1980) *Biochim. Biophys. Acta,* **600**, 238.
188. Casale,T.B., Friedman,M., Parada,N., Plekes,J. and Kaliner,M. (1984) *J. Membr. Biol.,* **79**, 33.
189. Kaevar,V., Szamel,M., Goppelt,M. and Resch,K. (1984) *Biochim. Biophys. Acta,* **776**, 133.
190. Adams,D.A., Freauff,S.J. and Erickson,K.L. (1985) *Anal. Biochem.,* **144**, 228.
191. Truding,R., Shelanski,M.L., Daniels,M.P. and Morell,P. (1974) *J. Biol. Chem.,* **249**, 3973.
192. Record,M., Laharrague,P., Fillola,G., Thomas,J., Ribes,G., Fontan,P., Chap,H., Corberand,J. and Douste-Blazy,L. (1985) *Biochim. Biophys. Acta,* **819**, 1.
193. O'Donnell,R.T. and Andersen,B.R. (1985) *Biochim. Biophys. Acta,* **814**, 307.
194. Chibber,R. and Castle,A.G. (1983) *Eur. J. Biochem.,* **136**, 383.
195. Arvan,P. and Castle,J.D. (1981) *J. Cell Biol.,* **95**, 8.
196. Boyd,C.A.R., Chipperfield,A.R. and Steele,L.W. (1979) *J. Dev. Physiol.,* **1**, 361.
197. Molitoris,B.A. and Simon,F.R. (1985) *J. Membr. Biol.,* **83**, 207.
198. Truman,P., Wakefield,S.J. and Ford,H.C. (1981) *Biochem. J.,* **196**, 121.
199. Cole,D.E.C. (1984) *Biochem. Biophys. Res. Commun.,* **123**, 223.
200. Khalfoun,B., Degenne,D., Arbeille-Brassart,B., Gutman,N. and Bardos,P. (1985) *FEBS Lett.,* **181**, 33.
201. Sixma,J.J. and Lips,J.P.M. (1978) *Thrombos. Haemostas. (Stuttgart),* **39**, 328,
202. Seiler,S. and Flesicher,S. (1982) *J. Biol. Chem.,* **257**, 13862.
203. Daniel,E.E., Grover,S.K. and Kwan,C.Y. (1982) *Fed.Proc.,* **41**, 2898.
204. Ljungstrom,M., Norberg,L., Olaisson,H., Wernstedt,C., Vega,F.V., Arvidson,G. and Mardh,S. (1984) *Biochim. Biophys. Acta,* **769**, 209.
205. Millette,C.F., O'Brien,D.A. and Moulding,C.T. (1980) *J. Cell Sci.,* **43**, 279.
206. Schneider,D., Brauer,D. and Falke,D. (1982) *Biochim. Biophys. Acta,* **685**, 94.
207. Haeffner,E.W., Kolbe,K., Schroeter,D. and Paweletz,N. (1980) *Biochim. Biophys. Acta,* **603**, 36.
208. Franke,W.W., Kapprell,H.P. and Mueller,H. (1983) *Eur. J. Cell Biol.,* **32**, 117.
209. Kensler,R.W. and Goodenough,D.A. (1980) *J. Cell Biol.,* **86**, 755.
210. Stevenson,B.R. and Goodenough,D.A. (1984) *J. Cell Biol.,* **98**, 1209.
211. Carlin,R.K., Grab,D.J., Cohen,R.S. and Siekevitz,P. (1980) *J. Cell. Biol.,* **86**, 831.
212. Somerville,R.A., Merz,P.A. and Carp,R.I. (1984) *J. Neurochem.,* **43**, 184.
213. Mitchell,R.D., Palade,P. and Fleischer,S. (1983) *J. Cell Biol.,* **96**, 1008.
214. Yoshinari,M., Taurog,A. and Krupp,P.P. (1985) *Endocrinology,* **117**, 580.
215. Glaumann,H., Ahlberg,J., Berkenstam,A. and Henell,F. (1986) *Exp. Cell Res.,* **163**, 151.
216. Montague,D.J., Peters,T.J. and Baum,H. (1984) *Biochim. Biophys. Acta,* **77**, 9.
217. Burnette,B. and Batra,P.P. (1985) *Anal. Biochem.,* **145**, 80.
218. Batra,S. (1982) *Acta Physiol. Scand.,* **114**, 91.
219. Warley,A. and Cook,G.M.W. (1976) *Biochem. J.,* **156**, 245.
220. Hutton,J.C., Penn,E.J. and Peshavaria,M. (1982) *Diabetologia,* **23**, 365.
221. Cameron,R.S. and Castle,J.D. (1984) *J. Membr. Biol.,* **79**, 127.
222. Fricker,L.D. and Snyder,S.N. (1982) *Proc. Natl. Acad. Sci. USA,* **79**, 3886.
223. Fishman,J.B. and Fine,R.E. (1987) *Cell,* **48**, 157..
224. Willems,M.F. and De-Potter,W.P. (1983) *J. Neurochem.,* **41**, 466.
225. Gordon-Weeks,P.R. and Lockerbie,R.O. (1984) *Neuroscience,* **13**, 119.
226. Hook,G.E.R. and Gilmore,L.B. (1982) *J. Biol. Chem.,* **257**, 9211.
227. Gonatas,J.O., Gonatas,N.K., Stieber,A. and Fleischer,B. (1985) *J. Neurochem.,* **45**, 497.

CHAPTER 2

Membrane fractions from plant cells

D.JAMES MORRÉ, ANDREW O.BRIGHTMAN and ANNA STINA SANDELIUS

1. INTRODUCTION

The aim of this chapter is to provide an overview of the problems and procedures involved in the isolation of membranes from plant cells. Isolated membranes are critical to the development of cell-free systems necessary for rapid advances in the understanding of complex physiological and biochemical phenomena (1).

The emphasis will be on newer developments especially the isolation of plasma membranes by two-phase partitioning and of plasma membranes and tonoplasts from the same homogenates by preparative free-flow electrophoresis. There is a growing body of information on marker enzymes and criteria for evaluation of fraction purity as well as preparation of balance sheets, and determinations of recovery. Indeed, the preparation of purified membranes from plants is rapidly approaching a level of experimental development that just a few years ago was possible only with rat liver (2).

Fractionation of plant cells is different from that of animal cells in three important respects.

(i)   The plant cell is surrounded by a cellulosic wall. The wall does not present a particular impediment to the preparation of membranes but is a major consideration in the development of homogenization procedures applicable to plant tissues and plant cells in culture. Homogenization methods must be severe enough to rupture the cell wall but not so severe that organelles are destroyed.

(ii)   In the centre of most plant cells is a large fluid-filled vacuole, surrounded by a membrane, the tonoplast or vacuole membrane, of surface area essentially equivalent to that of the plasma membrane. When the vacuoles are ruptured and reduced to small vesicles, most of the vacuolar contents are released thereby diluting the homogenization medium and also lowering the pH if the medium is not well buffered. Frequently the vacuolar contents include high concentration of tannins, or other secondary products, that prove deleterious to proteins and enzyme activities.

(iii)   The plastids form an organelle system in amount equal to or exceeding that of the mitochondria. In green tissues, the plastids (chloroplasts) contain chlorophyll. In normally non-green parts such as roots, the plastids lack chlorophyll but are, none-the-less, present in large numbers. They most often contain starch, so-called starch plastids or amyloplasts, and are the source of the large starch pellets released when root or storage tissues are homogenized and then centrifuged.

A favourite source of material for many types of growth experiments is so-called etiolated plants grown in darkness or semi-darkness. The plastids (etioplasts) derived from etiolated plants may contain starch and various prothylakoid membranes. Proplastids or pre-plastids are terms usually reserved for an early precursor stage in the development of any plastid line. Thus, no actively growing higher plant or algal tissue can be found where plastids are absent. The presence of plastids adds a degree of complexity to plant fractionation equal to or exceeding that resulting from the presence of mitochondria. Plastids exhibit a marked propensity to fragment and the fragments exhibit sedimentation or buoyant density characteristics different from those of intact plastids.

Fractionation of plant cells thus has these three added elements of complexity not encountered in animal cells: the cell wall, the vacuole and plastids. While lysosomes, *per se*, are not known in plants, the vacuolar system in plants is much more extensive than in animals. Mitochondria, nuclei, endoplasmic reticulum, Golgi apparatus, peroxisomes, lipid droplets, coated vesicles and secretory vesicles are common to both plants and animals (see Chapter 1). Because of the added complexity, it is essential that all procedures for preparation of membranes from plants include information on the distribution of tonoplast and plastids in addition to mitochondria, endoplasmic reticulum, Golgi apparatus and plasma membranes. This has been done in the past in only a few investigations and doubtless many membrane preparations have been used to generate experimental observations without an adequate accounting of all membranes present. This practice should no longer be regarded as permissible.

2. PREPARATION OF HOMOGENATES

Preparation of homogenates involves:

(i)   the medium in which the cell contents are to be released;
(ii)   the manner in which the cells are to be broken to release the contents; and
(iii)   the initial steps to remove cell walls, cell wall fragments and debris.

The outcome, the so-called total or filtered (cell-free) homogenate, is then utilized in subsequent steps. It is sampled in biochemical experiments for preparation of balance sheets and for the calculation of recoveries.

2.1 Homogenization media

Specific examples of isolation media will be given in the sections that relate to individual cell components. Media normally consist of an osmotically active solute (osmoticum) together with a buffer and materials to guard against degradative changes such as dextran, albumin, protease inhibitors (see Chapter 1 for classification) and sulphydryl protectants. Ions such as Mg^{2+} are essential for preparation of rough endoplasmic reticulum. Monovalent ions, sodium or potassium, or both, may be beneficial in some instances.

2.2 Methods of homogenization

Low-shear homogenization using a mechanized razor blade chopper (3) is one method of choice. Hand chopping, while useful for analytical purposes, is too time-consuming for preparative isolations. The Polytron homogenizer (4) gives a rotary scissors action. The speeds utilized are well below those generating sonic or cavitation effects.

38

The advantages of the Polytron in preparation of homogenates are its ease of application, reproducibility and the short homogenization times required. The yield of membranes is greater with the Polytron than with razor blade chopping. A disadvantage of the Polytron is that more fragments are generated since the cell contents are subjected to considerable mixing once the cells are broken. The latter is avoided with razor blade chopping. For razor blade chopping, 5 min at 50−60 chops per second is recommended for 10−20 g of starting material. With the Polytron, 2 min at 3000−4000 r.p.m. is usually sufficient. A mortar may also be used.

The ratio of tissue-to-medium should be high. If the Polytron is used, the entire amount of homogenization medium is added at the outset. If razor blade chopping is used, the homogenizing medium is added gradually. A ratio of 1 part tissue to 1 part medium (w/v) may be used but the ratio should not go lower than 1:2 for best morphological preservation.

The temperature during homogenization is usually maintained at about 4°C. If buffers are cold it is not essential to work in a cold room. For most purposes, homogenate temperatures should never exceed 25°C.

2.3 Removal of cell walls and debris

To remove cell walls, wall fragments and debris, some form of filtration is employed usually, this being preferable to centrifugation where fragments tend to float and large components such as nuclei are lost. A single layer of Miracloth (Chickopee Mills, New York), a porous cellulosic fabric, removes wall fragments and debris, but allows nuclei to filter through nearly quantitatively. A fine nylon mesh will serve the same purpose. Before filtering, the homogenate should be stirred and mashed with a flat spatula against the tube wall to extrude cell contents from broken cells. Likewise, the Miracloth and/or nylon mesh filtrate should be squeezed to collect as much liquid as possible for highest yields.

2.4 Stabilization of homogenates

To offset the problem of dilution by vacuolar content and the attendant deleterious effects of vacuolar products, the homogenization media may be supplemented with albumin or prepared in pre-centrifuged coconut water (milk) (3). Coconut milk is liquid endosperm from coconuts cleared by centrifugation for 1−3 h at 100 000 *g*. It contains no membranes, contributes little protein or extraneous enzyme activities to the fractions obtained and is an inexpensive and effective source of plant protoplasm (e.g. *Figure 2*).

Glutaraldehyde (0.1% v/v final concentration) added to freshly prepared homogenates (5, 6) or at any point thereafter in the procedure stabilizes the structure of organelles and provides an indication of morphology at the time of glutaraldehyde addition. While glutaraldehyde-stabilized preparations resist further degradation during purification, they have only limited utility for enzymatic studies. Usually some form of extrapolation method is required for analysis (7). Glutaraldehyde-stabilized preparations are useful for chemical analyses where cross-linking of amino-, SH- and hydroxyl-groups are not a problem (i.e. determination of total N or P) and for analyses of *in vivo* kinetics of isotope incorporation (8, 9).

3. GENERAL APPROACHES

The general approach for isolation of plant membranes, once a suitable homogenate has been prepared, is to concentrate in fractions the organelle or cell component in question. Preparations in useful yield and quantity, free from contamination of all other cell components and in a form functionally appropriate to the experiments being conducted are the objective. To effect the separation, any or all properties whereby the various cell components differ may be employed to advantage. For the most part, these include size, shape, density, surface charge and chemical or enzymatic differences in the external membrane surface, either natural or artificially imposed. For plant cells it is likely that in order to effect a complete fractionation, some combination of techniques involving several different properties will be necessary.

The following definitions as proposed by deDuve (10) will be used. *Cell component*, rather than organelle, denotes any organized region of the cytoplasm composed of different kinds of macromolecules. The term organelle, while acceptable for components possessing a membrane and that exist as discrete entities (e.g. mitochondria, plastids, nuclei, peroxisomes), is subject to wide variations in current usage. For more extensive cellular structures such as endoplasmic reticulum, plasma membrane, tonoplast and Golgi apparatus, the term cell component seems more appropriate. Cell constituent denotes the individual kinds of molecules (e.g. lipids, proteins, sugars, nucleic acids, etc.) of which cell components are formed.

A *marker* is defined as a cell constituent, naturally occurring or imposed, that is found mainly in one cell component or in a particular domain of a cell component and is retained when that cell component is removed from the normal positional relationship of the intact cell (11). When originally proposed by deDuve (12) the definition included the proviso that some markers might be appropriate only for cell components of certain cell types or for certain plant or animal species.

In the sections that follow, the fundamental principles of the three main approaches to plant cell fractionation are described briefly. Historical development and details can be found in the general references provided for each technique (see also Chapter 1).

3.1 Differential and density gradient centrifugation

Centrifugation methods have been used widely to prepare membranes and cell components. The two principle properties that contribute to centrifugal behaviour are sedimentation rate (s) and density (rho). Anderson has defined the concept of s−rho space (13) whereby organelles of either differing size (sedimentation rate) or density or both could be separated (see Figure 2, Chapter 1). In contrast, organelles and components of similar sizes and densities cannot be separated by centrifugation methods alone. Practically, three types of centrifugation methods are used to effect the separation of membranous cell components.

3.1.1 *Differential centrifugation*

With differential centrifugation, cell components with the greatest mass sediment first (i.e. whole cells and nuclei) and the smallest (i.e. microvesicles), last. Both stratification (where sufficiently great combinations of time and speed are used to sediment more than one component) or some separation (time and speed to remove principally only

Table 1. Forces sufficient to sediment cell components from plants.

Cell component	$\times g_{max}$	*Time (min)*	*g/min*
Whole cells	500	5	2500
Nuclei	500	7	3500
Chloroplasts	2000	2	4000
Mitochondria	6000	15	90 000
Golgi apparatus (dictyosomes)	10 000	20	200 000
Plasma membrane-tonoplast vesicles	16 000	20	320 000
Peroxisomes	17 500	20	350 000
Endoplasmic reticulum vesicles (microsomes)	17 500	20	350 000
Coated vesicles (microvesicles)	20 000	20	400 000

Values are approximate and are for the cell components suspended in 0.25 M sucrose and centrifuged in a standard tube (9 cm vertical path length).

one component) may be achieved. Centrifugation protocols are manipulated in terms of force × time (*g*.min). Approximate forces needed to sediment different cell components from plants are given in *Table 1.*

Separations by differential centrifugation from total homogenates are never complete. Older schemes that give so-called nuclei, mitochondria and microsome fractions by this method yielded, at best, crude mixtures. Differential centrifugations, however, do provide either partially enriched starting fractions for other methods or aid final separations of mixtures achieved by other techniques, for example free-flow electrophoresis. When using fractions as the starting material for a subsequent separation, an important consideration is to avoid a tightly-packed pellet. Centrifugation on a sucrose 'cushion' or 'mini-gradient' avoids this problem (e.g. *Figure 8*).

3.1.2 *Gradient centrifugation*

The gradient material usually is sucrose although other materials such as Renografin, Ficoll, Percoll and Metrizamide have found specific uses (see also Chapter 1). Concentrations are expressed as percents, molarity or density, the latter at a given temperature (see Appendix I). For sucrose, the least ambiguous designation is molarity. Gradients can be continuous (linear or non-linear) or discontinuous. The discontinous or 'step' gradient offers advantages for preparative cell fractionation in being easily prepared and in serving as a reproducible source of membranes in useful yield and purity. Gradients, especially step gradients, need not be centrifuged to equilibrium.

In order to minimize preparation time and to maximize preservation of structure and functional activity, as little as 30 min but not more than 90 min are commonly used. *Figure 1* shows a continuous gradient for analytical work for plant homogenates and *Figure 2* shows a discontinous gradient for preparative work derived from information generated with continuous gradients. Standard analytical sucrose gradient procedures, (for example ref 15), have been most useful in assignment studies but have only limited value for preparative work because of overlap with contaminating membranes.

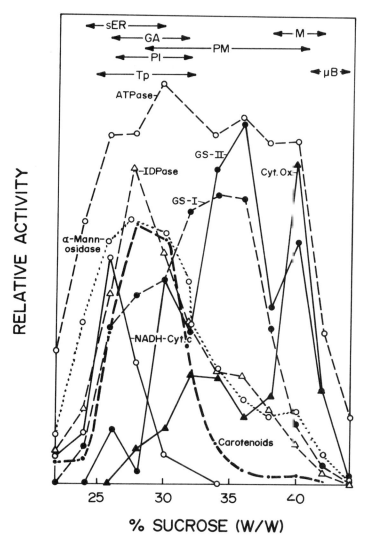

Figure 1. A continuous sucrose gradient separation of a total homogenate of etiolated zucchini hypocotyls (*Cucurbita pepo*), that illustrates the relative complexity of fractionation of plant homogenates. Large particles were first removed by filtration through nylon mesh and by an initial centrifugation at 4000 *g* for 10 min. Marker enzymes and putative cell components include α-mannosidase for tonoplast (Tp), NADH-cytochrome *c* (NADH-Cyt. c) reductase for smooth endoplasmic reticulum (sER) as well as glucan synthetase-I (GS-I) and IDPase (nucleoside diphosphatase, IDP as substrate) for Golgi apparatus (GA), glucan synthetase-II (GS-II) for plasma membrane (PM), carotenoids for plastids (P1) and cytochrome oxidase (Cyt. Ox.) for mitochondria (M). Peroxisomes or microbodies (μB) would be expected to band to the right of mitochondria. Modified from ref. (14). Original data courtesy of Dr G.F.E.Scherer, Botanical Institute, University of Bonn, FRG.

3.1.3 *Rate-zonal centrifugation*

The principle of rate-zonal centrifugation is to centrifuge in a medium of appropriate density or of different densities to retard preferentially one cell component relative to another thereby increasing the efficiency of differential centrifugation. An example would

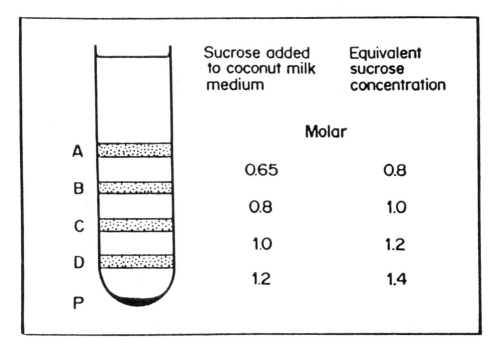

	Sucrose added to coconut milk medium	Equivalent sucrose concentration
	Molar	
A		
	0.65	0.8
B		
	0.8	1.0
C		
	1.0	1.2
D		
	1.2	1.4
P		

Figure 2. Diagram of a discontinuous sucrose and coconut water (milk) gradient used to separate cell components of soybean homogenates. The coconut water is used as a source of plant proteins to help stabilize membranes and enzymatic activities and is equivalent in osmotic value to about 0.2 M sucrose. For gradients prepared without the coconut water, the equivalent sucrose concentration is also given.

be in the separation of surface membranes from mitochondria from fraction D of the gradient shown in *Figure 2*. When layered over 1.25 M sucrose and centrifuged for 10 min at 10 000 *g*, mitochondria form a pellet whereas the surface membranes remain in the supernatant. The latter are then diluted 1:1 with buffer and collected by differential centrifugation.

3.2 Aqueous two-phase partition

Phase separations involve mixing of membranes with a mixture of polymers that themselves will separate into different phases. The procedure was first developed by Albertsson (16, 17), and takes advantage of the different surface properties of membranes. Since different membranes and perhaps even the same membranes from different species may possess different surface charges, it is usually necessary to evaluate a series of polymer and salt solutions. It is important to prepare these solutions very carefully, for polymers vary from batch to batch, and they need to be tested each time they are purchased (see also Chapter 1).

To effect the separation, the phase system containing the membranes is inverted and returned upright 20–40 times and then centrifuged at 4°C at low speed for short times in a swinging bucket rotor to resolve the polymers into two phases (*Figure 3*). The phases can be re-partitioned or washed to yield further purification or collected directly.

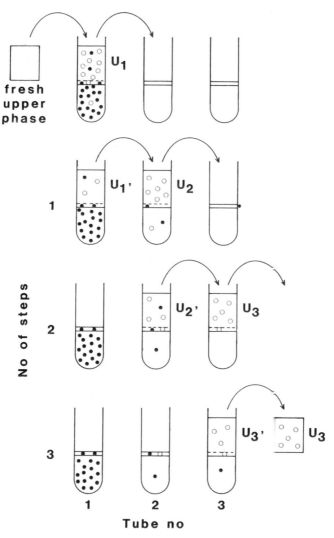

Figure 3. Purification of plasma membranes by aqueous two-phase separation. The mixture of plasma membranes (○) and other membranes (●) is suspended in a pre-weighted phase mixture to give the phase system of the desired final composition. The tube is well mixed by inversions (40 times) and allowed to settle. Separation of the two phases is facilitated by a low-speed centrifugation in a swinging bucket rotor. Then 90% of the upper phase (U_1) is removed (to dashed line) and re-partitioned twice with fresh lower phase (steps no. 1 and 2) to increase the purity, giving fraction U_3. To increase the yield of plasma membranes, the lower phases may be re-extracted with fresh upper phase (steps 1−3) to give U_3'. Fresh upper and lower phase are obtained from a bulk phase system prepared separately. For further details, see Section 7.2. From Larsson (18).

3.3 Preparative free-flow electrophoresis

In this technique developed by Hannig and co-workers (19), a mixture of components to be separated is introduced as a fine jet into a separation buffer moving perpendicular to the flux lines of an electrical field (see also Chapter 1). Membranes bearing different electrical charge-densities migrate different distances across the separation

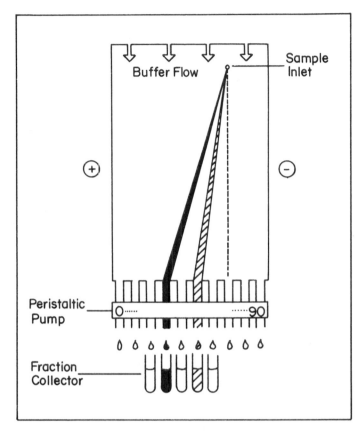

Figure 4. Principles of preparative continuous free-flow electrophoresis. Re-drawn and reprinted with permission from Bender and Hobein, Munich, FRG. From Morré *et al.* (20).

chamber and thus may be resolved (*Figure 4*). This technique has recently permitted preparation of plasma membranes and tonoplast from a single homogenate of either etiolated stems (21) or green leaves (22). In addition, mitochondria and endoplasmic reticulum vesicles were separated (23). The latter is important since mitochondria are frequent contaminants of endoplasmic reticulum fractions.

4. IDENTIFICATION AND CRITERIA FOR PURITY

Among the first tasks of cell fractionation is to identify the cell component(s) isolated. This may be done by assay of biochemical markers or from characteristic morphological features. Usually a combination of the two approaches is most desirable. Initially, identification should correlate with localization to a particular component within the intact cells. In mammalian cells, the compartment containing latent acid phosphatase was first isolated and later identified by enzyme cytochemistry and shown to be associated with morphologically identifiable lysosomes. For the Golgi apparatus, the opposite strategy was used. Golgi components were first isolated on the basis of their characteristic morphology (stacks of cisternae = dictyosomes) (24, 25) and only later correlated with

Table 2. Cell components from plants and criteria that permit identification upon isolation and simultaneous *in situ* identification.

Cell component	Criterion	Correlates with	Reference
Plasma membrane	Thick (~10 nm) membranes that stain with PTA at low pH	N-1-naphthylphthalamic binding	(30) (31)
Tonoplast	Thick (7−9 nm) membranes unreactive with PTA at low pH	Anion-stimulated, nitrate-inhibited ATPase	(21, 22 32, 33)
Mitochondria	Thin (6 nm) membranes Cristae with electron transport particles	Succinate dehydrogenase Cytochrome oxidase other mitochondrial activities	(34) (35)
Chloroplasts	Thin (6 nm) membranes Thylakoids	Chlorophyll	(36, 34)
Amyloplasts	Thin (6 nm) membranes reactive with $KMnO_4$	Monogalactosyldiglyceride synthetase Carotenoids Digalactosyl- and mono-galactosyldiglycerides	(37) (38)
Golgi apparatus	Both thick and thin membranes; cisternae organized into stacks	Latent IDPase, both isolated and *in situ* by cytochemistry	(5, 24, 26, 39)
Endoplasmic reticulum	Thin membranes (6 nm) with ribosomes attached	NADPH-cytochrome *c* reductase (NADH-cyto-chrome *c* reductase)	(35)
Peroxisomes	Single bounding membrane Dense content	Catalase and various H_2O_2-producing enzymes	(40)
Nuclei	Thin (6 nm) membranes the two membranes are joined at pores, with chromatin associated with the innermost of the two membranes	DNA	(41)
Coated vesicles	Clathrin coats	Clathrin (180−190 kd) band on SDS−PAGE	(42, 43, 44, 208)

specific sugar transferases (26−29). With plants, the various criteria for identification of cell components are listed in *Table 2*.

An important criterion, especially for identification of surface membranes (plasma membrane and tonoplast) is membrane thickness measured from electron micrographs (*Table 3, Figure 5*). Tonoplast and plasma membranes, as well as some membranes from the mature Golgi apparatus face, are thick (7−10 nm). Two classes of thick membranes are resolved on the basis of reactivity with phosphotungstic acid (PTA) at low pH. Plasma membranes are reactive and tonoplast membranes are unreactive with this contrasting method. All other membranes are both thin (5−6 nm) and unreactive with PTA at low pH. By projecting electron microscope negatives at magnifications approaching 1 000 000 times, these different classes of membranes are thus identified.

Correlative biochemical markers are established in one or both of two ways. In the

Table 3. Membrane dimensions in plants tissues as a criterion for their identification in isolated fractions.

Cell component	Membrane thickness, nm ± Standard deviation				
	Soybean hypocotyl[a]	Spinach leaf[b]	Onion stem[c]	Pythium aphanidermatum[d]	
	in situ	in situ	in situ	in situ	isolated
Plasma membrane	10.1 ± 0.7	10.5 ± 0.3	9.3	9.2 ± 0.4	9.6 ± 0.45
Tonoplast membrane	7.2 ± 0.8	8.1 ± 0.9	8.0	9.3 ± 0.5	9.7 ± 0.45
Endoplasmic reticulum	5.7 ± 0.7	6.3 ± 0.8	5.3	6.0 ± 0.1	6.0 ± 0.1
Nuclear envelope outer membrane leaflet	5.8 ± 0.8	6.8 ± 0.6			
Inner membrane leaflet	6.5 ± 0.6	5.8 ± 0.7			
Average both leaflets			5.6	6.1 ± 0.5	6.1 ± 0.45
Mitochondria outer membrane	5.0 ± 0.6	6.4 ± 0.8	5.5	6.0 ± 0.2	6.4 ± 0.3
inner membrane	6.2 ± 0.7	6.0 ± 0.9	6.0	6.2 ± 0.2	6.2 ± 0.4
Etioplast outer envelope membrane	5.0 ± 0.6				
Inner envelope membrane	6.1 ± 0.8				
Thykaloid	8.0 ± 1.0				
Chloroplast outer envelope membrane		5.1 ± 0.8			
inner envelope membrane		6.3 ± 0.8			
Grana thylakoid[e]		13.0 ± 1.1			
Peroxisome	5.5 ± 0.6	7.0 ± 0.4			
Golgi apparatus	6–9	6.6 ± 1.8	5–9	7–9[f]	7–9[f]

[a]From Sandelius et al. (21)
[b]From Auderset et al. (22)
[c]From unpublished data. Glutaraldehyde–osmium tetroxide fixation.
[d]From Powell et al. (33)
[e]Both appressed membranes together.
[f]Golgi apparatus equivalent. Includes secretory vesicles.

first, pure fractions of the cell component in question should contain the marker of highest specific activity. With an absolute marker (see below), recovery and balance sheet experiments should indicate that all or most of the marker co-isolates with the fraction in question. Secondly, in a series of fractions of increasing content of both marker and cell component, the two should correlate (e.g. *Figure 6*).

Constituents correlated in the manner outlined above are called markers. They may be *absolute markers* (restricted to that cell component) as with electron transport chain components of mitochondria and chloroplasts and chlorophyll of chloroplasts. Alter-

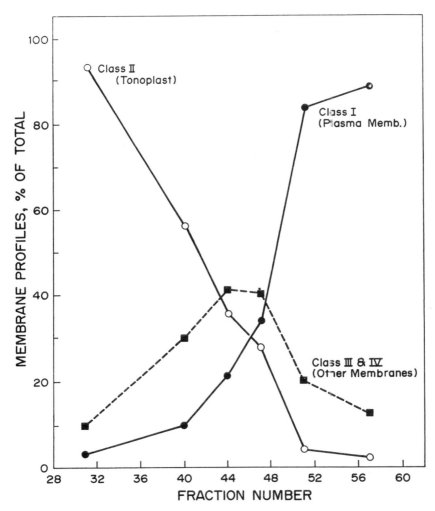

Figure 5. Distribution of class I (plasma membrane), class II (tonoplast) and class III (other) membranes by free-flow electrophoresis of green spinach leaves (23). The starting fraction was a total microsome pellet from which the bulk of the mitochondria and chloroplasts had been removed by differential centrifugation. The different classes of membranes were quantified by measurements of membrane thickness in electron micrographs, and defined as follows: Class I. Thick (7−11 nm) membranes also stained with PTA at low pH according to Roland *et al.* (31). Class II. Thick (7−11 nm) membranes not stained with PTA at low pH. Class III. Thin (5−7 nm) membranes and not stained with PTA at low pH. Class IV. Very thick (12−14 nm) membranes consisting of appressed membranes of grana thylakoids.

natively, they may be *operational markers* (concentrated in that cell component but present in lesser concentrations elsewhere) as with glucan synthetase II which is concentrated in plasma membrane but may be present also in Golgi apparatus. Markers are important not only to identify the particular cell component in question but to assess the purity of the preparation by allowing qualitative or quantitative estimates of levels of contaminating cell components. To be quantitative, one determines specific activities of marker enzymes in pure cell fractions either by direct assay or by extrapolation.

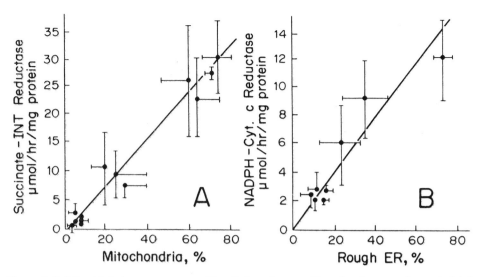

Figure 6. Relationship between enzyme specific activity and composition of soybean membranes isolated from sucrose gradients. (**A**) Succinate-INT reductase and mitochondria. (**B**) NADPH-cytochrome *c* reductase and rough endoplasmic reticulum. The membrane composition of the fractions was determined by morphometry. Results are ± standard deviations. From Morré *et al*. (23).

Examples of this type of information are given for etiolated hypocotyls of soybean in *Table 4*. A complete listing of markers for plant cell components is given in the reviews by Quail (34) and Bowles *et al*. (52).

The minimum requirements necessary to identify and establish the purity of a cell fraction should include at least two independent marker criteria for the principal component (e.g. glucan synthetase II and PTA at low pH for plasma membrane; the latter is quantified by morphometry) and a representative electron micrograph of the fraction plus a table listing specific activities of each of the markers summarized in *Table 4* or an equivalent activity to include all major cell components present in the homogenate. Markers for tonoplast and plastids, in addition to the usual markers for mitochondria, endoplasmic reticulum, plasma membrane and Golgi apparatus, must be evaluated. Although quantitatively minor, attention should be given to peroxisome distribution (see Section 13). Such evaluations are frequently complicated by the widespread unspecific distribution of catalase in plant cells. The electron microscopy should be quantified and accompanied by high quality electron micrographs to illustrate not so much the purity of the fraction (fields may be selected) but the morphological characteristics used for identification.

For quantification of electron micrographs, the following simple method is recommended as described originally by Loud (53) and applied subsequently to analysis of isolated fractions (54).

(i)   Prepare a simple transparent overlay with lines 1 cm apart as a photocopy transparency from an inked master copy.

(ii)   Analyse micrographs at a final magnification of about ×35 000. Count the intercepts of membranes with lines on the overlay to ascertain the total membranes present and then each cell component to be analysed.

Table 4. Examples of marker enzymes and constituents for cell components of etiolated hypocotyls of soybean.

Cell component	Marker enzyme	Sp. act.[a]	Units	References[b]
Endoplasmic reticulum[c]	NADPH-cytochrome c reductase	15.3	μmol cytochrome c reduced h/mg protein	(35, 45)
Golgi apparatus[d]	Latent IDPase (pH 7.2)	47	μmol P_i/h/mg protein	(39, 46)
Plasma membrane	Glucan synthetase II (1 mM UDPG)	585	nmol [14C]glucose incorporated/h/mg protein	(46, 47)
Tonoplast	NO3-inhibited, Cl-stimulated ATPase	0.55	μmol/h/mg protein	(21, 48)
Mitochondria[e]	Succinate-INT-reductase	37.5	μmol INT reduced/ h/mg/protein	(35, 49)
Etioplasts[f]	Carotenoids	80	ng/mg protein	(38)
Peroxisomes	Catalase	–	–	(40)

[a]For purified fraction calculated from regression analyses of fractons of known organelle content determined independently from morphometry.
[b]The first reference gives the assay method, the second the specific activity data. The actual calculations are detailed in Morré and Buckhout (50).
[c]Includes nuclear envelope.
[d]As an alternative, glucan synthetase I (1.5 μM UDPG) may be used, specific activity 0.2 nmol [14C]glucose incorporated from UDP-[14C]glucose into alkali-insoluble glucan/h/mg protein (46, 47).
[e]Cytochrome oxidase is one of several other alternatives, specific activity 1.13 μmol O_2/min/mg protein (50, 51).
[f]Content of monogalactosyl- and diagalactosyldiglycerides may be used as an alternative; specific content exceeding 1 mg/mg phospholipid (38).

(iii)   Express the results as intercepts per 100 total intercepts.

More sophisticated procedures and theoretical considerations of morphometric analyses also are available (55).

5. RECOVERY AND BALANCE SHEETS

Recovery and balance sheets are part of an accounting system for cell fractionation used to establish the validity and specificity of markers, to estimate the efficiency of the isolation procedure (yield), and in assignment studies to validate subcellular locations of specific markers. The basic procedure for a specific cell constituent or enzyme activity is to record the volume of both the starting homogenate and each final resuspended fraction, and to determine protein content and activity for a measured portion of each. The latter provides the specific activity used subsequently for the determination of absolute recovery. The total for each of the fractions should equal approximately the amount present in the total homogenate. A departure of more than $10-20\%$ from 100% indicates either an error or possible enzyme activation or inactivation.

To determine absolute recoveries, it is necessary to relate the isolated fractions to the composition of the intact cell as follows.

(i)   Construct whole cell montages from electron micrographs of median cross-sections from well fixed cells. With vacuolate plant cells, at least 10 cells are required. Micrographs of only portions of cytoplasm will give a bias even when photographed at random.

50

Table 5. Distribution of cell components within cortical cytoplasm of etiolated soybean hypocotyl by morphometry (50).

Cell component	% Total membrane	Protein (mg) per 10 g fresh weight
Nuclear envelope	3.9	0.6
Endoplasmic reticulum	30.6	4.5
Golgi apparatus	12.3	1.8
Plasma membrane	13.1	1.9
Tonoplast	10.3	1.5
Mitochondria	17.2	2.5
Etioplasts	12.2	1.8
Microbodies	0.2	0.03
Supernatant	–	37.4
Total homogenate	–	52.0

(ii)   Quantify the micrographs by the line intercept method (53) and calculate the relative abundance of each cell component as intercepts/100 intercepts (% of total membrane).

(iii)   Relate this to protein, by preparing a homogenate in which greater than 90% of the cells are broken; make sure that recovery of organelles is as complete as possible for determination of total protein.

(iv)   Centrifuge the homogenate at 100 000 g for 90 min to prepare total membranes as a pellet.

(v)   Determine the protein content of the total particulate fraction and relate this to the starting fresh weight of tissue. From these values the milligrams of protein per unit fresh weight will be known for each of the cell components as summarized in *Table 5* for membranes of etiolated soybean hypocotyls.

(vi)   Recoveries, are then determined as follows. Estimate the specific activity (units/unit protein) for the constituent in question in the total homogenate and in the purified fraction. With highly enriched reference fractions, one assumes initially that the fractions are pure. Alternatively, the purity is calculated based upon marker criteria. To calculate total activity, multiply the specific activity of the homogenate by the amount of homogenate protein. To calculate the total activity localized in any particular cell component, multiply the specific activity of the constituent for that cell component by the amount of protein for that particular cell component present in the cell and divide this value by fraction purity. If the amount present in the total homogenate and in the particular cell component are the same, the constituent analysed may be regarded as an absolute marker for that particular cell constituent.

Example. The specific activity of succinate-2-p-iodophenyl-3-p-nitrophenyl-5-phenyl-2H-tetrazolium chloride (succinate-INT) reductase of soybean total homogenate is 0.21 μmol of INT reduced/h/mg protein. The total protein of a soybean homogenate is 52 mg/10 g fresh weight (*Table 5*) or 52 × 2.1 = 109.2 μmol of INT reduced/h/10 g fresh weight. The specific activity of purified mitochondria for this same activity is 37.5 mol of INT reduced/h/mg protein (*Table 4*). Mitochondria account for 2.5 mg

51

of protein/10 g fresh weight (*Table 5*) with a fraction purity of 90% yielding a recovery of $37.5 \times 2.5 \div 0.9 = 104.2$ μmol/h/10 g fresh weight. The total activity present in the homogenate (109 μmol/h/mg/ protein) and that calculated for the total mitochondria (104 μmol/h/mg protein) are sufficiently close to consider this enzyme activity as an absolute marker for mitochondria.

Other examples of this approach are available for soybean (50) and rat liver (56).

6. MORPHOLOGY AND MORPHOMETRY

Part of the complete characterization of any membranous cell fraction should involve examination with the electron microscope. The light microscope, also, may be useful, for example to monitor cell and/or nuclear breakage.

The primary purpose of the electron microscope evaluation is to verify the composition of the fraction by direct observation and to quantify the membrane composition of isolated fractions. For this purpose, it is necessary to have sufficiently good morphological preservation so that the majority of the membranes are identifiable. Identification may be based on characteristic appearance (nuclei, dictyosomes, rough endoplasmic reticulum, peroxisomes, mitochondria, plastids), membrane thickness and differential staining with PTA at low pH (tonoplast, plasma membrane, endoplasmic reticulum lacking ribosomes) (see Section 4) or cytochemical localization of a particular enzymatic activity [e.g. latent inosine diphosphatase (IDPase) for Golgi apparatus].

For morphological quantitation to be valid, sampling should be representative of the preparation as a whole. A random sample of a homogeneous pellet is sufficient, but if it is stratified, micrographs must be evaluated across the pellet along the axis of centrifugation.

6.1 Fixation procedures

For plant membranes phosphate-buffered (0.1 M) glutaraldehyde (1−2.5%) at pH 7.2 for 1 h or longer gives consistently good results and is followed by post-fixation in 1% osmium tetroxide 1 h or overnight in the same buffer. Samples may remain in glutaraldehyde indefinitely and processing continued at any time. With PTA staining at low pH dehydration in acetone is preferred to dehydration in ethanol and the times for each change are shortened to 5 min or less. Embedding in Epon (57) gives satisfactory results. Embedding in Spurr Epon (58) may interfere with staining of plasma membranes with PTA at low pH.

6.2 Negative staining

Another approach to morphological evaluation uses negative staining.

(i)   Mix a drop of the sample resuspended in distilled water or dilute buffer (concentrated salts or sucrose interferes) on a carbon-shadowed, colloidion-coated, electron microscope grid together with a drop of 1% PTA, adjusted to pH 7.0 with potassium hydroxide.

(ii)   After mixing by drawing up and down in a Pasteur pipette or capillary, remove the excess fluid by slow absorption onto filter paper or the torn edge of a paper towel.

(iii) Examine the dried grid in the electron microscope immediately.

With plant membranes, structure can be stabilized by addition of dilute glutaraldehyde prior to preparation of the grids.

The advantage of negative staining is that information is obtained immediately and purification procedures can be monitored at critical steps; for example, to verify in two-phase separations whether or not additional purification cycles are necessary to remove mitochondria. A second advantage is that negative staining permits mitochondrial fragments to be distinguished from electron-transport particles. Intact Golgi apparatus membranes also exhibit a characteristic morphology distinguishable from plasma membrane and tonoplast in negatively stained preparations. However, tonoplast and plasma membrane have not been be distinguished in negatively stained preparations.

7. PLASMA MEMBRANES AND TONOPLAST

Studies of membrane dynamics, transport, growth and other aspects in plants are facilitated by procedures for isolation of tonoplast and plasma membranes in high purity from the same homogenates.

Earlier methods for the isolation and identification of plasma membranes from plant cells (59, 46) used PTA at low pH to specifically and characteristically stain plasma membrane vesicles in electron microscope sections of fixed and embedded pellets of membrane fractions (31). Subsequently sucrose gradients (47, 60−62) or aqueous two-phase systems consisting of dextran T500 and polyethylene glycol (PEG) 3350 (63−67), or isoelectric focusing (68) were used.

7.1 Plant plasma membrane markers

Two markers, a K^+-stimulated, vanadate-inhibited increment in Mg^{2+}ATPase activity (69) and a glucan synthetase exhibiting a high K_m for UDP-glucose (47), the so-called glucan synthetase II activity (70), have been associated with plasma membranes of plants. Additionally, a fraction of the plant plasma membrane binds the antagonist of polar auxin transport, N-1-naphthylphthalamic acid (NPA) (71). The latter marker was shown to correlate with the content of membranes stained specifically with PTA at low pH for a series of membrane fractions from oat coleoptiles (*Figure 7*) (30).

Where suitable markers do not exist, artificial markers (imposed labels) may be introduced by chemical linkage or by specific binding. In plants, the use of such markers has been restricted to the plasma membrane of protoplasts where surface labelling by means of enzymatic iodination, impermeable reactants such as diazotized sulphanilic acid or lectins have been attempted (72). These, however, have not achieved uniform success primarily due to lack of unequivocal evidence for specificity of surface labelling. Lanthanum compounds have been used as an electron-dense imposed label for identification of plasma membrane of protoplasts (73).

7.2 Plasma membrane isolation by aqueous two-phase partitioning

The procedure described here for hypocotyls of etiolated soybean seedlings is similar to that shown in *Figure 3* (18, 63) (see also Chapter 1, Table 8).

(i)   To isolate plasma membranes, centrifuge a filtered homogenate at 6 000 *g* for 10 min to remove the bulk of the mitochondria. Discard the pellet and centrifuge the supernatant for 30 min at 60 000 *g*.

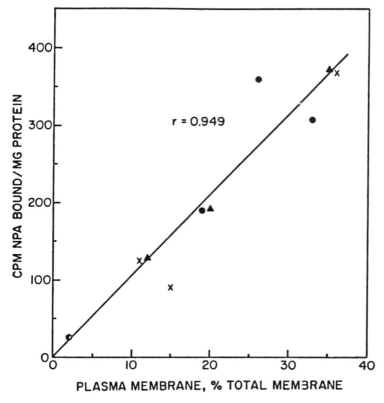

Figure 7. The correlation between N-1-naphthylphthalamic acid (NPA) bind ng and staining with PTA at low pH for membrane fractions from oat (*Avena sativa*) coleoptiles. From Lembi *et al.* (30).

(ii)   Suspend the pellet in 5 mM potassium phosphate, pH 6 8, and use this as starting material for the two-phase partition.

(iii)   Construct a 16 g two-phase system of the following composition: resuspended microsomal membranes (obtained from about 40 g of tissue), 6.4% (w/w) dextran T500 (Pharmacia), 6.4% (w/w) PEG 3350 (Fisher), 0.25 M sucrose and 5 mM potassium phosphate, pH 6.8.

(iv)   Mix the contents of the two-phase system by inversion of the tubes (40 times). Separate the two phases by centrifugation in a swinging bucket rotor at 150 *g* for 5 min.

(v)   Re-wash the upper phase, enriched in plasma membranes, twice against a fresh lower phase and separate the phases after each set of 40 time inversions by centrifugation (*Figure 3*).

(vi)   Repartition a fresh upper phase first against the original lower phase, followed by the two lower phases that were used to wash the original upper phase.

(vii)   Combine the two upper phases containing plasma membranes, dilute with 5 mM potassium phosphate, pH 6.8, and 0.25 M sucrose and centrifuge at 80 000 *g* for 30 min.

7.3 Isolation and identification of tonoplast

For the isolation of tonoplast, different approaches have been followed. Low-shear automated tissue slicing methods (74, 75) or preparation of protoplasts followed by controlled osmotic lysis (76−80), mechanical breakage (81) or by treatment with polybases are used to isolate vacuoles (82). These are identified due to the large size by light microscopy and are sedimented at very low centrifugal forces, for example at unit gravity (82). Natural pigments contained within the vacuole contents (74, 76) and neutral red, supplied exogenously (79, 81, 82) have been used to identify vacuoles. Yields generally have been low (below 0.5%) for both the mechanical and protoplast procedures (74,83).

While protoplasts provide, at present, one of the few approaches to the isolation of vacuoles from plant tissues (84), the mechanical, tissue slicing method has the advantage of minimizing changes which may occur during isolation of protoplasts. However, the tissue slicing procedure is largely restricted to firm tissues such as storage roots.

The lytic functions of plant vacuoles render them functionally equivalent to mammalian lysosomes (85), and vacuole preparations have been examined for hydrolase activities. Estimates of the purity of the preparations are based primarily on the absence of markers for other compartments (34, 80, 81, 84, 86). There is also evidence for the localization of the cyanogenic glucoside, dhurrin, in the vacuole of sorghum (79, 87) and the presence of proteinase inhibitor I (83) and malate (82) in vacuoles of other species. All of the constituents, as well as the α-mannosidase (80, 82, 88, 89) are content markers and are lost when the vacuoles are ruptured. Based on theoretical consideratons of the change in surface to volume ratios as vacuoles are broken progressively into smaller vesicles (90) and direct measurements of enzymatic activities (91), 97% of the content activities are solubilized during tissue homogenization of vacuolate plant cells whereas only 3% remain with the isolated, sealed vesicles.

The search for a membrane-associated marker for tonoplast vesicles was greatly aided by the assignment of an anion-stimulated, nitrate-inhibited, Mg^{2+}-ATPase to the tonoplast. This was based first on studies with latex vesicles of *Hevea brasilianensis* (92, 93) and subsequently with isolated vacuoles from red beet storage roots (94). Nevertheless, the absolute specificity of this assignment has been questioned (18). Our studies of etiolated soybean hypocotyls in which highly purified plasma membrane and tonoplast fractions from the same homogenate by free-flow electrophoresis (see Section 7) have been compared (21), support the location of an anion-stimulated, nitrate-inhibited Mg^{2+}-ATPase in the tonoplast.

7.4 Transport-competent (sealed) tonoplast vesicles

Tightly sealed vesicles of human erythrocyte membranes are separated from leaky ones by buoyancy differences in high polymer density gradients (95). Large molecules such as dextran will not permeate the membrane of sealed vesicles and yet they generate a low osmotic pressure so that the vesicles do not collapse. The density of sealed vesicles is determined mainly by the content within the vesicles (i.e. vesicles sealed in a medium of density 1.03 g/ml will equilibrate above a solution of 1.06 g/ml). On the other hand,

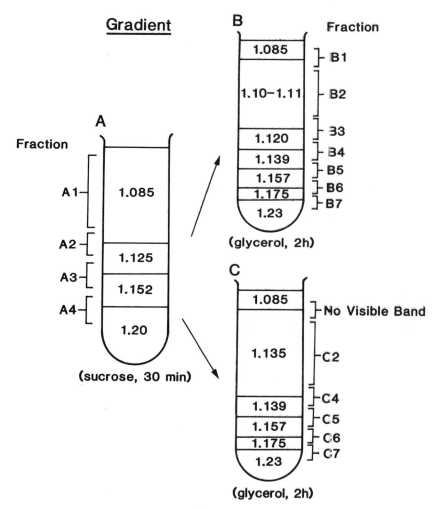

Figure 8. A sequential sucrose and glycerol gradient procedure for isolation of enriched plasma membrane and tonoplast fractions from plant stems. (Section 7.5). From Scherer (14, 99).

leaky vesicles will be dependent on the respective membrane density (1.10−1.20 g/ml). Transport-competent vesicles enriched in tonoplast membrane have been prepared using a dextran density cushion (96). Renografin gradients also give transport-active vesicles. These vesicles appear to have a higher specific activity of transport than vesicles prepared by differential centrifugation alone or vesicles purified by sucrose gradient fractionation as predicted from the above considerations (97, 98).

7.5 Consecutive sucrose and glycerol gradient centrifugation for preparation of tonoplast and plasma membranes

Enrichment of fractions in either tonoplast or plasma membrane vesicles active in ATP-dependent proton transport has been achieved by means of a preparative procedure involving two consecutive centrifugation steps: a sucrose step gradient and an isopycnic glycerol density gradient (14, 99) applicable for example, to etiolated plant stems.

(i)   Chop the tissue finely and then grind it using a mortar and pestle in ice-cold medium containing 8% ethanolamine (v/v), 20 mM ethylenediaminetetraacetic acid (EDTA), 0.4 M sodium β-glycerophosphate (grade III, Sigma) 2 mM dithioerythritol and 0.15 mM nupercaine or tetracaine [these prevent rapid enzymatic breakdown of membrane lipids following tissue homogenization (100)].

(ii)   Filter the homogenates through nylon cloth and centrifuge for 10 min at 4000 *g*.

(iii)   Load the supernatant onto a sucrose step gradient (*Figure 8*) and centrifuge for 30 min at 90 000 *g*.

(iv)   Collect fractions A2 and A3 from the sucrose gradient and re-load these onto glycerol gradients and re-centrifuge. Prior to loading, adjust fraction A2 to 0.15 M sucrose using a 1.2 M sucrose stock solution in homogenization buffer.

(v)   Prepare the glycerol gradients in Beckman SW 28.1 (17 ml) buckets or equivalent and centrifuge for 2 h at 90 000*g*.

Based on extensive characterization of marker activities, fraction B1 is enriched in tonoplast whereas fractions B5 + B6 and fractions C5 + C6 are enriched in plasma membranes (14, 99).

7.6 Free-flow electrophoresis method for separation of plasma membrane and tonoplast

Free-flow electrophoresis (*Figure 4*) affords a procedure whereby highly purified fractions of both plasma membrane and tonoplast are prepared from the same homogenate. Fractions collected from nearest the point of sample injection (farthest from the anode) are enriched in plasma membranes while fractions nearest the anode (farthest from the point of sample injection) are enriched in tonoplast (*Figure 5*). Other membrane cell components (dictyosomes, endoplasmic reticulum, plastids and mitochondria) are located in fractions intermediate between plasma membrane and tonoplast. A generalized procedure follows (21).

(i)   Prepare homogenates from 35−50 g of tissue (e.g. soybean hypocotyl segments), filter (nylon cloth) and centrifuge for 10 min at 6000 *g*.

(ii)   Centrifuge the resulting supernatant for 30 min at 60 000 *g*.

(iii)   Resuspend the pellets in electrophoresis chamber buffer (see below), re-centrifuge for 30 min at 60 000 *g* and finally resuspend the pellet in electrophoresis chamber buffer using approximately 1 ml per 10 g of starting fresh weight of tissue.

The electrophoresis medium (electrophoresis chamber buffer) contains 0.25 M sucrose, 2 mM KCl, 10 μM $CaCl_2$, 10 mM triethanolamine and 10 mM acetic acid adjusted to pH 7.5 with NaOH. The electrode buffer contains 100 mM triethanolamine and 100 mM acetic acid adjusted to pH 7.6 with NaOH.

(iv)   Carry out the separations in a continuous free-flow electrophoresis unit (VaP-5 produced by Bender and Hobein, Munich, FRG, or equivalent) under conditions of constant voltage (800 V/9.2 cm field), 165 ± 10 mA, buffer flow 1.7 ml/fraction/h, sample injection 2.7 ml/h, and constant temperature of 6°C. Monitor the distribution of membranes in the separation from the absorbance at 280 nm.

(v)   Collect the membranes from individual or pooled fractions by centrifugation at 80 000 *g* for 30 min.

8. MITOCHONDRIA

Functional mitochondria have been isolated from a variety of tissue sources (101). The general considerations given at the beginning of the chapter apply to mitochondria as well as to other organelles, and special considerations apply for their isolation from green tissue (102). To isolate mitochondria from plant homogenates, fatty acid-free bovine serum albumin (BSA) and quinone scavengers such as polyvinylpyrrolidone (PVP) are added. Potato (*Solanum tuberosum*) tubers, cauliflower (*Brassica oleracea* var *italica*) buds, and spadices of skunk cabbage (*Symplocarpus foeidus*) as well as tubers of Jerusalem artichoke (*Helianthus tuberosus*) are excellent sources of plant mitochondria although mitochondria can be isolated readily from virtually any plant tissue including maize (*Zea mays*) roots, pea (*Pisum sativum*) seedlings, spinach (*Spinacia oleracea*) leaves, mung bean (*Phaseolus aureus*) seedlings and beet (*Beta vulgaris*) and turnip (*Brassica napus*) roots.

(i)   Homogenize the tissues in two volumes of a medium containing 0.3 M mannitol (or sucrose), 4 mM cysteine, 1 mM EDTA, 30 mM Mops buffer, pH 7.8 and 0.1% de-fatted BSA with or without 0.6% (w/v) PVP.

(ii)   Filter the homogenates (cheese cloth, nylon mesh, Miracloth) and centrifuge for 3−4 min at 4000 *g* to remove starch and plastids.

(iii)   Centrifuge the supernatant from this initial centrifugation at 10 000 *g* for 20 min to pellet mitochondria.

(iv)   Mitochondria are purified from the differential pellets by centrifugation in gradients of sucrose (103−106). Resuspend the pellet in 4−6 ml of media and layer over a continuous 33−60% (w/w) sucrose gradient. Centrifuge at 100 000 *g* for 3 h (Beckman SW-28 rotor or equivalent) Mitochondria are located at a density of 1.19 g/ml and lysosomes (glyoxisomes) at 1.25 g/ml (193). Alternatively, Percoll (104, 105, 107−111), dextran (112, 113), Ficoll (114), or an aqueous two-phase polymer system (115, 116) may be used. Percoll gradients provide preparations capable of high rates of substrate oxidation with ADP:O_2 ratios approaching the theoretical values (117−119).

(v)   To assess the purity, measure cytochrome oxidase (120) or NADH-succinate oxidoreductase (49) activities.

Procedures that subfractionate mitochondria into outer and inner membranes, have been described (121), although there are no markers for the outer mitochondrial membrane of plants. Monoamine oxidase, used for this purpose in animal cells, is a soluble enzyme in plants (50). Sub-mitochondrial particles with membranes in an inside-out configuration when compared with intact mitochondria are useful to localize the components of the respiratory chain. These are prepared by sonication followed by differential centrifugation to remove unbroken mitochondria (122).

Additional marker activities (e.g. fumarase, cytochromes) and criteria of mitochondrial integrity (respiratory control, oxidative phosphorylation) and assays of membrane configuration have been described (103).

9. ENDOPLASMIC RETICULUM

There are several procedures for preparation of endoplasmic reticulum from various plants including castor bean endosperm (123), soybean hypocotyl (35, 124) and cress roots (125, 126). The procedure for isolation from castor beans (123) is as follows.

(i)   Chop 20 endosperm halves from 4-day old seedlings using a razor blade. The medium is 8 ml of 150 mM Tricine (pH 7.5) 10 mM KCl, 1 mM MgCl$_2$, 1 mM EDTA and 13% (w/w) sucrose.

(ii)   Filter (nylon cloth), adjust the volume to 10 ml and remove the debris by centrifuging for 10 min at 270 g .

(iii)   Apply the supernatant to a continuous 6−60% sucrose gradient dissolved in 150 mM Tricine, 10 mM KCl and 1 mM EDTA (pH 7.5).

(iv)   Centrifuge for 4 h at 100 000 g (Beckman SW-28 or equivalent). Collect the endoplasmic reticulum at a sucrose density of 1.12 g/ml (see Appendix for Tables). This is a crude fraction contaminated by other membranes (see Section 10).

Identification is based usually on the presence of ribosomes associated with the membranes, a situation favoured by incluson of 5−50 mM Mg^{2+} in the isolation medium. Marker enzymes used for endoplasmic reticulum include NADPH cytochrome c reductase, mannosyl transferase (127) and terminal enzymes of phospholipid biosynthesis (128) such as CDP-choline cytidyl transferase (123). The latter occurs also in Golgi apparatus (7, 129). NADH-cytochrome c reductase, although distributed among other cell components, has also been used as a marker for endoplasmic reticulum. For the mannosyl transferase reaction which occurs probably exclusively in endoplasmic reticulum of animal cells (130), the chloroform−methanol soluble product must be identified as a polyisoprenol derivative by several tests including its chromatographic behaviour on DEAE−cellulose (acetate form) and the release and identification of free mannose upon mild acid hydrolysis (127). Glucose-6-phosphatase does not occur in plants, and hydrolysis of glucose-6-phosphate in plant homogenates is due to an unspecific soluble phosphatase.

Although smooth endoplasmic reticulum occurs rather more rarely in plants than in animals, plants with oil-producing glands contain extensive areas of tubular endoplasmic reticulum lacking ribosomes (131). However, in most plant cells, smooth endoplasmic reticulum occurs only as local segments within larger cisternae with attached ribosomes.

A further method for separating rough endoplasmic reticulum from mitochondria with similar sizes and equilibrium densities involves the removal of ribosomes from the endoplasmic reticulum by extraction with EDTA and high KCl thereby lowering the density of the membranes. Alternatively, a shift to higher densities by addition of magnesium (132) has been used to aggregate endoplasmic reticulum. Another approach to resolution of endoplasmic reticulum and mitochondria uses free-flow electrophoresis (23).

10. GOLGI APPARATUS

Golgi apparatus fractions have been prepared from plant sources using continuous (133, 134) or discontinuous (24, 50) sucrose gradients as well as by rate-zonal centrifugation (26). A simple procedure for preparation of a Golgi apparatus-enriched fraction from

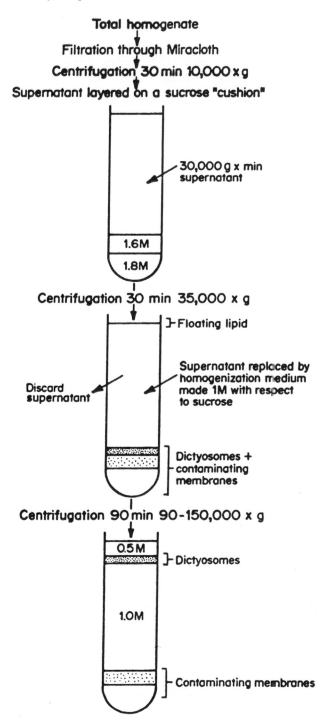

Figure 9. A simple discontinuous sucrose gradient procedure for Golgi dictyosome isolation. From Morré and Buckhout (50).

soybean stem is shown in *Figure 9*. Dextran is added to the isolation medium (1% final concentration) and it is necessary to work fast as the dictyosomes continually unstack as the isolation proceeds (24). Identification is based on the characteristic morphology (fenestrated cisternae) (3) and density in sucrose gradients (1.12 g/ml) as well as the presence of a latent IDPase activity (26, 39) and a glucan synthetase I activity (70), originally described as having a high K_m for UDP-glucose (47). The procedures for estimation of latent IDPase activity vary. The original description was of a latency overcome by storage of the fractions for several days at low temperature (26). This type of latency seems more specific than the more usual type of structure-linked latency defined by rupture of membranes with detergent treatment (39). Although there are many parallels between the two IDPase activities (135), the detergent-stimulated activity usually yields a somewhat broader subcellular distribution than that appearing upon cold storage.

At present there are no Golgi apparatus fractions from plants in which purity above 50% has been established. The composition of a Golgi apparatus fraction prepared from soybean hypocotyl (*Figure 9*) and characterized by morphometry and assay of enzyme markers was: 45% Golgi apparatus, 15% plasma membrane, 4% tonoplast, 1% mitochondria, 18% endoplasmic reticulum, and 10% plastids with 7% of the protein unaccounted for (50).

Secretory vesicles derived from Golgi apparatus were isolated from pollen tubes by a combined size discrimination (nucleopore filter) — centrifugation method (136). Subsequently, a method based on a two-step centrifugation on a discontinuous sucrose gradient was developed for vesicles from germinating pollen (137) and was used to demonstrate β-glucan synthetase activity associated with the vesicle membranes (138 − 140). More recent methods for putative secretory vesicle fractions from other plant sources have been described (141 − 143) but the fractions are poorly characterized.

11. PLASTIDS

No general procedure for isolation of plastids can be given, for the method employed depends both on the choice of tissue and on the developmental stage of the plastids to be isolated. The following will however exemplify methods for isolation of different kinds of plastids and plastid subfractions.

11.1 Chloroplasts

To isolate functional chloroplasts, the tissues of choice are spinach and peas. A detailed account of chloroplast isolation procedures, given by Leegood and Walker (144), is briefly summarized here.

The osmoticum most commonly employed for chloroplast isolation is 0.33 M sorbitol, although a variety of sugars (e.g. sucrose) and sugar alcohols (e.g. mannitol) may be used. Similarly, the choice of buffers is not critical, although pyrophosphate, Tris, Mes and Hepes are used widely. Small amounts of $MgCl_2$ and $MnCl_2$ may be included, as well as inorganic phosphate and/or EDTA to prevent aggregation. Isoascorbate and BSA may be added as protective agents. To increase the yield, the leaves are exposed to a high intensity illumination for 20−30 min prior to homogenization.

(i)   Disrupt de-ribbed leaves in three parts of grinding medium (see above) to one

(ii)   part of leaves with either a Waring-type blendor (full speed for 3−5 sec) or by means of a Polytron (low speed for up to 3 sec). Avoid long homogenization times as they tend to increase the proportion of broken chloroplasts.

(ii)   Filter the homogenates (Miracloth, nylon mesh or cheese cloth) and centrifuge at 6000−8000 g in a swing-out rotor for up to 90 sec. Lower centrifugation forces, for example 1000−2000 g for 20−60 sec, are also commonly used (36).

(iii)   Discard the supernatant and wash the pellet surface with a small quantity of medium to remove the upper layer of broken chloroplasts.

(iv)   Resuspend the remainder of the pellet, enriched in intact chloroplasts, in a small amount of medium by gentle swirling or with a soft paint brush.

The chloroplast fraction obtained requires further purification for functional or assignment studies. Chloroplasts purified by centrifugation on sucrose gradients may lose much of the capacity to fix CO_2. Methods that combine sedimentation velocity and isopycnic separations (145, 146) or that utilize rate separations (147) have been more successful. Percoll gradients yield chloroplasts with good retention of photosynthetic activity (148−151). A combination gradient of Percoll, PEG and Ficoll has also been employed successfully (152). Chloroplast fractions of very high purity can be obtained through purification of crude chloroplast fractions by preparative free-flow electrophoresis (153, 154) or by aqueous polymer two-phase partition (155, 156). These techniques have not been extensively used, although a purification step based on membrane surface properties is needed to remove plasma membrane-enclosed chloroplasts from the chloroplast fraction (155, 156).

The range of plants from which functional chloroplasts are isolated has been extended markedly by the use of protoplasts as starting material (144). However, it should be noted that protoplast isolation *per se* has been reported to affect several physiological properties, for example, protein synthesis (157), lipid and fatty acid metabolism (158, 159) and O_2 evolution and CO_2 fixation (160).

The most critical indicator of chloroplast functional integrity is the intactness and the efficiency of carbon assimilation (144). To test for intactness, any substrate prevented from entering the chloroplast by the envelope may be used. Among the simplest and most widely used is ferricyanide as a Hill oxidant. Competence in CO_2 assimilation is provided by measurements of photosynthesis (161).

The most frequently used marker for chloroplasts is chlorophyll (162), but note that this marker does not discriminate between intact plastids and free thylakoids. Other commonly used markers include stroma-localized enzymes, such as those specific to the reductive pentose phosphate pathway, for example $NADP^+$-dependent glyceraldehydephosphate dehydrogenase and ribulose biphosphate carboxylase (144).

11.2 Chloroplast subfractions

Functionally active thylakoids are best isolated from intact chloroplasts, as thylakoids prepared directly from homogenates may be degraded during isolation. The isolated chloroplasts are suspended in a hypotonic medium (163) and the thylakoids released are collected by differential centrifugaiton. Thylakoid vesicles of different orientations (inside-out, stroma-side-out) can be obtained by Yeda press treatment of stacked thylakoids followed by aqueous polymer two-phase partition (164). Markers for thylakoids include both membrane constituents and photosystem reactions.

Purification of chloroplast envelopes began with the work of Mackender and Leech (165) and procedures have now been described for their isolation in highly purified form from intact chloroplasts by gentle osmotic shock followed by centrifugation in a discontinous sucrose gradient (86, 166−168). If chloroplasts are purified on a Percoll gradient, they should be washed prior to disruption, or two mixed envelope membrane fractions may be obtained (169).

To separate the inner and outer envelope membranes, chloroplasts are infiltrated with hypertonic buffer and disrupted either by a freeze−thaw cycle (170), by French press treatment (171) or by homogenization in a ground-glass homogenizer (172). The two envelope membranes are resolved from each other and from other chloroplast constituents after discontinuous sucrose gradient centrifugations. For marker enzymes of chloroplast envelope membranes, see Keegstra and Yousif (172).

11.3 Etioplasts

Etioplasts are usually formed when tissues, in which plastid development normally leads to formation of chloroplasts (e.g. leaves, stems), are kept in darkness.

Etioplasts are isolated by procedures similar to those used for chloroplasts, with modifications due to differences in size and density. The procedure summarized here for isolation of etioplasts from dark-grown wheat leaves, utilizes purification of a crude etioplast fraction by Percoll gradient centrifugation (173).

(i)   Harvest leaves under dim green light.
(ii) Homogenize approximately 35 g of leaf tissue in 200 ml of isolation medium (0.50 M sucrose, 10 mM Hepes, 20 mM Tes, adjusted to pH 7.6 with KOH) using a Waring blendor for 2×5 sec at full speed.
(iii) Filter the homogenate through eight layers of cotton gauze and one layer of nylon cloth (mesh size 15 μm) and rinse the residue with 2×20 ml of isolation medium.
(iv) Centrifuge the filtrate for 6 min at 1000 g_{max} in a fixed-angle rotor (Sorvall, SS-34).
(v)   Resuspend the crude etioplast pellets in isolation medium and load onto gradients made of Percoll suspended in isolation medium (25% v/v) resting on a cushion of 67% (v/v) Percoll.
(vi) Centrifuge for 20 min at 1000 g_{max} (Sorvall, swing-out bucket rotor, HB-4), and collect the etioplasts from the interface between 25 and 67% Percoll.
(vii) Wash the etioplasts with isolation medium by centrifuging for 15 min at 1000 g_{max} (Sorvall, HB-4 rotor). Repeat this centrifugation step twice.

Levels of carotenoids and mono- and digalactosyldiacylglycerols are used to monitor relative etioplast enrichment (38). Fraction purity can be determined from morphometric analysis applied to preparations for transmission electron microscopy, where etioplast membranes are intensely stained selectively with aqueous permanganate fixation (38).

11.4 Etioplast subfractions

An etioplast envelope fraction can be isolated by procedures developed for chloroplast envelopes (174). After separation of the envelope membranes from the etioplasts, the inner membranes, prolamellar bodies and prothylakoids can then be separated from each other either by homogenization in a ground-glass homogenizer (175, 176) or by homogenization followed by sonication (173, 177). The prolamellar body and pro-thylakoid fractions are then resolved by continuous or discontinuous sucrose gradient centrifugation. Sonication will yield a very pure fraction of prolamellar bodies (177), with the drawback that promellar body fragments will contaminate the prothylakoid fraction. The degree of contamination increases with increased force of sonication (173).

11.5 Amyloplasts

Amloplasts occur in non-green tissues of plants and contain one to several starch-grains per plastid. Amyloplast fractions have been isolated from a variety of sources, including potato tubers (37), suspension cultures of soybean (178) and sycamore (179), maize endosperm (180), and pea roots (181). To isolate amyloplasts, the tissue is disrupted either mechanically (37, 181) or enzymatically (178–180), followed by collection of the amyloplasts by sedimentation at either unity gravity (37, 180) or by a low-speed centrifugation (178, 179, 181). Amyloplast fractions obtained by centrifugation may contain starch-grains devoid of envelope membranes. Plastids with multiple starch grains (e.g. amyloplasts of pea roots) tend to be more stable to centrifugation forces than amyloplasts with one starch-grain only (180).

A procedure for isolating amyloplasts from roots of dark-grown pea is as follows (181).

(i)   Slice approximately 20 g of roots with a razor blade in one volume of 50 mM Tricine-NaOH pH 7.9, 330 mM sorbitol, 1 mM EDTA, 1 mM $MgCl_2$, 0.1% BSA, and then gently macerate using a mortar and pestle.

(ii)   Filter the homogenate through six layers of muslin and centrifuge for 1 min at 200 g to remove intact cells, nuclei and cell debris.

(iii)   Transfer 10–15 ml aliquots of the resulting supernatant to centrifuge tubes, and underlay with 10 ml of 50 mM Tricine-NaOH pH 7.9, 330 mM sorbitol, 0.1% BSA and 10% (v/v) freshly dialysed Percoll.

(iv)   Centrifuge for 5 min at 4000 g in a swinging bucket rotor to sediment the amyloplasts. Re-suspend the amyloplast pellet carefully.

Markers for stroma-containing amyloplasts include nitrite reductase, glutamate syn-thase, glucose-6-phosphate dehydrogenase and gluconate-6-phosphate dehydrogenase; for descriptions of enzyme assays, see Emes and England (181). To test for intactness of the plastids obtained, latency and resistance to trypsin digestion may be assayed (178, 181).

The fate of ruptured amyloplast membranes is unknown in most tissue fractionation schemes. Carotenoids and galactolipids (37, 38), as well as the capability of galac-tolipid biosynthesis (37) are appropriate markers for membranes of amyloplasts.

11.6 Chromoplasts

Chromoplasts are formed from chloroplasts or amyloplasts by re-formation of internal membranes and are characterized by a high proportion of non-photosynthetic, mainly carotenoid, pigments. Several plant tissues, for example petals from *Narcissus pseudonarcissus* (182–184) and *Tropaeolum majus* (185), and fruits from *Capsicum annuum* (186) and *Lycopersicum esculentum* (187) have served as sources for chromoplasts. Generally, the tissue is homogenized mechanically and a crude chromoplast fraction is isolated from other cell components by differential and/or gradient centrifugations.

Chromoplast localized enzymes studied are membrane-associated, such as the enzymes involved in galactolipid (183, 185) and β-carotene (184) synthesis.

11.7 Plastoglobuli

Plastoglobuli are lipid-rich droplets, normally osmiophilic in sections for transmission electron microscopy, that are present in most kinds of plastids (188−191). Due to their low density, plastoglobuli usually collect at the surface of sucrose density gradients after centrifugations of ruptured plastids. The procedure for isolating plastoglobuli from crude fractions to chloroplasts or chromoplasts is as follows (188):

(i)   Suspend the plastids in buffer (1 mM phosphate buffer pH 7.6, 0.1 mM DTE) and sonicate for 5 min at 20 kHz. Alternatively, French press treat the suspension at 50 MPa.

(ii)   Centrifuge the resulting suspension for 60 min at 150 000 g in a fixed-angle rotor.

(iii)   Collect the slightly turbid plastoglobuli zone at the top and the agglomeration of plastoglobuli on the side of the tube and add to a solution of 5% Ficoll in phosphate buffer pH 7.6.

(iv)   In a centrifuge tube, prepare a discontinuous floating gradient of plastoglobuli suspension (bottom) overlayed with 2.5% Ficoll in phosphate buffer pH 7.6 and fill the tube with 1 mM phosphate buffer pH 7.6. Centrifuge gradient for 90 min at 250 000 g in a swinging bucket rotor.

(v)   Collect plastoglobuli from the top of the tube and repeat washing procedure (steps iv and v) at least six times.

(vi)   To prevent aggregation of plastoglobuli, briefly sonicate the final plastoglobuli suspension prior to use.

For composition of plastoglobuli, see references 188−191.

12. NUCLEI AND NUCLEAR ENVELOPE

Nuclei are usually prepared from filtered homogenatess obtained under low shear conditions (razor blade chopper, Polytron, mortar and pestle) by low-speed centrifugation (500 g, 10 min) followed by purification by centrifugation over 2.3 M sucrose. The nuclei pass through the sucrose during a 60 min centrifugation at 60 000 g, whereas other cell membranes do not enter the sucrose (192).

Nuclear preparations may be monitored by light and electron microscopy for content and purity. One criterion for intactness is that both inner and outer leaflets of the nuclear envelope are preserved. There are no enzymatic markers known for nuclei (see also Chapter 1). Lamella and tubules of endoplasmic reticulum connected to the outer leaflet of the nuclear envelope are the most common source of contamination. Acceptable yields from a homogenate would be 20−25% of the total DNA as intact nuclei. Details of the isolation and characterization of plant nuclei are provided by Dunham and Bryant (193).

12.1 **Nuclear envelope**

Nuclear envelopes from plants have been characterized (41). Their enzymatic and chemical composition is similar if not nearly identical to that of rough endoplasmic reticulum. Intact nuclear envelopes or nuclear envelope fragments can be identified in electron microscope sections or in negatively stained preparations by the presence of characteristic pore complexes that connect inner and outer membrane leaflets.

A standard method of preparation of nuclear envelopes from plant and animal cells is given below.

(i)   Dilute the sucrose-isolated nuclei with distilled water, causing nuclei to swell rapidly and the envelope to rupture releasing chromatin and nucleoli.

(ii)   After gentle sonication and low-speed centrifugation (3 min, 300 g) to remove debris and intact nuclei, purify the nuclear envelopes by layering the sonicated suspensions over a 62% (w/v) sucrose cushion and centrifuging for 30 min at 3000 g.

(iii)   Nucleoli and chromatin sediment whereas the envelope membranes collect on the cushion from which they are collected and then pelleted by centrifugation (194, 195).

12.2 **Chromatin and nucleoli**

To separate nucleoli from chromatin fragments, nuclear preparations are lysed with Triton, or sonicated, and then centrifuged at 2000 g in 30% glycerol. Nucleoli pellet whereas chromatin remains in the supernatant (196).

13. MICROBODIES

Microbodies (peroxisomes, glyoxysomes) are $0.5-1.5$ μm in diameter and spherical. Common features include a single bounding membrane, a dense stroma or granular matrix which stains cytochemically for catalase, and an equilibrium density of $1.24-1.26$ g/ml in sucrose density gradients (197, 198). Spinach leaves are often used to isolate leaf peroxisomes, while castor bean endosperm is a commonly used source of glyoxysomes.

Due to their single bounding membrane, microbodies are fragile during isolation. The method of tissue homogenization should therefore be gentle and repeated centrifugations and resuspensions avoided. The following general method can be used for the isolation of glyoxysomes from castor bean endosperm (199).

(i)   Chop endosperm tissue with a razor blade in one ml per gram tissue of homogenizing buffer: 0.45 M sucrose, 150 mM Tricine-KOH pH 7.5, 0.1 mM $MgSO_4$, 0.1 mM KCl, 0.05 mM $CaSO_4$, 0.1% BSA.

(ii)   Filter homogenate through nylon mesh.

(iii)   Layer $15-20$ ml of filtered homogenate on top of a linear gradient of $15-60$% (w/w) sucrose in 2.5 mM Tris-Mes pH 7.2 and centrifuge for 1 h at 80 000 g in a swinging bucket rotor.

(iv)   Collect the glyoxysome fraction from the gradient (at around 1.24 g/ml) and dilute it slowly with 10 volumes of dilution buffer (13.5% (w/w) sucrose, 0.1 mM $MgSO_4$, 0.1 mM KCl, 0.05 mM $CaSO_4$, 10 mM Tris-Mes pH 6.8, 0.05% BSA).

(v)   Pellet glyoxysomes by centrifugation at 10 000 g for 10 min.

Cellular debris, nuclei and intact plastids can be removed from the endosperm homogenate by a low speed centrifugation (150−500 g for 10 min) and, if necessary, the microbodies in the resulting supernatant concentrated by a 10 min centrifugation at 10 000 g prior to sucrose density gradient centrifugation (200).

In addition to catalase, leaf peroxisomes contain enzymes of the glycolate pathway. Glyoxysomes of fat storing tissues contain enzymes of β-oxidation of fatty acids and of the glyoxylate cycle (197−204).

Other organelles seldom contaminate microbody fractions obtained by density gradient centrifugation. There is, however, no obvious method to detect low levels of contamination of other membrane fractions by microbodies except from morphological examination of the pellets. The usual markers for identifcation of microbodies in purified fractions, catalase and peroxidase, while concentrated in microbodies, are widely distributed among plant fractions.

14. COATED VESICLES

Coated pits and coated vesicles are present in plant cells (205). Ultrastructurally, both are recognized by the presence of a clathrin coat on the cytoplasmic membrane surface. Both exocytotic (205) and endocytotic (206, 207) functions have been suggested.

The following method describes isolation of coated vesicles from protoplasts of suspension cultured soybean (208).

(i)   Re-suspend washed protoplasts in three volumes of homogenization buffer [100 mM 1,4-piperazine ethanesulphonic acid (Pipes) pH 6.9, 2 mM DTE, 2 mM EDTA, 1 mM MgCl$_2$, 0.30 M sorbitol] and homogenize with 50 strokes of a chilled Dounce homogenizer using a loose fitting pestle.

(ii)   Centrifuge homogenate for 10 min at 1000 g_{max}.

(iii)   Apply the supernatant from (ii) to a linear gradient of 15−60% (w/w) sucrose in a gradient buffer (45 mM Mes pH 6.5, 1 mM DTE, 1 mM EDTA, 0.1 mM MgCl$_2$) and centrifuge for 3 h at 140 000 g_{av} in a vertical rotor, using slow acceleration to 10 000 r.p.m., and slow braking.

(iv)   Collect 2 ml fractions from the bottom of the gradient. Dilute the coated vesicle-enriched fractions (48−52% sucrose w/w) with two volumes of gradient buffer, and centrifuge for 70 min at 75 000 g_{max} to pellet the coated vesicles.

(v)   To purify the coated vesicle-enriched fraction, dilute the 48−52% sucrose fractions with a gradient buffer to 10% (w/w) sucrose and layer onto a linear 15−25% (w/w) sucrose gradient. Centrifuge for 90 min at 16 000 g_{av} and collect 1.5 ml fractions from the bottom. Coated vesicles are enriched in the 16−18% sucrose fractions.

Coated vesicles of soybean protoplasts have a major polypeptide of a relative molecular weight of 180−190 kd (42−44, 208), and exhibit Glucan synthetase I activity (209).

15. CONCLUDING COMMENTS

It is now possible to isolate all cell components from plant cells but not necessarily from the some homogenate. Despite this limitation, it seems important that studies are not carried out using poorly or incompletely defined and characterized subcellular fractions. All major cell components should be accounted for (including tonoplast and plastids) and the electron microscope morphology of a representative portion of the pellet should be illustrated. If possible, the entire pellets should be sampled from a representative number of preparations and quantified by morphometry. Additionally, balance sheets should be constructed whenever possible for all markers and constituents measured, together with estimates of yields and recoveries. Only in this manner can the full potential of fractionation techniques applied to plant cells be realized.

16. REFERENCES

1. Morré,D.J., Gripshover,B., Boss,W.J. and Tuite,P.J. (1984) *Annals of the Phytochemistry Society of Europe.* Boudet,A.M., Alibert,G., Marigo,G. and Lea,P.J. (eds), Clarendon Press, Oxford, Vol. **64**, p. 247.
2. Morré,D.J. (1973) In *Molecular Techniques and Approaches in Developmental Biology.* Chrispeels,M.J. (ed.), John Wiley, New York, p. 1.
3. Morré,D.J. (1971) *Methods Enzymol.*, **22**, 130.
4. Mollenhauer,H.H., Morré,D.J. and Kogut,C. (1969) *Exp. Mol. Pathol*, **11**, 113.
5. Cunningham,W.P., Morré,D.J. and Mollenhauer,H.H. (1966) *J. Cell Biol.*, **28**, 169.
6. Morré,D.J., Mollenhauer,H.H. and Chambers,J.E. (1965) *Exp. Cell Res.*, **38**, 672.
7. Morré,D.J., Nyquist,S. and Rivera,E. (1970) *Plant Physiol.*, **45**, 800.
8. Morré,D.J. (1970) *Plant Physiol.*, **45**, 791.
9. Kappler,R., Kristen,U. and Morré,D.J. (1986) *Protoplasma*, **132**, 38.
10. deDuve,C. (1967) *Science*, **189**, 186.
11. Morré,D.J., Cline,G.B., Coleman,R., Evans,W.H., Glaumann,H., Headon,D.R., Reid,R., Siebert,G. and Widnell,C.C. (1979) *Eur. J. Cell Biol.*, **19**, 231.
12. deDuve,C. (1964) *J. Theor. Biol.*, **6**, 33.
13. Anderson,N.G., Harris,W.W., Barber,A.A., Rankin,C.T.,Jr. and Candler,E.L. (1966) *Natl. Cancer Inst. Monogr.*, **21**, 253.
14. Scherer,G.F.E. (1984) *Planta*, **160**, 348.
15. Chadwick,C.M. and Northcote,D.H. (1980) *Biochem. J.*, **186**, 411.
16. Albertsson,P.A. (1958) *Biochim. Biophys. Acta*, **27**, 328.
17. Albertsson,P.A., Andersson,B., Larsson,C. and Akerlund,H.E. (1982) *Methods Biochem. Anal.*, **28**, 115.
18. Larsson,C. (1985) In *Modern Methods of Plant Analysis, New Series.* Linskins,H.F. and Jackson,J.F. (eds), Springer-Verlag, Berlin/Heidelberg, Vol. **1**, p. 85.
19. Hannig,K. and Heidrich,H.-G. (1977) In *Cell Separation Methods. Part IV. Electrophoretic Methods.* Bloemendal,H. (ed.), Elsevier/North Holland Biomedical Press, New York, p. 93.
20. Morré,D.J., Creek,K.E., Matyas,G.R., Minnifield,N., Sun,I., Baudoin,P., Morré,D.M. and Crane,F.L. (1984) *BioTechniques*, **2**, 224.
21. Sandelius,A.S., Penel,C., Auderset,G., Brightman,A., Millard,M. and Morré,D.J. (1986) *Plant Physiol.*, **81**, 177.
22. Auderset,G., Sandelius,A.S., Penel,C., Brightman,A., Greppin,H. and Morré,D.J. (1986) *Plant Physiol.*, **68**, 1.
23. Morré,D.J., Lem,N.W. and Sandelius,A.S. (1984) In *Structure, Function and Metabolism of Plant Lipids.* Siegenthaler,P.S. and Eichenberger,W. (eds), Elsevier Science Amsterdam, p. 325.
24. Morré,D.J. and Mollenhauer,H.H. (1964) *J. Cell Biol.*, **23**, 295.
25. Morré,D.J., Hamilton,R.L., Mollenhauer,H.H., Mahley,R.W., Cunningham,W.P., Cheetham,R.D. and LeQuire,V.A. (1970) *J. Cell Biol.*, **44**, 492.
26. Ray,P.M., Shininger,T.L. and Ray,M.M. (1969) *Proc. Natl. Acad. Sci. USA*, **64**, 605.
27. Morré,D.J., Merlin,L.M. and Keenan,T.W. (1969) *Biochem. Biophys Res. Commun.*, **37**, 605.
28. Schachter,H., Jabbal,I., Hudgin,R.L., Pinteric,L., McGuire,E.J. and Roseman,S. (1970) *J. Biol. Chem.*, **245**, 1090.
29. Fleischer,B., Fleischer,S. and Ozawa,H. (1969) *J. Cell Biol.*, **43**, 59.
30. Lembi,C.A., Morré,D.J., Thomson,K.St. and Hertel,R. (1971) *Planta*, **99**, 37.
31. Roland,J.C., Lembi,C.A. and Morré,D.J. (1972) *Stain Technol.*, **47**, 195.
32. Marin,B., Cretin,H. and D'Auzac,D. (1982) *Physiol. Veg.*, **20**, 333.
33. Powell,M.F., Bracker,C.E. and Morré,D.J. (1983) *Protoplasma*, **111** 87.
34. Quail,P.H. (1979) *Annu. Rev. Plant Physiol.*, **30**, 425.

35. Lem,N. and Morré,D.J. (1985) *Protoplasma*, **128**, 14.
36. Jensen,R.G. and Bassham,J.A. (1966) *Proc. Natl. Acad. Sci. USA*, **56**, 1095.
37. Fishwick,M.J. and Wright,A.J. (1980) *Phytochemistry*, **19**, 55.
38. Hurkman,W.J., Morré,D.J., Bracker,C.E. and Mollenhauer,H.H. (1979) *Plant Physiol.*, **64**, 398.
39. Morré,D.J., Lembi,C.A. and VanDerWoude,W.J. (1977) *Cytobiologie*, **17**, 165.
40. Vigil,E.L. (1973) *Sub-Cell Biochem.*, **2**, 237.
41. Phillip,E., Franke,W.W., Keenan,T.W., Stadler,J. and Jarasch,E. (1976) *J. Cell Biol.*, **68**, 11.
42. Pearse,B.M.F. (1975) *J. Mol. Biol.*, **97**, 93.
43. Robinson,D.G. (1985) *Plant Membranes*. Wiley Interscience, New York.
44. Mersey,B.G., Fowke,K.C., Constabel,F. and Newcomb,E.H. (1982) *Exp. Cell Res.*, **141**, 459.
45. Hodges,T.K. and Leonard,R.T. (1974) *Methods Enzymol.*, **32**, 392.
46. Hardin,J.W., Cherry,J.H., Morré,D.J. and Lembi,C.A. (1972) *Proc. Natl. Acad. Sci. USA*, **63**, 3146.
47. VanDerWoude,W.J., Lembi,C.A., Morré,D.J., Kidinger,J.A. and Ordin,L. (1974) *Plant Physiol.*, **54**, 333.
48. Scherer,G.F.E. (1984) *Planta*, **161**, 395.
49. Pennington,R.J. (1969) *Biochem. J.*, **80**, 649.
50. Morré,D.J. and Buckhout,T.J. (1979) In *Plant Organelles*. Reid,E. (ed.), Ellis Horwood, Chichester, p. 117.
51. Cooperstein,S.A. and Lazarow,A. (1951) *J. Biol. Chem.*, **189**, 665.
52. Bowles,D.J., Quail,P.H., Morré,D.J. and Hartman,G.C. (1979) In *Plant Organelles. Methodological Surveys B: Biochemistry*. Reid,E. (ed.), Ellis Horwood, Chichester, Vol. 9, p. 207.
53. Loud,A.V. (1962) *J. Cell Biol.*, **15**, 481.
54. Morré,D.J., Yunghans,W.N., Vigil,E.L. and Keenan,T.W. (1974) In *Methodological Developments in Biochemistry. Subcellular Studies*. Reid,E. (ed.), Longman/London, Vol. 4, p. 195.
55. Blouin,A., Bolender,R.P. and Weibel,E. (1977) *J. Cell Biol.*, **72**, 441.
56. Croze,E. and Morré,D.J. (1984) *J. Cell Physiol.*, **72**, 441.
57. Luft,J.M. (1961) *J. Biophys. Biochem. Cytol.*, **9**, 409.
58. Spurr,A.R. (1969) *J. Ultrastruct, Res.*, **26**, 31.
59. Hodges,T.K., Leonard,R.T. and Bracker,C.E. (1972) *Proc. Natl. Acad. Sci. USA*, **69**, 3307.
60. Anderson,R.L. and Ray,P.M. (1978) *Plant Physiol.*, **61**, 723.
61. Nagahashi,G., Leonard,R.T. and Thomson,W.W. (1978) *Plant Physiol.*, **61**, 993.
62. Pierce,W.S. and Hendrix,D.L. (1979) *Planta*, **146**, 161.
63. Kjellbom,P. and Larsson,C. (1984) *Physiol. Plant.*, **62**, 501.
64. Lundborg,T., Widell,S. and Larsson,C. (1981) *Physiol. Plant.*, **52**, 89.
65. Uemura,M. and Yoshida,S. (1983) *Plant Physiol.*, **73**, 586.
66. Widell,S., Lundborg,T. and Larsson,C. (1982) *Plant Physiol.*, **70**, 1429.
67. Yoshida,S., Uemura,M., Niki,T., Sakai,A. and Agusta,L.V. (1983) *Plant Physiol.*, **72**, 105.
68. Griffing,L.R. and Quantro,R.S. (1984) *Proc. Natl. Acad. Sci. USA*, **81**, 4804.
69. Hodges,T.K. (1976) In *Transport in Plants II. Part A. Cells*. Luttge,U. and Pitman,M.G. (eds), Springer-Verlag, Berlin/Heidelberg/New York, p. 260.
70. Ray,P.M. (1977) *Plant Physiol.*, **59**, 594.
71. Hertel,R. (1974) In *Membrane Transport in Plants*. Zimmerman,W. and Dainty,J. (eds), Springer-Verlag, Berlin/New York, p. 859.
72. Hall,J.L. (1983) In *Isolation of Membranes and Organelles from Plant Cells*. Hall,J.L. and Moore,A.L. (eds), Academic Press, New York, p. 55.
73. Taylor,A.R.D. and Hall,J.L. (1978) *Protoplasma*, **96**, 613.
74. Leigh,R.A. and Branton,D. (1976) *Plant Physiol.*, **58**, 656.
75. Leigh,R.A., Branton,D. and Marty,F. (1979) *Plant Organelles, Methodological Surveys B. Biochemistry*. Reid,E. (ed.), Ellis Horwood, Chichester, p. 69.
76. Wagner,C.R. and Siegelman,H.W. (1975) *Science*, **190**, 1298.
77. Butcher,H.C., Wagner,G.J. and Seigelman,H.W. (1977) *Plant Physiol.*, **59**, 1098.
78. Lin,W., Wagner,G.J., Siegelman,H.W. and Hind,G. (1977) *Biochim. Biophys. Acta*, **405**, 110.
79. Saunders,J.A. and Conn,E.E. (1978) *Plant Physiol.*, **61**, 154.
80. Boller,T. and Kende,H. (1979) *Plant Physiol.*, **63**, 1123.
81. Nishimura,M. and Beevers,H. (1978) *Plant Physiol.*, **62**, 44.
82. Buser,C. and Matile,P. (1977) *Z. Pflanzenphysiol.*, **82**, 462.
83. Walker-Simmons,M. and Ryan,C.A. (1977) *Plant Physiol.*, **60**, 61.
84. Wagner,C.R. (1983) In *Isolation of Membranes and Organelles from Plant Cells*. Hall,J.L. and Moore,A.L. (eds), Academic Press, New York/London, p. 83.
85. Marty,M.F. (1978) *Proc. Natl. Acad. Sci. USA*, **75**, 852.
86. Douce,R., Holtz,R.B. and Benson,A.A. (1973) *J. Biol. Chem.*, **248**, 7215.

87. Saunders,J.A., Conn,E.E., Lin,C.H. and Stocking,C.R. (1977) *Plant Physiol.*, **59**, 647.
88. Matile,P. (1975) *The Lytic Compartment of Plant Cells*. Springer-Verlag, Vienna/New York.
89. Matile,P. (1966) *Z. Naturforsch.*, **21**, 871.
90. Twohig,F. (1974) MS Thesis, Purdue University.
91. Twohig,F., Morré,D.J. and Vigil,E.L. (1974) *Proc. Indian Acad. Sci.*, **83**, 86.
92. Marin,B., Marin-Lanza,M. and Komor,E. (1981) *Biochem. J.*, **198**, 365.
93. Marin,B. (1983) *Planta*, **157**, 324.
94. Walker,R.R. and Leigh,R.A. (1981) *Planta*, **153**, 140.
95. Steck,J.L. and Kant,J.A. (1974) *Methods Enzymol.*, **31**, 172.
96. Thom,M. and Komar,E. (1984) *FEBS Lett.*, **173**, 1.
97. Sze,H. (1983) *Biochim. Biophys. Acta*, **732**, 596.
98. Sze,H. (1985) *Annu. Rev. Plant Physiol.*, **36**, 175.
99. Scherer,G.F.E. (1984) *Z. Naturforsch.*, **37C**, 550.
100. Scherer,G.F.E. and Morré,D.J. (1978) *Plant Physiol.*, **62**, 933.
101. Laties,G.G. (1974) *Methods Enzymol.*, **31**, 589.
102. Leaver,C.J., Hack,E. and Forde,B.G. (1983) *Methods Enzymol.*, **97**, 476.
103. Moore,A.L. and Proudlove,M.O. (1983) In *Isolation of Membranes and Organelles from Plant Cells*. Hall,J.L. and Moore,A.L. (eds), Academic Press, New York/London, p. 153.
104. Jackson,C. and Moore,A.L. (1979) In *Plant Organelles. Methodological Surveys B. Biochemistry*, Reid,E. (ed.), Ellis Horwood, Chichester, Vol. 9, p. 173.
105. Tolbert,N.E. (1974) *Methods Enzymol.*, **31**, 734.
106. Day,D.A. and Hanson,J.B. (1977) *Plant Sci. Lett.*, **11**, 99.
107. Gardeström,P. and Edwards,G.E. (1963) *Plant Physiol.*, **74**, 24.
108. Jackson,C., Dench,J.E., Hall,D.O. and Moore,A.L. (1979) *Plant Physiol.*, **64**, 150.
109. Goldstein,A.H., Anderson,J.O. and McDaniel,R.G. (1980) *Plant Physiol.*, **66**, 488.
110. Goldstein,A.H., Anderson,J.O. and McDaniel,R.G. (1981) *Plant Physiol.*, **67**, 594.
111. Hrubec,T.C., Robinson,J.M. and Donaldson,R.P. (1985) *Plant Physiol.*, **77**, 1010.
112. Koeppe,D.E., Cox,J.R. and Gruenwald,P.J. (1978) In *Plant Mitochondria*. Ducet,G. and Lance,C. (eds), Elsevier, Amsterdam, p. 419.
113. Solomos,T., Malhora,S.S. and Spencer,M. (1973) *Plant Physiol.*, **59**, 807.
114. Pitt,D. and Galpin,M. (1973) *Plant Physiol.*, **109**, 233.
115. Larsson,C. and Andersson,B. (1979) In *Plant Organelles. Methodological Surveys B. Biochemistry*. Reid,E. (ed.), Ellis Horwood, Chichester, Vol. 9, p. 35.
116. Fisher,D. (1981) *Biochem. J.*, **196**, 1.
117. Bergman,A., Gardeström, P. and Ericson,I. (1980) *Plant Physiol.*, **66**, 442.
118. Gardeström,P., Bergman,A., Ericson,I. and Sahlstrom,S. (1981) In *Proceedings of the 5th International Congress on Photosynthesis*. Akoyunoglou,G. (ed.), p. 633.
119. Gardeström,P., Bergman,A. and Ericson,I. (1980) *Plant Physiol.*, **65**, 389.
120. Sun,I. and Crane,F.L. (1969) *Biochim. Biophys. Acta*, **172**, 417.
121. Bowles,D.J., Schnarrenberger,C. and Kauss,H. (1976) *Biochem. J.*, **150**, 375.
122. Møller,I.M., Bergman,A., Gardeström,P., Ericson,I. and Palmer,J.M. (1981) *FEBS Lett.*, **126**, 13.
123. Lord,J.M., Kasagawa,T., Moore,T.S. and Beevers,H. (1973) *J. Cell Biol.*, **57**, 659.
124. Williamson,F.A., Morré,D.J. and Jaffee,M.J. (1975) *Plant Physiol.*, **56**, 738.
125. Buckhout,T.J., Heyder-Caspers,L. and Sievers,A. (1982) *Planta*, **156**, 108.
126. Buckhout,T.J. (1983) *Planta*, **159**, 84.
127. Lord,J.M. (1983) In *Isolation of Membranes and Organelles from Plant Cells*. Hall,J.L. and Moore,A.L. (eds), Academic Press, New York/London, p. 119.
128. Bowden,L. and Lord,T.M. (1975) *FEBS Lett.*, **49**, 369.
129. Montague,M.J. and Ray,P.M. (1977) *Plant Physiol.*, **59**, 225.
130. Creek,K.E., Morré,D.J., Silverman-Jones,C.S., Shidoji,Y. and De Luca,L.M. (1983) *Biochem. J.*, **210**, 541.
131. Schnepf,E. (1972) *Biochem. Physiol. Pflanzen*, **163**, 113.
132. DePierre,J.W. and Dallner,G. (1975) *Biochim. Biophys. Acta*, **415**, 411.
133. Chanson,A., McNaughton,E. and Taiz,L. (1984) *Plant Physiol.*, **76**, 498.
134. Chanson,A. and Taiz,L. (1985) *Plant Physiol.*, **78**, 232.
135. Nagahashi,J. and Kane,A.P. (1982) *Protoplasma*, **112**, 167.
136. VanDerWoude,W.J., Morré,D.J. and Bracker,C.E. (1971) *J. Cell Sci*, **8**, 331.
137. Engles,F.M. (1973) *Acta Bot. Neerl.*, **22**, 6.
138. Engles,F.M. (1974) *Acta Bot. Neerl.*, **23**, 81.
139. Hespler,J.P.F.G., Veerkamp,J.H. and Sassen,M.M.A. (1977) *Planta*, **133**, 303.
140. Hespler,J.P.F.G. (1981) *Acta Bot. Neerl.*, **30**, 1.

141. Ray,P.M., Eisinger,W.R. and Robinson,D.G. (1976) *Ber. Dtsch. Bot. Ges.*, **89**, 121.
142. Taiz,L., Murry,M. and Robinson,D.G. (1984) *Planta*, **758**, 34.
143. Binari,L.L.W. and Racusen,R.N. (1983) *Plant Physiol.*, **76**, 26.
144. Leegood,R.C. and Walker,H. (1983) In *Isolation of Membranes and Organelles from Plant Cells.* Hall,J.L. and Moore,A.L. (eds), Academic Press, New York/London, p. 185.
145. Rocha,V. and Ting,I.P. (1970) *Arch. Biochem. Biophys.*, **140**, 398.
146. Miflin,B.J. and Beevers,H. (1974) *Plant Physiol.*, **53**, 870.
147. Leech,R.M. (1964) *Biochim. Biophys. Acta.*, **79**, 637.
148. Morgenthaler,J.-J., Price,C.A., Robinson,J.M. and Gibbs,M. (1974) *Plant Physiol.*, **54**, 532.
149. Mills,W.F. and Joy,K.W. (1980) *Planta*, **148**, 75.
150. Mourioux,G. and Douce,R. (1981) *Plant Physiol.*, **67**, 470.
151. Bertrams,M., Wrage,K. and Heinz,E. (1981) *Z. Naturforsch.*, **36**, 62.
152. Takabe,T., Nishimura,M. and Akazawa,T. (1979) *Agric. Biol. Chem.*, **43**, 2137.
153. Klofat,V.W. and Hannig,K. (1967) *Z. Physiol. Chem.*, **348**, 739.
154. Dubacq,J.-P. and Kader,J.-C. (1978) *Plant Physiol.*, **61**, 465.
155. Larsson,C. and Albertsson,P.-A. (1974) *Biochim. Biophys. Acta*, **245**, 425.
156. Larsson,C., Andersson,B. and Roos,G. (1977) *Plant. Sci. Lett.*, **8**, 291.
157. Reusink,A.W. (1978) *Physiol. Plant.*, **44**, 48.
158. Webb,M.S. and Williams,J.P. (1984) *Plant Cell Physiol.*, **25**, 1541.
159. Webb,M.S. and Williams,J.P. (1984) *Plant Cell Physiol.*, **25**, 1551.
160. Morris,P., Linstead,P. and Thain,J.F. (1981) *J. Exp. Bot.*, **32**, 801.
161. Walker,D.A. (1980) *Methods Enzymol.*, **69**, 94.
162. Arnon,D.I. (1949) *Plant Physiol.*, **24**, 1.
163. Reeves,S.G. and Hall,D.O. (1980) *Methods Enzymol.*, **69**, 85.
164. Andersson,B., Sundby,C., Akerlund,H.-E. and Albertsson,P.-A. (1985) *Physiol. Plant.*, **65**, 322.
165. Mackender,R.O. and Leech,R.M. (1970) *Nature*, **228**, 1347.
166. Block,M.A., Dorne,A.-T., Joyard,J. and Douce,R. (1983) *J. Biol. Chem.*, **258**, 13272.
167. Poincelot,R.P. (1980) *Methods Enzymol.*, **69**, 121.
168. Douce,R. and Joyard,J. (1979) *Adv. Bot. Res.*, **7**, 1.
169. Haas,R., Siebertz,H.P., Wrage,K. and Heinz,E. (1980) *Planta*, **148**, 238.
170. Cline,K., Andrews,J., Mersey,N., Newcomb,E.H. and Keegstra,K. (1981) *Proc. Natl. Acad. Sci. USA*, **78**, 3595.
171. Douce,R. and Joyard,J. (1982) In *Methods in Chloroplast Molecular Biology.* Edelman,M., Hallick,R. and Chua,H.-H. (eds), Elsevier, Amsterdam, New York, p. 239.
172. Keegstra,K. and Yousif,A.E. (1986) *Methods Enzymol.*, **119**,
173. Sandelius,A.S. and Selstam,E. (1984) *Plant Physiol.*, **76**, 1041.
174. Bahl,J. (1977) *Planta*, **136**, 21.
175. Wellburn,A.R. and Hampp,R. (1979) *Biochem. Biophys. Acta*, **547**, 380.
176. Lutz,C. (1978) In *Chloroplast Developments.* Akoyunoglou,G. and Argyroudi-Akoyunoglou,J.H. (eds), Elsevier/North Holland, Amsterdam, p. 481.
177. Ryberg,M. and Sundqvist,C. (1982) *Physiol. Plant.*, **56**, 125.
178. MacDonald,F.D. and ap Rees,T. (1983) *Biochim. Biophys. Acta*, **755** 81.
179. Macherel,D., Viale,A. and Akazawa,T. (1986) *Plant Physiol.*, **80**, 1041.
180. Echeverria,E., Boyer,C., Liu,K.-C. and Shannon,J. (1985) *Plant Physiol.*, **77**, 513.
181. Emes,M. and England,S. (1986) *Planta*, **168**, 161.
182. Liedvogel,B., Sitte,P. and Falk,H. (1976) *Cytobiology*, **12**, 155.
183. Liedvogel,B. and Kleinig,H. (1976) *Planta*, **129**, 19.
184. Kreutz,K., Beyer,P. and Kleinig,H. (1982) *Planta*, **154**, 66.
185. Liedvogel,B., Kleinig,H., Thompson,J.A. and Falk,H. (1978) *Planta*, **191**, 303.
186. Camara,B., Bardat,F. and Moneger,R. (1982) *Eur. J. Biochem.*, **127**, 255.
187. Iwatsuki,N., Moriyama,R. and Asahi,T. (1984) *Plant Cell Physiol.*, **25**, 763.
188. Steinmuller,D. and Tevini,M. (1985) *Planta*, **163**, 201.
189. Hansmann,P. and Sitte,P. (1982) *Plant Cell Reports*, **1**, 111.
190. Tevini,M. and Steinmuller,K. (1985) *Planta*, **163**, 91.
191. Dahlin,C. and Ryberg,H. (1986) *Physiol. Plant.*, **68**, 39.
192. Kuehl,L. (1964) *Z. Naturforsch.*, **19b**, 525.
193. Dunham,V.L. and Bryant,J.A. (1983) In *Isolation of Membranes and Organelles from Plant Cells*, Hall,J.L. and Moore,A.L. (eds), Academic Press, New York and London, p. 237.
194. Franke,W.W. (1974) *Phil. Trans. Ry. Soc. London B*, **268**, 67.
195. Franke,W.W. (1966) *J. Cell Biol.*, **31**, 619.
196. Lin,C.-Y., Guilfoyle,T.J., Chen,Y.M. and Key,J.L. (1975) *Plant Physiol.*, **56**, 850.

197. Vigil,E.L. (1973) *Sub. Cell Biochem.*, **2**, 237.
198. Tolbert,N.E. (1980) In *The Biochemistry of Plants*. Stumpf,P.K. and Conn E.E. (eds), Academic Press, New York, p. 359.
199. Mettler,I.J. and Beevers,H. (1980) *Plant Physiol.*, **66**, 555.
200. Huang,A.H.C., Trelease,R.N. and Moore,T.S.Jr (1983) *Plant Peroxisomes*. Academic Press, New York.
201. Baker,A.L. and Tolbert,N.E. (1966) *Methods Enzymol.*, **9**, 338.
202. Tolbert,N.E., Oeser,A., Kisaki,T., Hageman,R.H. and Yamazaki,R.K. (1968) *J. Biol. Chem.*, **243**, 5179.
203. Cooper,T.G. and Beevers,H. (1969) *J. Biol. Chem.*, **244**, 3507.
204. Cooper,T.G. and Beevers,H. (1969) *J. Biol. Chem.*, **244**, 3514.
205. Newcomb,E.H. (1980) In *Coated Vesicles*. Ockleford,C.D. and Whyte,A. (eds), Cambridge University Press, Cambridge, p. 55.
206. Tanchak,M.A., Griffing,L.R., Mersey,B.G. and Fowke,L.C. (1984) *Planta*, **162**, 481.
207. Joachim,S. and Robinson,D.G. (1984) *Eur. J. Cell Biol.*, **34**, 212.
208. Mersey,B.G., Griffing,L.R., Rennie,P.J. and Fowke,L.C. (1985) *Planta*, **163**, 317.
209. Griffing,L.R., Mersey,B.G. and Fowke,L.C. (1986) *Planta*, **167**, 175.

CHAPTER 3

Immunological methods applicable to membranes

ELAINE M.BAILYES, PETER J.RICHARDSON and J.PAUL LUZIO

1. INTRODUCTION

Immunological techniques are widely used in the study of membrane structure and function. Their development has been stimulated by the ability to obtain specific antibodies to membrane antigens even when the latter have not been fully purified or characterized. It is now common for immunological methods to be used for the identification, intracellular localization, quantitation, topographical analysis and purification of membrane components. Antibodies may also be used in the isolation of membrane fractions, to modulate biological function and as probes in the molecular cloning of cDNA coding for membrane proteins. In the present chapter no attempt is made to describe general techniques for the detection of cell surface antigens, nor to describe the variety of labels and methods that have enabled the visualization of antibody—antigen complexes with both the light and electron microscope. Our aim is to describe the relative merits and practicalities of raising polyclonal or monoclonal antibodies to membrane components and their use in immunoblotting, the purification of membrane fractions or membrane macromolecules and as probes for molecular cloning.

2. THE PREPARATION OF POLYCLONAL AND MONOCLONAL ANTIBODIES TO MEMBRANE COMPONENTS

2.1 The relative merits of polyclonal and monoclonal antibodies

Although antibodies are used as tools in many biological studies, it is at best difficult and often impossible to predetermine their properties. Polyclonal antisera will contain antibodies of different specificity and affinity as well as antibodies of different class and type. In addition, unless the immunogen used is pure, polyclonal antisera are likely to contain antibodies to other constituents of the immunogen preparation. Even in the most favourable circumstances, specific antibodies are unlikely to comprise more than 5% of the total serum immunoglobulin, although if pure antigen is available these antibodies may be affinity purified.

Monoclonal antibodies may be prepared as pure chemical reagents even when the available antigen is impure. Indeed, as described below, monoclonal antibodies have proven extremely useful in the immunoaffinity purification of many minor membrane proteins. Most monoclonal antibodies prepared are of low affinity ($K_a < 0.5$ nM^{-1}) (1). Whilst low affinity antibodies are useful for some purposes, particularly immunoaffinity purification, high affinity monoclonal antibodies or affinity-purified polyclonal antibodies of high avidity are likely to be advantageous in techniques such as immunoaffinity purification of membrane fractions, immunoblotting and as probes for molecular cloning. In some circumstances, polyclonal antibodies may be more useful reagents since they recognize more than one antigenic site (epitope) on the macromolecule of interest. Monoclonal antibodies, by definition, will only recognize a single type of epitope, though this epitope may occur more than once on the antigen, for example on the dimer of the membrane enzyme 5′-nucleotidase (2).

The properties of a section of protein causing it to be immunogenic have been the subject of much interest in recent years. Studies, especially by Atassi and co-workers (3) have suggested that polyclonal antisera raised to native proteins contain antibodies only to certain defined regions of the protein. They generated overlapping peptide fragments representing the entire sequence of the proteins lysozyme, myoglobin, haemoglobin and bovine serum albumin and showed that a limited number of fragments could absorb all the antibodies from polyclonal antisera raised to the intact molecules. Epitopes were classified into two types: either 'continuous', consisting of a single, peptide bond-linked, section of primary sequence such as is found in all five sites in myoglobin; or 'discontinuous', formed by the juxtaposition in the secondary or tertiary structure of distant sections of primary sequence, such as all three sites in lysozyme. Both types of epitope, which tend to occur once every 5−10 kd, are at the surface of the molecules, sometimes in prominent loops. They are limited in size, typically covering 6−7 amino acid residues at the protein surface, though they are not necessarily the most hydrophilic zones. An alternative approach to defining epitopes on the surface of proteins pursued by Lerner and co-workers (4) has provided more hope for the prospect of preparing antibodies of predetermined specificity raised against chemically synthesized peptides. In experiments in which antisera were raised to overlapping synthetic peptides representing the total sequence of an influenza virus haemagglutinin (HA1), 18 out of 20 sera reacted with the intact native protein. From the known structure of this haemagglutinin, it was calculated that some part of each immunoreactive peptide would be exposed at the surface of the intact molecule. In general the antigenicity of a section of protein appears to be related to accessibility, hydrophilicity and mobility (5).

Polyclonal antisera are the end product of a complex series of events in the immune system which may include the suppression of lymphocyte clones producing antibodies to some parts of the immunogen. Since the production of monoclonal antibodies effectively freezes and immortalizes a stage in the development of antibody-producing clones, antibodies to unexpected parts of the antigenic molecule not represented in a polyclonal antiserum may be produced. Analysis of monoclonal antibodies prepared at different stages of the maturation of the immune response (6) may eventually result in the development of guidelines for improving the chances of obtaining antibodies of particular affinity and specificity.

2.2 Immunization for the preparation of polyclonal antibodies

It is not possible to recommend a single immunization schedule suitable for all antigens that will produce high titre antisera. Many techniques have been reported but were rarely based on comparative trials. It is worthwhile consulting a general experimental immunology handbook before commencing immunization with membranes or membrane macromolecules, since this will provide a useful guide both to routes of immunization (7) and the advisability of using an adjuvant (8). Membrane proteins may require solubilization in detergent before immunization and excess detergent may interfere with the production of adjuvant emulsions. Alternatives to using solubilized proteins include immunization with gel slices containing the protein from sodium dodecyl-sulphate — polyacrylamide gel electrophoresis (SDS — PAGE) gels (9) (though the protein may be electroeluted first)(see Chapter 6) or even strips from nitrocellulose blots (10). The raising of antibodies to lipid antigens has proved difficult, with most of the antisera successfully raised recognizing primarily the carbohydrate portion of glycolipids (11). A variety of immunization schedules have been used, in particular immunization with whole membrane fractions (12), whole cells (13) or purified glycolipid emulsified with protein (14, 15).

In this laboratory, when using membrane proteins as immunogens, it is usual to inject rabbits, guinea pigs or sheep at multiple intradermal sites with $10-100$ μg of protein per animal on each occasion of primary or secondary immunization. The primary injection is usually given using complete Freunds as adjuvant and boosts with incomplete Freunds as adjuvant. A minimum of 6 weeks elapses between injections and the animals are bled 10 days after second and subsequent injections. Collected blood is allowed to clot, and the serum is separated by centrifugation and heated at $56°C$ for 45 min, if required, to inactivate the complement pathway. Immunoglobulin fractions may be prepared by salt precipitation and column chromatographic methods as described in several experimental texts (e.g. 16) or by binding to Protein A — Sepharose (17).

2.3 The preparation of monoclonal antibodies

Monoclonal antibodies are generated by immortalizing the antibody production of individual B lymphocytes by fusion with an appropriate myeloma cell line. The resultant hybrid cells retain the *'in vitro'* growth characteristics of the myeloma and secrete the antibody encoded by the lymphocyte. A permanent source of a defined antibody is thus obtained (18).

The principles and protocols for the production of monoclonal antibodies have been described (19, 20). In this chapter it is only possible to outline the basic procedures involved and to give the protocol used in this laboratory (*Table 1*). Spleen cells from

Table 1. The preparation of monoclonal antibodies to membrane proteins.

A. *Immunization schedule*

The antigen is emulsified in Freunds complete adjuvant and $1-100$ μg protein per animal injected subcutaneously in a total volume of $0.2-0.5$ ml per mouse or up to 1 ml in the rat. The available mouse myelomas were derived from the BALB/c strain and the rat myelomas from the Lou strain. The use of alternative strains for immunization will require the breeding of appropriate hybrid animals for the production of ascites fluid. The amount of protein injected depends on the degree of purity of the antigen, 10 μg should be ample for a purified preparation. If whole cells are to be used as immunogen they should be washed free of protein-containing medium and injected into the peritoneum ($0.5 \times 10^6-1 \times 10^7$ cells in $0.5-1.0$ ml saline).

An intermediate injection of antigen in Freunds incomplete adjuvant can be given $3-4$ weeks after the first immunization but in many cases a satisfactory immune response can be obtained from a boost injection of antigen in saline, $6-8$ weeks after the primary immunization. In the mouse this injection can be i.p. or i.v. but in the rat the latter route is the only choice. Three days later, the animals should be tested for the presence of serum antibodies to select individuals for the fusions on the following day. Samples of $100-200$ μl of blood can be obtained by cutting off under light anaesthesia a small portion of the tail with a sharp scalpel blade.

Table 1. Continued

B. *Culture media*

1.   Grow myeloma and hybridoma cells at 37°C in Dulbecco's modified Eagle's Medium (DMEM) containing heat-inactivated fetal calf serum. 5–10% serum (DMEM 5–DMEM 10) is sufficient for well-established hybridomas and myeloma cells but new hybridomas require 20% serum.

2.   HAT and HT media can be bought complete or prepared from stock concentrates.
        100 × hypoxanthine/thymidine (HT): dissolve 136 mg of hypoxanthine and 39 mg of thymidine 100 ml of water either by heating up to 70°C or adding NaOH.
        100 × aminopterin (A): 1.76 mg/100 ml water.
        Both solutions can be filter sterilized or added to the medium before it is filter sterilized. HAT medium (containing 5 ml of 100 × HT and 5 ml of 100 × A per 500 ml) should be tested before the fusion for its ability to maintain hybrid cell growth and to kill myelomas.

3.   Prepare 50% PEG the day before the fusion by autoclaving 10 g of PEG 1500 for 20 min and adding 10 ml of DMEM (warmed to 37°C) whilst it is still hot. On the day of fusion adjust the pH by shaking in the air of the sterile cabinet until it is slightly alkaline, pink/orange by the phenol red dye.

C. *Myeloma cells*

1.   Grow up 2 × 10^7 myeloma cells per fusion in DMEM 10. Use P3-NS1-Ag 4-1 (NS-1) (no endogenous heavy chain production), X63/Ag 8.653 or NSO/1 (no endogenous chain production) for mouse fusions; for rat fusions choose Y3-Ag 1.2.3 (no endogenous heavy chain production).

2.   Maintain cultures in log phase for at least 1 week, at cell concentrations of $10-30 \times 10^4$/ml, before the fusions.

D. *Cell fusion*

1.   Kill the animal by cervical dislocation or CO_2 asphyxiation, swab or dip it in alcohol and remove the spleen under sterile conditions into a Petri dish containing 5 ml of DMEM 2.

2.   Collect as much blood as possible from the dead animal by cardiac puncture and retain the serum for assay.

3.   Remove any connective tissue from the spleen, snip the capsule, transfer to a round-bottomed tube and disperse the spleen cells into 10 ml of DMEM 2 using a loose-fitting Teflon homogenizer.

4.   Transfer the suspension to a 30 ml Universal tube, top up to 20 ml and centrifuge at 600–700 r.p.m. for 7 min at room temperature.

5.   Resuspend the spleen cell pellet in 21 ml of DMEM 2, remove 1 ml into 19 ml of DMEM to count the cells.

6.   Pellet spleen cells and myeloma cells, then resuspend 2 × 10^8 spleen cells (~1 mouse spleen or ½ rat spleen) and 2 × 10^7 myeloma cells together in 20 ml of DMEM and pellet again.

7.   Drain the cells well, tap them to disperse the pellet and place the tube in a beaker of water at 37°C.

8.   Add 1 ml of 50% PEG, (pre-warmed to 37°C) over 1 min, stirring them gently. Stir for 1 min, add 1 ml of DMEM (37°C) over 1 min, followed by 19 ml of DMEM (37°C) over the next 5 min, with stirring.

9.   Pellet the cells and resuspend them gently to 100 ml in DMEM 20 and plate out over four 96-well plates or four 24-well plates.

10.   Grow overnight in a CO_2 humidified 37°C incubator.

E. *HAT selection*

1.   On the following day, remove half of the supernatants by aspiration and replenish with HAT medium containing 20% fetal calf serum.

2.   Repeat this procedure for the next 2 days. Most of the cells will die during this time.

3.   Hybridoma growth should be clearly visible within a week and the cell supernatants can be assayed for specific antibodies.

Aminopterin is a very effective enzyme inhibitor and therefore the hybridomas must not be transferred directly to DMEM 20 after HAT selection has killed all the unfused myelomas. HT medium is used to 'wean' the cells off HAT medium.

F. *Cloning*

Transfer cells selected by the screening assay to 24-well plates, grow to confluence and clone them by limit dilution or on semi-solid agar. Agar cloning is the method of choice for rat cell lines; limit dilution is satisfactory for most mouse cell lines.

(i) Limit dilution

Feeder cells are required to support hybridoma cell growth at the low cell concentrations required for this cloning method.

1. Prepare a suspension of spleen cells from a non-immune animal (see Fusion protocol) in 20 ml of DMEM 20 per spleen (mouse) or prepare peritoneal macrophages by washing the peritoneum with 10 ml of DMEM 20.
2. Count the hybridoma cells and dilute down to 10 and 20 hybridoma cells per ml in 20 ml of DMEM 20 containing HT and 2 ml of feeder cell suspensions and disperse over a 96-well plate.
3. About 1 week later screen the plates visually for single colony growth and take the supernatants from such wells for assay. Grow up several positive wells, freeze stocks and re-clone. A monoclonal cell line will be 100% positive upon re-cloning.

(ii) Semi-solid agar

1. Prepare 1.1% (w/v) Difco Bacto agar by heating, cool and maintain at 45°C.
2. Add an equal volume of 2 × concentrated DMEM (at 45°C) and 1/10 volume fetal calf serum.
3. Add 2 ml per 35 mm diameter Petri dish and leave until set.
4. Prepare six 6-fold dilutions from the positive cell well in 1 ml of DMEM 10, add 1 ml of the agar solution and transfer to the Petri dishes containing the set agar.
5. Grow for 1 week in a humidified CO_2 incubator by which time the individual colonies can be picked off using a Pasteur pipette and transferred to liquid culture.

G. *Freezing cells*

1. Pellet the cells (growing exponentially) and resuspend to 2×10^6 cells/ml in cold DMEM 20 containing 10% DMSO.
2. Place 1 ml aliquots into cryotubes, transfer to an expanded polystyrene box in a −70°C freezer for 1 day and then transfer to a liquid N_2 freezer.
3. Thaw the cells rapidly by immersing in a 37°C water bath, add 10 ml of DMEM 20, gently pellet the cells and resuspend them in a few ml of DMEM 20 and culture them.
4. A sample vial from each frozen batch should be thawed and checked for viability and sterility.

H. *Ascites fluid production*

1. Two weeks or more before immunizing the animals with antibody-producing cells, prime the animals with an i.p. injection of pristane (tetramethylpentadecane), 0.5 ml per mouse or 1.5 ml per rat.
2. Grow the hybridomas exponentially to obtain approximately 5×10^6 cells per animal.
3. Before injecting the cells into the peritoneum, wash the cells in serum-free medium.
4. Ascites fluid production usually begins within 1−2 weeks. The animals can be left until swollen with fluid and then killed, yielding 5 ml per mouse, 40 ml per rat, or the animals can be tapped on alternate days by inserting a wide gauge needle into the peritoneum. By this method 15−20 ml of ascites fluid per mouse is obtainable.
5. Centrifuge the ascites fluid to remove the cells.

I. *Antibody purification from ascites fluid*

1. Add an equal volume of saturated ammonium sulphate and collect the precipitated protein by centrifugation.
2. Re-dissolve in an equal volume of 100 mM potassium phosphate buffer, pH 8, and repeat the precipitation, re-dissolving in 10 mM potassium phosphate buffer, pH 8.
3. Dialyse against the 10 mM buffer and apply to a DE-52 ion-exchange column equilibrated in the same buffer. The antibody is found in the first major protein peak and can be concentrated by Amicon filtration using, for example, a PM30 membrane. The yield of antibody varies with cell line but ≥5 mg protein per ml ascites fluid should be obtainable.
4. Elute the anitbody using a gradient of 10−200 mM potassium phosphate.

an immunized animal are fused with myeloma cells using 50% polyethylene glycol and the cells are placed in medium containing hypoxanthine, aminopterin and thymidine (HAT medium). Aminopterin, a folic acid antagonist, inhibits the *'de novo'* nucleotide synthetic pathway. The alternative 'salvage' pathway requires hypoxanthine and thymidine but the myeloma cells used in fusions lack an essential enzyme in this pathway, hypoxanthine guanine phosphoribosyl transferase, and are killed in HAT medium. Spleen cells possess this enzyme but do not grow in cell culture. Therefore only hybrid cells are capable of growth in HAT medium. A typical fusion using 2×10^7 myeloma cells and 2×10^8 spleen cells will give rise to perhaps 10^3 hybrid cell lines. The hybrids are grown in culture for approximately 1 week before testing the spent medium for antibodies of interest. Chosen lines are then cloned twice, usually by limit dilution, and stocks of the cell lines frozen down. Preparative amounts of antibodies are obtained by growing as ascites fluid in animals of an appropriate strain. The antibody can be purified from the ascites fluid by ammonium sulphate fractionation and ion-exchange chromatography (16). A method used to prepare monoclonal antibodies is given in *Table 1*.

2.4 Screening assays for monoclonal antibodies

A good screening assay is crucial to the success of a fusion. The chosen assay(s) should be capable of detecting the appropriate antibodies at concentrations of less than 1 μg/ml in $50-100$ μl of spent medium. One fusion will generate approximately $100-400$ wells for assay depending on whether the fusion has been plated out over four 24-well plates or four 96-well plates. The use of 24-well plates will provide $1-2$ ml of spent medium for assay per well, but has the disadvantage that slow growing positive clones are more likely to be overgrown by more vigorous negative clones. To screen the 96-well plates, it is usually time and cost effective to pool the medium from four or six wells for assay and to re-assay the individual wells of positive pools.

The assay method chosen will depend on the nature, amount and purity of the antigen available. Many of the screening assays used in this laboratory have been based on an immunoadsorbent of sheep anti-mouse IgG antibodies coupled to aminocellulose (*Tables 2* and *3B, Figure 1*). The sheep was immunized with the monoclonal antibody MOPC 21 IgG from the cell line P3-X63/Ag 8. Typically a coupling ratio of 0.3 mg IgG/mg cellulose is achieved and 0.25 mg of this immunoadsorbent has the capacity to bind all of the mouse immunoglobulin in 100 μl of spent medium or 100 μl of a 10^{-3} dilution of mouse serum. Alternatively the anti-mouse (or rat) immunoglobulin can be adsorbed onto 96-well plates (*Table 3A*). The binding capacity is at least 10-fold lower than the cellulose immunoadsorbent but the assay procedure may be considered less tedious. Either method of immobilizing the antibodies may be used in a simple binding assay if the antigen is pure enough and suitable for radiolabelling (see *Table 3B* for an example of a suitable assay protocol). The antigen need not be purified if it can be pre-bound to a radiolabelled ligand (21), although this method will not detect antibodies to the ligand-binding site.

Screening assays for enzymes can be devised which do not require prior purification of the enzyme. Non-inhibitory antibodies or partially inhibiting antibodies can be detected by incubating the immobilized antibody with a dilute enzyme solution and, after washing

Table 2. The preparation of cellulose immunoadsorbents.

A. *Preparation of 'Aminocellulose'*

1. Dissolve 0.5 g of sodium acetate in 2 ml of water, and 1.4 g of *m*-nitrobenzyloxymethyl pyridinium chloride (BDH) in 18 ml of ethanol.
2. Mix the two solutions, add 5 g of cellulose (Whatman CC41) and stir to a slurry.
3. Heat to 70°C in a Petri dish until dry (~30 min), and then for a further 40 min at 125°C.
4. Wash three times with 200 ml of benzene in a sintered glass funnel and suck dry. Wash with 1 litre of water.
5. Resuspend in 150 ml of 200 g/l sodium dithionite and mix at 55−60°C for 30 min.
6. Wash three times with 200 ml of water and twice with 200 ml of 30% acetic acid, and again with water until there remains no smell of H_2S.
7. Dry in a desiccator at 20°C. If lumpy, powder the 'aminocellulose' and store desiccated at 4°C, where it is stable for several months.

B. *Protein preparation*

For anti-(IgG) immunoglobulin from serum or ascites, use an appropriate ammonium sulphate cut (between 40% and 50%) well dialysed against a buffer solution.

C. *Preparation of diazocellulose*

1. Dissolve 1.5 g of cupric chloride in a minimum amount of water (5 ml) to form a green solution, and add *fresh* 1 M NaOH in excess (~75 ml) with constant stirring.
2. Wash the blue precipitate twice with 100 ml of water in a Buchner funnel through two layers of Whatman 41 filter paper until the pH of the wash is less than 9. Partially dry on the funnel.
3. Scrape off the precipitate and dissolve it in 40 ml of fresh 0.88 ammonia solution (deep blue) to form a saturated solution. Stir well for at least 15 min.
4. Dissolve 0.5 g of the powdered 'aminocellulose' in 40 ml of this ammoniacal cupric hydroxide tion. Stir well for at least 15 min.
5. Centrifuge (~5000 *g* for 5 min) to remove any excess cupric hydroxide and/or any remaining lumps of 'aminocellulose'.
6. Decant into 1500 ml of water to give a light blue solution. Add 10% H_2SO_4 until the solution is colourless or very pale blue and the pH <4, when the 'aminocellulose' flocculates to form a white precipitate.
7. Leave to stand for 30 min to settle and then syphon off as much water as possible.
8. Wash the precipitate four times with cold water (100 ml aliquots), centrifuging between washes. If the cellulose becomes difficult to spin down add a drop or two of 1 M HCl.
9. Cool at 4°C. The cellulose may turn pink.
10. Suspend the 'aminocellulose' in 50 ml of cold fresh 2 M HCl. Add 2 ml of 1% w/v sodium nitrite and test for free oxidant using starch iodide paper (Whatman) which turns black. Then mix for 20 min at 4°C.
11. Add excess solid urea until the starch iodide test is almost negative and then wash at 4°C three times with water and twice with 0.2 M borate buffer[a]. Centrifuge between washes.
12. Suspend in a small volume (25 ml) of 0.2 M borate buffer. To test for completeness of diazotization add some to a little β-naphthol in water—it should go bright orange.

D. *Coupling*

1. Immediately after preparing, add an equal quantity of diazocellulose (usually 100 mg per 100 mg protein) to the protein in borate buffer (~15 ml total).
2. Mix for 48 h at 4°C in the dark.
3. Centrifuge and retain the supernatant which can be used again for coupling.
4. Wash the immunoadsorbent three times in 0.2 M borate buffer.
5. Resuspend the immunoadsorbent at 5 mg/ml in 0.2 M borate buffer.
6. Measure the amount of protein bound to the immunoadsorbent by the method of Lowry. Usually 200−300 μg protein is bound per mg cellulose.

The various steps in making cellulose immunoadsorbents are shown in *Figure 1*.

[a]Borate buffer 0.2 M borate: 12.4 g of boric acid, 14.9 g of KCl per litre adjusted to pH 8.2 with NaOH.

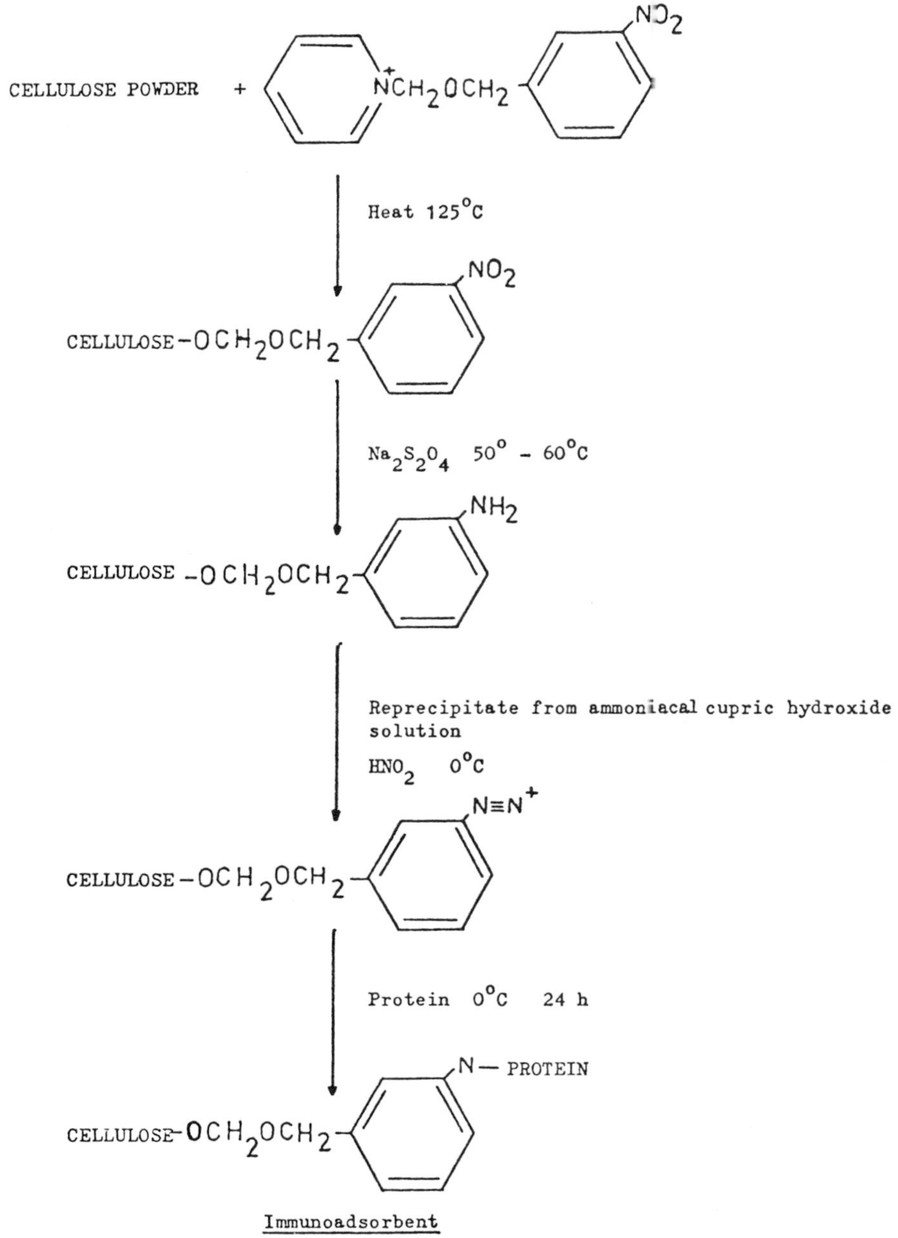

Figure 1. General scheme for preparation of cellulose-based immunoadsorbent. For full details, see Appendix II.

to remove unbound enzyme, adding enzyme assay cocktail. This method is particularly suitable for colorimetic assays performed on 96-well plates since the absorbance can be read rapidly on a Titertek Multiscan (Flow Laboratories); alternatively and even more quickly, positive wells can be determined by visual inspection. This technique has been used to detect monoclonal antibodies to the isoenzymes of alkaline phosphatase (22, 23).

It is inadvisable to test for antibodies solely by the effect of spent medium on the antigen assay because the spent medium, containing fetal calf serum and cellular debris, may interfere non-specifically with the assay system. For example, in the course of raising monoclonal antibodies to rat liver 5'-nucleotidase, we found that spent medium produced an apparent inhibition of 5'-nucleotidase due to further metabolism of the assay product [^{3}H]adenosine by cytosolic enzymes, unless a cold adenosine 'trap' was included in the assay cocktail (Bailyes, 1979, unpublished observation). Inhibitory as well as non-inhibitory antibodies can be detected by using the immunoadsorbent/monoclonal antibody complex to remove enzyme activity from the applied solution. This method should be applicable for any antigen for which there is a specific assay.

A major alternative to the methods described above is to coat a solid surface such as a 96-well plate with the antigen, incubate with spent medium and add labelled anti-mouse or anti-rat immunoglobulin or labelled protein A. This method requires larger amounts of purified antigen and solubilized membrane antigens may not bind to the surface very well as detergents present may inhibit adsorption (20). For example, placental alkaline phosphatase was found to bind poorly to the 96-well plates and a satisfactory assay could only be obtained by pre-coating the wells with rabbit anti-placental alkaline phosphatase (24). Cell membranes, whole cells and glycolipids can also be attached to 96-well plates and used as the basis for an assay (*Table 3*). Assays involving whole cells must be done at less than 4°C to prevent endocytosis or capping of added antibodies.

The labelled reagent in these assays can be enzymatic (peroxidase, alkaline phosphatase or β-galactosidase) or radioactive. The enzyme-linked immunosorbent assays (ELISA) are less tedious than the radiolabelled assays because the coloured assay products can be determined rapidly by inspection or by the Multiskan plate reader, whereas the wells in radiolabelled assays have to be cut out and loaded individually onto a gamma counter. However, endogenous enzyme activity in crude membrane preparations or cells can interfere with the ELISA and the working range of the assay is not as great.

In a radiolabelled assay the choice is between iodinated protein A or iodinated anti-rat or anti-mouse immunoglobulin. Rat IgG, mouse IgM and mouse IgG$_1$ do not react or react poorly with protein A (20) so iodinated anti-rat or mouse immunoglobulin is probably the best choice.

Immunofluorescence using the fluorescence-activated cell sorter (20) which is useful for detecting antibodies against sub-populations of cells and cytotoxicity assays (25) for complement-fixing antibodies against cell surface antigens are beyond the scope of this article.

3. IMMUNOBLOTTING, INCLUDING EPITOPE ANALYSIS

The combination of SDS−PAGE to fractionate membrane proteins, and the transfer of these proteins onto nitrocellulose allowing subsequent detection with antibodies, has greatly improved the immunological analysis of membrane proteins. The technique known as immunoblotting or 'Western' blotting (26, 27) has effectively superseded earlier methods of immunochemical characterization of gel-separated proteins such as

Table 3. Screening assays for monoclonal antibodies to membrane components.

A. *Methods used to coat 96-well plates*

(i) Soluble proteins

1. Incubate each well with 100 μl of $10-100$ μg protein/ml in dilute buffer of near neutral pH, e.g. PBS (pH 7.5), 50 mM potassium phosphate (pH 6.5), for several hours at room temperature or overnight at 4°C. Alternatively, the protein can be dried down in wells by leaving the plates in the airflow of a laminar flow cabinet overnight.
2. Remove the protein solution and wash the wells $2-3$ times with, e.g. $1-10$ mg/ml BSA or 5 mg/ml gelatin in the buffer (e.g. PBS) to block all protein-binding sites.

(ii) Membranes

Membranes can be bound directly to the plates as described for soluble proteins or attached via poly-L-lysine.

1. Incubate each well at 37°C for 1 h with 50 μl of poly-L-lysine (Sigma No. P-1399; mol. wt $150-300$ kd) at $1-2$ mg/100 ml in buffer, e.g. PBS, pH 7.5.
2. Remove and wash once with buffer.
3. Add 25 μl of the membrane suspension at $60-120$ μg protein/ml and incubate for 1 h at 4°C or centrifuge the membranes at 1000 g for 20 min at 4°C (e.g. Hereus Christ Minifuge GL).
4. Remove the liquid and wash with buffer.
5. Fill the wells with buffer containing 1% BSA and incubate for 1 h at 4°C to block remaining sites. The plate can be stored at -20°C in this state.

(iii) Whole cells

Adherent cells such as fibroblasts can be grown directly onto the assay plates. Other cells can be used without attachment if a suitable centrifuge is available. Otherwise cells can be attached via poly-L-lysine essentially as described for cell membranes. Use 10^5-10^6 cells per well. Glutaraldehyde fixation can also be used to attach the cells although this treatment may destroy certain antigenic sites[a].

1. Add 10^6 cells per well and centrifuge the plate (100 g, 5 min).
2. Carefully immerse the plate in a 1 litre glass beaker containing freshly prepared 0.25% glutaraldehyde in PBS at 4°C for 5 min.
3. Wash in PBS containing 1 mg/ml BSA.

(iv) Glycolipids

Glycolipids can be adsorbed onto the wells by adding the lipids in a volatile solvent, e.g. methanol, ethanol. Evaporate the solvent either by air drying or a nitrogen stream and wash with a buffered protein solution to block excess sites.

B. *Assay protocols*

(i) Labelled antibody and antigen-coated plates

1. Coat a 96-well plate with antigen as described in part A.
2. Incubate the wells with $50-100$ μl of spent medium plus 50 μl of buffer containing ≥ 1 mg/ml BSA for 1 h at room temperature.
3. Wash the plate $2-3$ times with buffer.
4. Incubate the plate with an appropriate dilution of labelled antibody, e.g. [125I]anti-immunoglobulin (10 000 c.p.m. per well) or enzyme-conjugated anti-immunoglobulin for 1 h at room temperature.
5. Wash the plate $2-3$ times with buffer.
6. Cut out wells to determine the bound radioactivity or add enzyme $-$ substrate cocktail and determine the positive wells by visual inspection or using the Multiscan instrument.

(ii) Radiolabelled antigen and immunoglobulin-coated plates

1. Iodinate the protein to incorporate ≤ 1 mol of ^{125}I per mol of antigen Dilute to 200 000 c.p.m./ml in a suitable buffer containing ≥ 1 mg/ml BSA and an appropriate detergent, e.g. 0.1% Triton X-100.
2. Incubate $50-100$ μl of spent medium plus 50 μl of buffer containing ≥ 1 mg/ml BSA and an appropriate detergent, in wells pre-coated with sheep anti-mouse (or rat) immunoglobulin for 1 h at room temperature.

3.   Wash twice with buffer, removing the liquid by aspiration or vigorous tapping.

4.   Add 100 μl samples of iodinated antigen prepared in step 1. Incubate for 1−4 h at room temperature or overnight at 4°C.

5.   Wash twice, cut out the wells, and determine the bound radioactivity with a gamma counter.

(iii) <u>Radiolabelled antigen and cellulose immunoadsorbent</u>

1.   Iodinate the protein to incorporate ≤ 1 mol of ^{125}I per mol of antigen. Dilute to 200 000 c.p.m./ml in a suitable buffer containing ≥ 1 mg/ml BSA and an appropriate detergent for the membrane protein. Aliquot out 100 μl samples.

2.   Incubate with 100 μl of spent medium for 1−4 h at room temperature or overnight at 4°C.

3.   Add 50 μl of 5 mg/ml cellulose sheep anti-mouse (or rat) immunoglobulin immunoadsorbent and incubate for 1 h at room temperature.

4.   Wash twice with 2 ml of the assay buffer, pelleting the immunoadsorbent by centrifugation. Decant and measure the bound radioactivity. The blank controls bind less than 100 c.p.m. Higher values suggest a problem with antigen aggregation and different detergents and/or high salt concentrations in the assay buffer should be tried.

[a]See reference 20.

crossed immunoelectrophoresis (28). Immunoblotting is an improvement on these techniques since it does not require diffusion of antibodies into a gel or precipitation of the immune complex. Instead, the proteins are electrophoretically transferred and immobilized on the surface of a filter, and the probing antibodies have direct access to the immobilized protein on this surface. Thus, the technique retains the resolution of SDS−PAGE and also has great sensitivity, for it is often possible to detect less than 1 ng of a membrane protein by using a high titre antibody. The methodology and applications of immunoblotting proteins have been reviewed (29−33). Most laboratories have developed their own modifications either of electrophoretic transfer, choice of membrane filter support, protein blocking agents, antibody incubation conditions or detection (radioactive or enzymatic) systems. However, protocols that have been found generally applicable are presented in *Table 4*.

A number of specific variations of basic immunoblotting have been developed. Thus, if it is difficult to identify the exact nature of a protein antigen amongst a mixture of proteins separated by SDS or two-dimensional PAGE, the gel may be fixed and stained with Coomassie blue before electrophoretic transfer onto nitrocellulose. Whilst fixed and stained protein will only transfer at low efficiency out of the gel, almost complete transfer to the nitrocellulose filter may be achieved if the gel is simply incubated for 2 h at room temperature in electrophoresis buffer containing SDS prior to electrophoretic transfer (34). The blue-stained pattern will then appear on the nitrocellulose membrane so that subsequent incubation with first antibody, peroxidase-labelled second antibody and development with diaminobenzidine results in a brown stain that can be observed through a blue filter and exactly corresponds with a blue-stained protein.

A further useful technique in the analysis of membrane proteins has been the use of the immobilized protein on the filter for the immunoaffinity purification of antibodies (35). This technique may be extended to establish whether two protein bands on a gel contain the same epitopes, by eluting antibody bound to one protein band and then reacting it with the second. Polypeptides derived by proteolytic modification of a larger protein, perhaps during membrane isolation, may be identified in this way (2).

Whilst both polyclonal and monoclonal antibodies may be used in immunoblotting, problems may arise with the latter as probes. Thus epitopes may be destroyed during

Table 4. Immunoblotting of membrane proteins.

A. *Gel electrophoresis*

In principle many gel and buffer systems may be used prior to blotting. In practice any standard slab gel SDS−PAGE system with a polyacrylamide concentration $\leq 12\%$ w/v total monomer allows good subsequent transfer to nitrocellulose. See also Chapter 6.

B. *Transfer*

1. Immediately after completion of SDS−PAGE package the gel into a sandwich ready for transfer. The sandwich consists of:
    a perforated plastic leaf;
    a sheet of Scotchbrite (or foam rubber);
    2 sheets of Whatman 3MM filter paper;
    the gel slab;
    a sheet of nitrocellulose (Schleicher & Schull, 0.45 μm, BA 85, 20 cm square trimmed to gel size);
    2 sheets of Whatman 3MM filter paper;
    a sheet of Scotchbrite (or foam rubber);
    a perforated plastic leaf.
2. Soak each layer of the sandwich in transfer buffer and gently roll each layer on to the previous layer with a roller to ensure the absence of air bubbles. 0.22 μm nitrocellulose filters are preferable to 0.45 μm filters for some membrane proteins.
3. Place the completed sandwich in a plastic electrophoretic transfer tank such that the gel is towards the negative electrode and the nitrocellulose towards the positive electrode. A suitable transfer tank, the Trans-blot Cell with power supply, is available from BioRad, Richmond,CA, USA or Hoeffer Scientific Instruments, San Francisco, CA, USA. These tanks take approximately 3 litres of transfer buffer[a].
4. Carry out electrophoretic transfer overnight at room temperature at approximately 6 V/cm (distance measured between electrodes), i.e. approximately 40 V in the BioRad apparatus.
    More rapid transfer may be achieved using a semi-dry blotting apparatus (Sartorius GmbH).

C. *Staining, blocking, washing, antibody incubation and development*

(i) Immunoperoxidase staining

The nitrocellulose filter can be stained with Ponceau S (0.2% w/v in 3% w/v trichloroacetic acid; Serva) for 5 min and washed with water to allow visualization of protein standards. The position of these standards must be marked since the stain is lost during subsequent washes.

Wash, block and develop the filter as follows using PBS containing 10% v/v horse serum and 0.1% v/v Triton X-100 (solution A) unless otherwise stated:

1. Wash the filter three times for 15 min each in solution A.
2. Incubate the filter with first antibody for 60 min. Ideally use an IgG fraction at $1-20$ μg/ml.
3. Wash the filter three times for 15 min each in solution A.
4. Incubate the filter with horseradish peroxidase-labelled second antibody for 60 min (1:2000 dilution antibody supplied by TAGO Inc. Burlingame, CA, USA)
5. Wash the filter three times for 15 min each in solution A.
6. Wash the filter twice with PBS, 5 min each.
7. Develop the filter in 0.1 M Tris-HCl, pH 7.4, containing 0.5 mg/ml 3,3'-diaminobenzidine tetrahydrochloride and 1:5000 H_2O_2.

The volume of washes depends on the size of filter used and may vary from 2 ml for a a single gel track to 25 ml for a whole filter.

Alternatives to horse serum as a blocking agent include gelatin (0.5% w/v), haemoglobin or powdered milk.

(ii) ^{125}I-protein A staining

1. Iodinate protein A or purchase from Amersham Int., or New England Nuclear.
2. Transfer proteins to nitrocellulose filters and stain for 1 min in 0.3% Amido-Black dissolved in H_2O, methanol, acetic acid (50/40/10). Wash filters in PBS.

3.   Incubate filters with antiserum/antibody in 1% BSA in PBS for 60 min at 37°C or overnight at 4°C. Dilution of antibody depends on titre (1:20 to 1:1000).
4.   Wash filters three times with PBS.
5.   Incubate filters with ^{125}I-protein A (1×10^6 c.p.m.) for 45 min at room temperature.
6.   Wash filters extensively in 0.1% Triton X-100 in PBS, dry and expose to KODAK-X-OMAT AR films.

[a]The transfer buffer contains 12 g of glycine and 2.5 g of Tris base per litre. Methanol may be added to improve transfer but its inclusion is not always necessary.

sample processing prior to PAGE [e.g. by reduction (36)] and cannot be reconstituted. It is thought that proteins bind to nitrocellulose by hydrophobic interactions (37) and in general it is expected that the immunoblotting technique allows the more hydrophilic exterior parts of the proteins to adopt their correct epitopic conformations. In fact, proteins may recover their functional properties during blotting as has been shown for the enzymic activity of 2'-3' cyclic nucleotide 3'-phosphodiesterase (38), bungarotoxin binding of the acetylcholine receptor (39), immunoreactivity of antibodies (40) and Ca^{2+} binding of certain proteins (41). The ability to recover correct antigenic conformation on blotting is also true of some proteolytic or chemically cleaved fragments of proteins as well as the intact protein. This has allowed epitope mapping by immunoblotting of both monoclonal and polyclonal antibody interactions with a variety of proteins (42). Epitope mapping in this way is a particularly powerful technique when used for antibodies which modify the function of proteins of known primary sequence, since it can lead to testable hypotheses concerning topography (e.g. across a membrane) and biological activity (43).

Immunoblotting of glycolipid antigens has also been developed. In this case glycolipids are separated on t.l.c. plates and then transferred to nitrocellulose (44). Alternatively, the glycolipid antigens can be detected directly on the plates after immobilization using polyisobutylmethacrylate (45, 46).

4. THE IMMUNOLOGICAL PURIFICATION OF SUBCELLULAR FRACTIONS

In contrast to fractionation techniques dependent on differences in physical parameters between organelles (see Chapters 1 and 2), immunological techniques for subcellular fractionation rely on biological differences (i.e. the expression of different antigens). The most popular are immunoaffinity techniques which offer the prospect of fast purification in high yield and these may be particularly valuable in situations where only a small amount of cell material is available (e.g. tissue culture cells). They also have the advantage of being applicable in physiological salt solutions and do not expose the organelles to unusual media or to osmotic and electric shock. Whilst it is some time since an immunoaffinity approach to subcellular fractionation was first discussed (47) and successful methods for particular organelle isolation were reported (48−50), it is only recently that immunoaffinity techniques have been used widely and successfully in a variety of biological systems (*Table 5*). The criteria for establishing successful immunoaffinity subcellular fractionation techniques have been reviewed (51, 52) and require the investigator to be concerned with antibody requirements and the nature of the solid support.

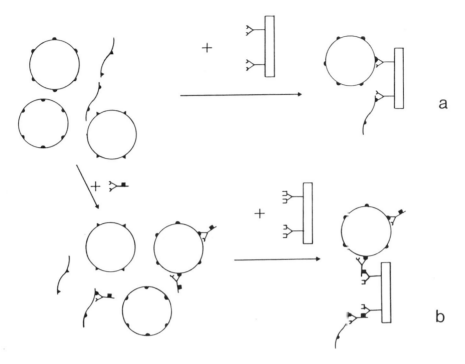

Figure 2. Schematic representation of direct (**a**) and indirect (**b**) immunoaffinity techniques for the preparation of subcellular organelles and membranes. In the direct method (**a**) the homogenate or partially purified subcellular fraction is reacted with the anti-organelle antibody immobilized to give a solid phase immunoadsorbent. In the indirect method (**b**) the free anti-organelle antibody is reacted with the homogenate (or fraction), unbound antibody washed away and the organelles then isolated on an immunoglobulin binding immunoadsorbent. In the diagram (**b**) this immunoadsorbent is shown as an anti-immunoglobulin antibody coupled to a solid phase and capable of reacting with an antigenic site (■) on the F_c region of the anti-organelle antibody. The diagram shows that both direct and indirect immunoaffinity techniques will select only those membranes and organelles showing the correct antigenic topography, ignoring those with other antigens or those that are 'inside out'.

4.1 Antibodies for subcellular fractionation

Immunoaffinity techniques for subcellular fractionation can be divided into the use of either direct immunoadsorbents, where anti-organelle antibody is itself covalently bound to the solid phase, or indirect imunoadsorbents where an antibody-binding reagent is covalently bound to the solid phase providing an immunoadsorbent capable of isolating the antibody−organelle complex (*Figure 2*). In practice, direct immunoadsorbents have rarely proved successful and it has been argued, by analogy with affinity chromatography, that an indirect technique is required either to provide a more flexible spacer arm or for steric reasons due to the difficulties inherent in the binding of two large particles. The antibody-binding reagent coupled to the solid support may be either an anti-immunoglobulin second antibody (polyclonal or monoclonal) or protein A from *Staphylococcus aureus* since this binds to the F_c portion of many immunoglobulin molecules (63). Occasionally, fixed, heat-inactivated *S. aureus* bacteria are themselves used as the solid support (60, 61). The indirect technique makes it possible to standardize the solid phase for use in the preparation of many organelles using different antibodies.

The choice of first and second antibodies may be dictated by availability, although it is often possible to incorporate some element of design into the choice. In principle the first antibody should be specific to the organelle or membrane fragment of interest, and be of high avidity. In practice, polyclonal antibodies prepared to isolated organelles are unlikely to be specific due to the impurity of the organelle preparations and because common antigens exist on different membranes. This may not prove an impossible obstacle since, for example, in isolating plasma membrane a polyclonal antiserum can be reacted with intact cells before homogenization so that only the plasma membrane is antibody-labelled (49). Performing such an operation in the cold prevents endocytosis of antibody (64) and the kinetics of antibody binding with fast 'on rates' and slow 'off rates' militates against the subsequent redistribution of antibody.

Polyclonal anti-organelle antibodies may be prepared to specific antigens (protein, lipid or carbohydrate) and/or be immunoaffinity purified on the antigen before use. Such purification considerably improves the specificity of the reagent and removes unwanted immunoglobulin which will interefere with the use of the solid-phase antibody-binding immunoadsorbent. Removal of the non-bound immunoglobulin after the first antibody step normally necessitates washing, but the use of a suitable dilution of affinity-purified antibody may preclude the need for this. The use of affinity-purified anti-organelle antibodies also allows the possibility of a single purification step since they may be pre-bound to the anti-immunoglobulin immunoadsorbent before incubation with the cell homogenate or partially purified subcellular fraction (58).

A modification of using a polyclonal antiserum directed at an endogenous component of an organelle is to alter the antigenic structure of the organelle. In the case of the plasma membrane this has been achieved by incorporating a viral protein antigen (56) and then isolating the membranes by reaction with anti-viral protein antibodies and anti-IgG immunoadsorbent. Viral protein is incorporated into the membrane simply by lowering the extracellular pH, and it then diffuses two-dimensionally in the plane of the phospholipid bilayer. It is therefore possible to show that indirect immunoaffinity isolation of the membrane is dependent on the antigen density being greater than 50 molecules/μm^2. This may be a general rule (54, 56) and is of importance in selecting an antigen for preparation of the anti-organelle antibody. It may also allow immunoaffinity isolation techniques to be used to prepare membranes and organelles that are particularly rich in an antigen that is present at lower concentration elsewhere.

High affinity monoclonal antibodies also are likely to be of use for subcellular fractionation. It is not known whether the fact that monoclonal antibodies each react with only one antigenic site affects their general use as first antibodies for immunoaffinity isolation, though it should be possible to overcome any difficulties by using mixtures of monoclonal antibodies. A more serious objection to their use is the problem of a suitable screening protocol with which to select antibody-secreting hybridomas. Proof that an antibody is directed to a specific organelle may require the use of immuno-fluorescence microscopy or even immunoelectron microscopy which can be time consuming, but simpler protocols using binding assays to organelle and membrane fractions have been developed and may be satisfactory (*Table 3*). In the future, antibodies of predetermined specificity prepared against synthetic polypeptide epitopes or from a molecular cloning approach may provide good anti-organelle antibodies for subcellular fractionation.

The anti-immunoglobulin antibody coupled to the solid phase for indirect immunoaffinity purification should be of high avidity, react with many immunoglobulin subclasses and be available in large amounts. Appropriately, selected monoclonal antibodies are particularly useful since essentially unlimited quantities of antibody can be produced from a single hybridoma cell line. For some species of polyclonal first antibody (e.g. sheep and rabbit) suitable mouse monoclonal second antibodies are available, and if mouse monoclonal first antibodies are used rat anti-mouse Ig monoclonal antibodies may be coupled to the solid support. *S. aureus* Protein A used on the solid support binds to the F_c portion of immunoglobulins, thus allowing directional single step immunoadsorbents to be prepared by reacting with affinity-purified polyclonal antibodies and stabilizing by chemical cross-linking (65).

In any specific immunoaffinity protocol for isolating an organelle or membrane fragment, the time course and dose−response curve for binding of first antibody and immunoadsorbent must be experimentally determined to find the most appropriate conditions for specific isolation in high yield in the shortest possible time.

4.2 Solid supports for immunoaffinity subcellular fractionation

The choice of solid support in the affinity purification of subcellular particles is determined by a number of criteria; the most important of these are high binding capacity, non-porous nature, adequate spacer arms, low non-specific binding, ease of separation from unbound particles and availability. It is recommended, for a number of reasons, that immunoadsorbents of high binding capacity should be used to isolate subcellular particles. For example, if an indirect approach is used there will be some free unbound first antibody present despite extensive washing. Such free antibody will then compete successfully with particle-bound antibody for binding sites on the immunoadsorbent. In addition, the binding of the immunoadsorbent and the subcellular particle is facilitated by the presence of potential binding sites over the entire surfaces of both.

The binding capacity of any immunoadsorbent is governed largely by the number of attachment sites for the IgG binding molecule (usually the second antibody) and the degree of inactivation suffered during covalent bonding to the matrix. The most favoured matrices for affinity purification of subcellular particles have been non-porous (e.g. Sepharose 6MB, polyacrylamide, cellulose) since these give the greatest surface binding capacities. Although porous supports, for example Sepharose 4B, give high total binding capacities, most of the binding sites are within the beads and so unavailable for the binding of relatively large subcellular organelles. The highest capacity (as determined by IgG binding) support reported (57) was obtained with a monoclonal second antibody coupled to cellulose (66−68). Cellulose immunomatrices are easy to use in the separation of bound and free particles since this can be achieved by low-speed centrifugation. Other solid supports may be separated by column-based methods or magnetism. Indeed, the latter may provide free-flow systems maintaining the magnetic particles in suspension and thus giving a significant reduction in non-specific binding (51).

Table 5. Immunoaffinity purification of subcellular fractions.

Organelle	Cell type/tissue	Adsorption efficiency[a]	Reference
Plasma membrane	Adipocyte, hepatocyte	8.8	49, 53, 54
Plasma membrane domains	Liver, kidney cells	14.6	55, 56
Right-side-out and inside-out plasma membrane vesicles	Erythrocyte	31	56
Cholinergic synaptosomes	Brain	85	57
Microsomes	Liver		50
Golgi	Liver	9.1	58
Lysosomes	Liver	30	59
Clathrin-coated vesicles	Brain		60, 61
Synaptic vesicles	Brain		62

[a]The adsorption efficiency is given as the ratio of specific:non-specific binding and refers only to the first cell type and reference indicated in each category. The efficiency is dependent on the solid support used. Further references to successful immunoaffinity subcellular fractionation and to adsorption efficiency may be found in (52).

Immunoaffinity purification of subcellular fractions can provide a far greater degree of purification than classical methods. The amount of contamination must be assessed, both as the non-specific binding of organelles to the matrix in the absence of the primary antibody, and also from the amount of a given contaminant which co-purifies with the particle. Non-specific binding to matrices is determined both by the quantity of the matrix used and the amount of protein already coupled. Although there are a number of ways of calculating the efficiency of any given adsorption process (51), the most informative is the ratio of specific to non-specific binding which illustrates the degree of purification achieved (52, *Table 5*). This ratio falls with increases in the ratio of matrix to particle used in the adsorption process. Consequently, high capacity immunoadsorbents tend to increase the efficiency of the process, by reducing the amount of matrix required.

4.3 The practicalities and versatility of immunoaffinity isolation of organelles

In general, it appears that indirect immunoaffinity techniques are most appropriate for subcellular fractionation. The first antibody should be specific, of high avidity and directed to an antigen present at high density. The immunoadsorbent should exhibit low levels of non-specific binding and be of high capacity using a high avidity second antibody. Incubation and washing conditions are also of importance and should be gentle to prevent disruption of organelles and vesiculation of membranes (51). A method for immunoaffinity isolation of cholinergic specific synaptosomes (nerve terminals) from rat brain (57) is described in *Table 6*. The selection of a suitable alternative first antibody and appropriate modification of the protocol can allow membranes and organelles from other tissues and/or species to be prepared.

The versatility of immunoaffinity approaches to subcellular fractionation can be gauged by reference to *Table 5*. These methods have been used to isolate a variety of organelle fractions, ranging from cholinergic synaptosomes present in the complex homogenates

Table 6. Immunoaffinity purification of cholinergic nerve terminals (See also Chapter 5, Table 5).

1.	Using 1 g of scraped rat cerebral cortex make a 10% tissue homogenate in 11% sucrose solution[a] by using 12 up-and-down strokes in a rotating (640 r.p.m.) Teflon/glass (loose-fitting) homogenizer. Keep at 4°C.
2.	Centrifuge at 1000 g for 10 min, take the supernatant and centrifuge at 18 000 g for 20 min. Adjust the pellet to 10 ml with the sucrose solution.
3.	Divide into two 5 ml portions. Incubate one with sheep anti-(Chol-1) serum (1:25 to 1:200 dilution), and the other with a similar dilution of non-immune sheep serum, for 40 min at 4°C with periodic shaking. Dilute to 30 ml with sucrose.
4.	Centrifuge at 18 000 g for 15 min (Sorvall SS-34, 14 000 r.p.m.).
5.	Wash the pellets three times with 30 ml of sucrose, centrifuging at above speed between washes.
6.	Resuspend the pellets in 3 ml of Percoll[b] and (see Chapter 1) centrifuge at 9000 g for 2 min. Take the top layer which contains nerve terminals, dilute with KRH + EDTA solution[c] and centrifuge again.
7.	Incubate 200 μl aliquots with 1−2 mg of immunoadsorbent[d] in a total volume of 500 μl. Ensure good mixing. Incubate for 40 min at 4°C with occasional swirling
8.	Separate the bound and free nerve terminals by addition of 2.5 ml of Percoll, mix and centrifuge at 9000 g for 2 min.
9.	Remove the supernatant. Wash the immunoadsorbent three times with 2.5 ml of Percoll and once with KRH solution[e]

[a]11% sucrose solution is 0.32 M sucrose in 1 mM EDTA, 5 mM Hepes, pH 7.4.
[b]Percoll solution is 45% Percoll in KRH + EDTA solution c (v/v).
[c]KRH + EDTA solution is as KRH[e] + 1 mM EDTA, pH 7.4, without Ca^{2+} and Mg^{2+}
[d]Immunoadsorbent is cellulose bearing monoclonal mouse anti-(sheep IgG) antibody ($\sim$0.3 mg mouse IgG/mg cellulose)
[e]KRH solution is 140 mM NaCl, 5 mM KCl, 2.6 mM CaCl$_2$, 2 mM MgCl$_2$, 5 mM glucose, 10 mM Hepes, pH 7.4.

of rat brain (57) to clathrin-coated vesicles (61). It is possible to prepare fractions within 2 h of adding antibody, and to select right-side-out (see *Figure 2*) or sometimes inside-out (56) vesicles which may be of importance in subsequent functional studies. It is usually difficult to separate membrane fractions from the immunoadsorbent without using extreme conditions, but elution is rarely necessary since most immunoaffinity-purified particles retain their normal biochemical functions while attached to the solid support.

5. THE IMMUNOLOGICAL PURIFICATION OF MEMBRANE COMPONENTS

5.1 Solubilization of membrane proteins

Membrane proteins are classified as extrinsic or intrinsic proteins according to the methods used to solubilize them (69). Characterization and purification in solution of intrinsic proteins requires that they are extracted from the membrane lipid bilayer and inserted into detergent micelles.

A wide variety of detergents of defined structure have become available to the membrane biochemist and these are discussed in Chapter 5. Early work on membrane protein solubilization circumvented the problem of obtaining amphiphilic proteins in aqueous solution by removing the hydrophobic part of the molecule by proteolysis (70) and characterizing the soluble hydrophilic fragment bearing biological activity. This dissection of a membrane protein into different domains is of particular interest when performed in parallel with detergent solubilization to obtain the intact molecule. However it is no longer regarded as the best approach to solubilization since not all membrane proteins are susceptible to proteolytic cleavage between their hydrophilic and hydrophobic domains. Organic solvents, such as butanol, can be used to extract some intrinsic membrane proteins but the precise action of these solvents is not clear. It has been suggested in the case of alkaline phosphatase that butanol treatment causes the activation of an endogenous phospholipase (71) since exogenous phospholipase C is capable of extracting alkaline phosphatase into aqueous solution.

5.1.1 Detergent properties

Detergents may be classified by their electric charge as non-ionic, zwitterionic, anionic or cationic, and by their overall structure as Type A or B (72). Type A detergents (e.g. octylglucoside, sarkosyl, Brij, Triton X and N series, Lubrol) have hydrophilic head groups and flexible hydrophobic tails and form micelles of molecular weight of the order of 20−90 kd into which the hydrophobic regions of membrane proteins are inserted. The critical micelle concentration (CMC) of the Type A detergents with non-ionic or zwitterionic head groups is low (<1 mM). Below this concentration the detergents exist in solution as monomers. The higher CMC of ionic detergents appears to be due to ionic repulsion of the head groups since increasing the ionic strength of the medium decreases the CMC and increases the micelle molecular weight. Type B detergents (e.g. bile salts and conjugated bile salts, CHAPS) form low molecular weight micelles and have high CMCs. These detergents may not interact with the membrane proteins as micelles but form a monolayer coat over the hydrophobic regions of the membrane proteins (73).

More comprehensive information on the properties of selected detergents is given in Chapter 5.

Charge-shift electrophoresis (74) (see Appendix V) can be used to demonstrate the presence or absence of a membrane-binding domain on the solubilized molecule. If the molecule has been solubilized by the binding of a neutral detergent micelle, addition of anionic detergent which will partition into the detergent micelle increases the protein's electrophoretic mobility (2). The absence of an electrophoretic shift suggests that the protein has not been solubilized by detergent binding, and therefore it lacks a membrane-binding domain.

5.1.2 Criteria of solubilization

To choose the best detergent to solubilize the protein of interest, a convenient plan of attack is to prepare a suitable membrane fraction, as a source of the protein, and store it in aliquots at −70°C so that all the experiments can be carried out on the same batch of membranes (see also Chapter 5).

(i) *Sedimentation at high g forces.* The effect of 0.2−3% w/v detergent on releasing the protein of interest from the membranes resuspended to 5 mg protein/ml, should be assessed at room temperature and 4°C. Incubate the detergents with the membranes for 30 min before centrifuging at 100 000 *g* for 60 min and assay the supernatant and an uncentrifuged sample for activity and total protein. These experiments can be performed in small plastic tubes, several of which can be floated in water-filled ultracentrifugation tubes for the centrifugation. Some proteins may require stabilizing agents to survive solubilization, for example glycerol, reducing agents, chelating agents and protease inhibitors (see Chapter 1) and these should be added if addition of detergent results in inactivation of the protein.

(ii) *Size-fractionation*. Mere non-sedimentation at high g forces is not a sufficient criterion to use in judging the extent of solubilization (75). Some estimate of the size of the released protein must be made, either by gel filtration or PAGE (see also Chapter 6). The advantages and disadvantages of choosing PAGE are discussed in (76) where full descriptions of several buffer systems in the pH range $5.2-10.9$ are also given. Suitable media for the initial trial gel filtration are Sepharose 6B (Pharmacia) or Ultrogel AcA22 (LKB) with exclusion limits of about 10^3 kd. Solubilized proteins should elute in the included volume of the column. The molecular weight of the solubilized protein can be determined by comparison with the R_f of standard proteins on gel filtration or in PAGE. The molecular weight values obtained for solubilized membrane proteins are however unlikely to be absolute, because of the effect of detergent binding and the fact that the axial ratios and partial specific volumes of membrane−detergent complexes differ from those of the soluble proteins used in calibration. The extent of this systematic error is unknown. Nevertheless, comparison of the size of the protein in different detergents is valuable in determining the degree of solubilization obtained.

The hydrophobic interactions of membrane protein with lipids and other proteins are amenable to detergent solubilization but hydrophilic interactions with other proteins are unlikely to be affected by detergent action. For example, in one study it was found that immunoaffinity-purified HLA antigens were substantially contaminated with cytoplasmic actin unless the membranes used were first pre-treated with a low ionic strength buffer containing ATP and dithiothreitol (77). If a similar problem is suspected with a given protein, buffers of varying ionic strength should be used in the initial solubilization.

Purification of detergent-solubilized membrane proteins using monoclonal antibodies is described in Section 5.2. In scaling up from pilot studies to preparative work, the optimum detergent concentration may need to be increased, since the detergent:lipid and detergent:protein ratios as well as the absolute detergent concentration affect the efficiency of solubilization. For example, 0.5% w/v Sulphobetaine 14 was sufficient to solubilize all the 5'-nucleotidase from plasma membranes at 3 mg protein/ml but at the higher protein and lipid concentrations in the preparative purification, 2% w/v was used for complete solubilization (78). There are reports that detergent action can disrupt the antigen/antibody bond (79−81). In such cases, the detergent used for solubilization need not be the same as that used for the immunoaffinity step. A change of detergent can be effected by gel filtration.

5.2 Immunoadsorbent purification of membrane proteins

Although some success was achieved using polyclonal antibodies for the purification of membrane proteins (82), only the advent of monoclonal antibodies led to the widespread use of immunoaffinity purification techniques. Most investigators have coupled their purified monoclonal antibody to a solid support using CNBr-activated Sepharose 4B or Affigel which are available from Pharmacia and Bio-Rad, respectively. Coupling to these supports usually achieves 1 − 10 mg protein/ml gel. Coupling to Ultrogel AcA34 (a polymer of agarose and acrylamide) has also been achieved with glutaraldehyde (83). An alternative approach is to couple antibody to derivatized cellulose (68, *Table 2*) which provides a high capacity immunoadsorbent, though without significant flow properties. It is therefore usual to use such cellulose immunoadsorbents in a batch rather than column protocol.

5.2.1 *Application of antigen to monoclonal antibody immunoadsorbent*

The application of the antigen to the immunoadsorbent can be done at 4°C or room temperature. Binding is faster at room temperature but such conditions are inadvisable if the antigen is unstable or subject to proteolysis. The antigen can be applied to the Sepharose columns at flow-rates up to 20 bed volumes/h. The effluent should be monitored for antigen activity and, if significant amounts of unbound antigen appear, the column capacity and/or flow-rate should be checked, although persistent non-binding of a proportion of the applied antigen may be due to heterogeneity of the antigen. Incubation of a cellulose immunoadsorbent with the antigen should be done with stirring either for 30 − 60 min at 20°C or 4 h at 4°C. A sample of the mixture should be centrifuged (2 min, Eppendorf centrifuge at 9000 g) and the supernatant checked to determine the degree of antigen binding.

Both techniques, column and batch, require extensive washing of the immunoadsorbent after antigen binding. It is advisable to pre-elute the support with the eluting buffer before equilibrating in the binding buffer and also to perform at least one high salt wash that does not elute the antigen.

5.2.2 *Elution of antigen from immunoadsorbents*

The conditions used to disrupt the antigen/antibody bond can be divided into three main classes: extremes of pH, chaotropic ions/high ionic strength and competitive ligands.

Many of the membrane proteins purified have been eluted at alkaline pH. Diethylamine (50 mM, pH 11.2 − 11.5), is widely used (*Table 7*). It is important to neutralize the eluted enzyme quickly; for example 1 ml column fractions can be collected into tubes containing 0.1 ml of 2 M Tris-HCl pH 6.8. When using cellulose, it is advisable to pass the eluate through a sterile filter to ensure that the solution is clear of fines. Alternatively, the eluate can be passed down a small gel filtration column. The use of acid pH is less popular but high ionic strength and chaotropic ions are major alternatives to alkaline pH. Sometimes, it may be acceptable to elute with harsh conditions to obtain antigen for such purposes as sequence analysis if the milder elution conditions do not give full yield (84). Occasionally, it is possible to use an elution condition specifically tailored for the antibody used. In purifying receptors, it may be possible to elute with ligand if the monoclonal antibody is competitive with ligand (85, 86). For example, in the purification of the Ca^{2+}/calmodulin-dependent cyclic nucleotide phosphodiesterase, the monoclonal antibody was selected for its specific binding to the calmodulin phosphodiesterase complex and for lack of binding to calmodulin or phosphodiesterase alone. Binding of the complex occurred in the presence of Ca^{2+} and the enzyme was eluted by chelation of Ca^{2+} with 2 mM EGTA (83).

Investigation of the properties of the monoclonal antibody can prove fruitful. The

Table 7. Conditions for elution of proteins from monoclonal antibody immunoadsorbents.

Elution conditions		Proteins purified
A. High pH	50 mM diethylamine pH 11.2−11.5 ±0.1−0.5% detergent	rat brain Thy-1 glycoprotein (88)
		epoxide hydrolase (89)
		T cell antigen receptor (90)
		HLA-A2, HLA-B7 antigens (77)
		rat thymocyte W3/13 antigen (91)
		rat OX2 antigen (92)
		common acute lymphoblastic leukaemia-associated antigen; transferrin receptor (65)
		Epstein−Barr virus membrane antigen (93)
		rat liver 5'-nucleotidase (78)
		human liver alkaline phosphatase (23)
	+ 0.14 M NaCl	human C3b inactivator (94)
	100 mM diethylamine pH 11.5 ±0.2−0.5% detergent	murine cell surface glycoprotein (95)
		Toxoplasma gondii membrane protein (96)
	0.6 M 2-amino-2-methyl-1-propanol pH 10.2	bovine cartilage alkaline phosphatase (97)
	0.2 M glycine/OH pH 10 0.5 M NaCl, 1% sodium cholate	*Torpedo* acetylcholine receptor (98)
	20 mM lysine/OH pH 11 0.1% Triton X-100	EGF receptor (84)
	100 mM triethylamine pH 10	murine calmodulin-binding protein (99)
	100 mM triethylamine pH 11.5 (0.5% sodium deoxycholate)	murine Fc receptor (100)
B. Low pH	0.05−0.1 M glycine-HCl pH 2.5−2.8 (0.1−0.2% detergent) +0.5 M NaCl	Electric organ acetylcholinesterases (101)
		EGF receptor (denatured) (84)
		human plasminogen activator (102)
	0.2 M acetic acid pH 2.5−3 0.15−0.3 M NaCl, 0.1% detergent	mouse submaxillary renin (103)
		urokinase (104)
		leukocyte interferon (105)
	PBS pH 3.25% glycerol 0.1% Emulgen 911	cytochrome P-450 (106)
	0.2 M NH$_4$ acetate pH 3	somatomedin-C/insulin-like growth factor 1 (107)
	9 M ethane-diol, 0.3M NaCl, 0.1 M citric acid pH 2	leukocyte interferon (108)
C. Chaotropic ions/high ionic strength	1 M NaCl pH 7.4, 0.1% detergent	H-2k^k antigen (109)
	2 M NaCl	rat muscle nicotinic acetylcholine receptors (110)
	3 M NaCl	liver alkaline phosphatase (111)
	2 M MgCl$_2$ pH 7.4	human complement component C9 (112)
	3 M MgCl$_2$ pH 6.8	tubulin−tyrosine ligase (113)
	4−8 M urea in PBS	pregnancy specific β1-glycoprotein (114)
	6 M urea	EGF receptor (denatured) (85)
	6 M guanidine-HCl 0.5% detergent	interleukin 2 receptor (115)
	50 mM Na-acetate 1 M KCl pH 5.5	DNA polymerase α (116)

epidermal growth factor receptor from A431 cells possesses the carbohydrate structure of the blood group A antigen and a monoclonal antibody raised to the receptor was shown to react with this blood group antigen. Elution of active receptor was obtained with the competing sugar, 0.25 M N-acetylgalactosamine (87).

5.2.3 *Choice of antibody and elution conditions*

If there is a choice of antibodies, those of lower affinity are probably the most suitable for immunopurification. Pilot scale studies of different elution conditions can be carried out using a sheep anti-mouse IgG cellulose immunoadsorbent (50 μl of 5 mg/ml) bound with the monoclonal antibody culture supernatant or ascites at 10^{-3} dilution (100 μl) and the antigen. It may be useful to radiolabel the antigen if the assay for bioactivity is tedious and to screen for appearance of label in the eluate. Successful conditions can then be tried for elution of biologically active antigen. After elution, it is advisable to perform gel filtration to remove any aggregated or denatured antigen or traces of unbound monoclonal antibody.

6. THE USE OF ANTIBODIES AS PROBES FOR THE SELECTION OF COMPLEMENTARY DNA CODING FOR MEMBRANE PROTEINS

A complete understanding of the structural properties of a protein requires a knowledge of its primary sequence. However, for minor membrane proteins it is difficult to purify sufficient protein to achieve a complete primary sequence by conventional protein chemistry techniques. An alternative approach is to sequence complementary DNA (cDNA) coding for the protein, though this in turn requires the availability of a probe capable of detecting the specific DNA. If short sections of primary sequence are available it is possible to synthesize oligonucleotide probes for cDNA screening and the criteria for suitable probes are well described (117). When no primary sequence information is available it is impossible to construct such probes, and since for most minor membrane proteins, no means are known of increasing the abundance of mRNA, screening of cDNA clones by hybrid-selected translation (118) is also inappropriate. An elegant solution for overcoming screening difficulties is to use specific antibodies to identify cDNA clones.

6.1 **Phage vectors**

A number of bacterial cloning vectors, both phage and plasmid, have been developed for use with antibody probes (119−124). One of the most widely used is the phage vector λgt11 (123, 125) in which the cDNA is inserted towards the 3' end of the structural gene coding for β-galactosidase. Recombinants can thus produce fusion proteins consisting of most of the β-galactosidase protein and a carboxy-terminal segment encoded by the cDNA. Production of the fusion protein can be induced chemically. Proteins released by the lysis of cells within plaques are immobilized on a nitrocellulose filter and the filter is then challenged with specific antibody in a manner analogous to that used in immunoblotting described in Section 3. Excellent descriptions of the experimental protocols necessary for construction and screening of cDNA libraries in λgt11 are available (125). The usefulness of λgt11 in cloning and sequencing cDNA

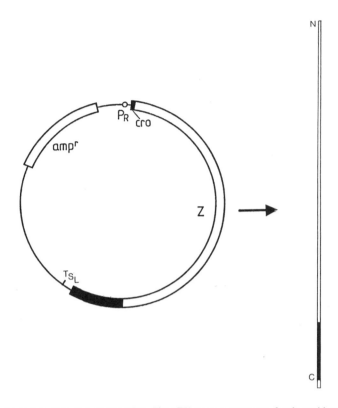

Figure 3. The bacterial expression vector pEX. The pEX vectors express a β galactosidase hybrid protein which is insoluble, forming dense inclusion bodies in the cell accounting for up to 25% of total *E. coli* protein. The plasmids contain a β lactamase gene, amp^r; a strong promoter, P$_R$; a cro-lac I − lac Z gene fusion, Z; an oligonucleotide linker containing cloning sites, L which in the diagram contains a short open reading frame cDNA (solid); an oligonucleotide containing translation stop codons in all reading frames, S and transcription terminator fragments, T from phage fd. The expression polypeptide corresspondong to the hybrid gene and cDNA insert is also shown.

coding for minor membrane proteins is exemplified by the recent determination of the structure of the human glucose transporter (126).

6.2 Plasmid vectors

A well controlled plasmid vector, pEX (124, *Figure 3*) is also available for expression cloning. This vector again employs insertion of cDNA towards the 3′ end of a gene coding for β-galactosidase, itself downstream from a strong promoter. pEX is available in three constructions allowing insertion of foreign cDNA in the correct reading frame. Insertion is into a linker containing a choice of restriction sites. Control of fusion protein synthesis is achieved by using the plasmids in an *Escherichia coli* strain producing a temperature-sensitive repressor so that fusion proteins are only made when the temperature is shifted to 42°C. The cells are grown at 30°C to allow plasmid amplification before the temperature shift.

Whichever expression vector is used, it is important to have tight control of fusion protein synthesized (by chemical induction in λgt11 or temperature induction in pEX),

Table 8. Colony blot procedure for screening pEX libraries.

A. *Growth and expression*

1. Pour 24 cm square plates with 200 ml of L-Broth containing 100 μg/ml ampicillin and 1.6% w/v agar. Dry the inverted plates overnight with their lids displaced.
2. Dilute the library to about 10^4 cells/ml in L-Broth, incubate for 15 min at 34°C and then spread 1.5 ml per plate. Grow for 20 h, at 30°C.
3. Displace the lid for 15 min. Carefully lay the dry nitrocellulose filter (Schleicher & Schull, 0.45 μm) onto the plate. Roll gently and mark the position with pin holes. Peel off the filter with colonies attached ($1-5 \times 10^4$) and lay (colonies uppermost) onto two sheets of Whatman 3MM paper soak-ed in L-Broth containing ampicillin. (Paper is rolled to remove air bubbles). Incubate for 2 h 42°C.

B. *Lysis and electrophoresis*

1. Place the filter (colonies uppermost) onto two sheets of Whatman 3MM paper previously soaked in 5% w/v SDS and rolled. Heat, in containers with lids, either in a conventional oven (95°C, 15 min) or in a microwave oven (45 sec, 600 W, then 60 sec, 200 W) until the colonies are translucent.
2. Arrange the filter in a blotting sandwich (as described in *Table 4* for immunoblotting proteins, but with no gel present) and with the colonies towards the negative electrode, electrophorese for 30 min, at 50 V. This ensures that colony protein is firmly attached to the nitrocellulose and removes excess SDS. Electrophoresis may be carried out in specially made tanks to accommodate large filters, or the filters may be cut to fit into conventional blotting apparatus.

C. *Antibody incubation, development and staining*

The filters are incubated as follows using PBS pH 7.4, containing 0.5% w/v gelatin (or 10% v/v horse serum) and 0.1% v/v Triton X-100 unless otherwise stated:

1. Wash the filters for 5 min with buffer.
2. Wash the filters for 20 min with 10 μg/ml DNase in the buffer.
3. Wash the filters twice for 5 min.
4. Place the filter between dry sheets of Whatman 3MM paper and roll with a roller to remove bacterial debris.
5. Incubate with first antibody for 60 min. Usually use 1 μg of IgG/ml buffer with *E. coli* protein pre-sent in the buffer to prevent non-specific binding to colonies.
6. Wash the filters three times for 5 min.
7. Incubate the filters with horseradish peroxidase-labelled second antibody for 60 min, (1:2000 dilu-tion antibody supplied by TAGO Inc, Burlinghame, CA, USA).
8. Wash the filters three times for 5 min.
9. Wash the filters twice with PBS for 5 min.
10. Develop in 0.1 M Tris-HCl, pH 7.4 containing 0.5 mg/ml 3,3′-diaminobenzidine tetrahydrochloride and 1:5000 H_2O_2.
11. Mark the positive colonies, then counter-stain with Ponceau S. Pick the positive colonies from the master plate.

The volume of washes used is $20-30$ ml for a 24 cm square filter.

in case any of the fusion proteins encoded by the foreign cDNA are lethal. Although pEX has been less widely used than λgt11 for screening cDNA libraries, the ease of working with plasmid DNA and the ability to produce fusion protein at up to 25% of total cell protein attest to its advantages. This may be particularly true when an expression vector is required to clone a specific piece of cDNA to obtain protein that may be used as immunogen to prepare antibodies of pre-determined specificity.

6.3 Screening

A colony blot procedure (127) for use in antibody screening of cDNA libraries in pEX is described in *Table 8*. The similarity of the protocol to that for immunoblotting proteins described in *Table 4* is apparent. In general, an antibody capable of detecting 10 ng of protein in a standard immunoblot procedure can be used as a probe in expression screening.

There is some dispute as to whether polyclonal or monoclonal antibodies are best for initial screening. An affinity-purified polyclonal antibody is an attractive probe, since one is then screening for the presence of any of several epitopes on the original protein. However, the chance of false positives due to polyclonal antibodies against any impurity in the immunogen will be amplified by expression screening since the antigenic impurity may be produced at the same concentration as the antigen of interest. A compromise is to have clones detected by both affinity-purified polyclonal antibodies and monoclonal antibodies (36, 124), though cocktails of monoclonal antibodies have also been used successfully (128). The investigator's confidence in positive clones may be strengthened by the success of colony purification and secondary screening, immunoblotting of the fusion protein, affinity purification of antibodies on the fusion protein with retroblotting on to the original antigen, as well as reaction with more than one antibody. However, confirmation of clones requires sequencing and comparison with at least some information concerning the primary structure of the protein. This information may include compositional analysis, sections of primary sequence and/or a knowledge of defined proteolytic cleavage sites.

6.4 Future prospects

Whilst the use of antibodies as probes for expression cloning can lead to false positives (129), there are an increasing number of successful examples. In future, expression vectors may prove to be particularly important in studying the topography of membrane proteins. Thus, domains of the protein can be expressed *in vitro* and used to raise antibodies of pre-determined specificity. These may then used to map protein topography across the membrane with respect to the DNA sequence. In addition, it will be possible to prepare antibodies to particular parts of a membrane protein for purposes such as the modification of function or to use as first antibodies in the immunoaffinity purification of specific organelles. Such pre-determined antibodies will undoubtedly prove important reagents in the future investigation of cell membrane structure and function.

7. ACKNOWLEDGEMENTS

We thank Professor C.N.Hales, Dr K.Siddle, Dr K.K.Stanley and Dr K.Howell for many useful and stimulating discussions. Experimental work by the authors reported in this article was supported by grants from the Medical Research Council, the Arthritis and Rheumatism Council, the British Diabetic Association, the Beit Foundation, the European Molecular Biology Organisation, St. John's College, Cambridge, IQ Bio Ltd., Cambridge and the Muscular Dystrophy Group of Great Britain and Northern Ireland.

8. REFERENCES

1. Siddle,K. and Soos,M. (1981) In *Monoclonal Antibodies and Developments in Immunoassay.* Albertini,A. and Ekins,R. (eds), Elsevier/North-Holland, Amsterdam, New York and Oxford, p. 53.
2. Bailyes,E.M., Soos,M., Jackson,P., Newby,A.C., Siddle,K. and Luzio,J.P. (1984) *Biochem. J.*, **221**, 369.

3. Atassi,M.Z. (1985) *Eur. J. Biochem.*, **145**, 1.
4. Lerner,R.A. (1984) *Adv. Immunol.*, **36**, 1.
5. Berzofsky,J.A. (1985) *Science*, **229**, 932.
6. Berek,C., Griffiths,G.M. and Milstein,C. (1985) *Nature*, **316**, 412.
7. Herbert,W.J. (1978) In *Handbook of Experimental Immunology*. 3rd edition, Weir,D.M. (ed.), Blackwell, Oxford, London, Edinburgh and Melbourne, Vol. 3, Appendix 4.
8. Herbert,W.J. (1978) In *Handbook of Experimental Immunology*. 3rd edition, Weir,D.M. (ed.), Blackwell, Oxford, London, Edinburgh and Melbourne, Vol. 3, Appendix 3.
9. Repasky,E.A., Granger,B.L. and Lazarides,E. (1982) *Cell*, **29**, 821.
10. Knudsen,K.A. (1985) *Anal. Biochem.*, **147**, 285.
11. Rapport,M.M., Graf,L. and Yariv,J. (1961) *Arch. Biochem. Biophys.*, **92**, 438.
12. Jones,R.T., Walker,J.H., Richardson,P.J., Fox,G.Q. and Whittaker,V.P. (1981) *Cell Tissue Res.*, **218**, 355.
13. Eisenbarth,G.S., Walsh,F.S. and Nirenberg,M. (1979) *Proc. Natl. Acad. Sci. USA*, **76**, 4913.
14. Geiger,B. and Smolarsky,M. (1977) *J. Immunol. Methods*, **17**, 7.
15. Young,W.W., Hakamori,S.-I., Durdik,J.M. and Henney,C.S. (1980) *J. Immunol.*, **124**, 199.
16. Weir,D.M., ed. (1978) *Handbook of Experimental Immunology*. Blackwell, Oxford, London, Edinburgh and Melbourne, Vols. 1−3.
17. Goding,J.W. (1978) *J. Immunol. Methods*, **20**, 241.
18. Kohler,G. and Milstein,C. (1975) *Nature*, **256**, 495.
19. Galfre,G. and Milstein,C. (1981) *Methods Enzymol.*, **73**, 3.
20. Goding,J.W. (1983) *Monoclonal Antibodies: Principles and Practice*. Academic Press, Inc.
21. Morgan,D.O. and Roth,R.A. (1985) *Endocrinology*, **116**, 1224.
22. Jemmerson,R. and Fishman,W.H. (1982) *Anal. Biochem.*, **124**, 286.
23. Bailyes,E.M., Siddle,K. and Luzio,J.P. (1985) *Biochem. Soc. Trans.*, **13** 107.
24. Slaughter,C.A., Coseo,M.C., Abrams,C., Cancro,M.P. and Harris,H. (1980) In *Monoclonal Antibodies, Hybridomas*. Kennett,R.H., McKearn,T.J. and Bechtol,K.B. (eds), Plenum Press, New York, London, p. 111.
25. Kennett,R.H. and McKearn,T.J. (1980) In *Monoclonal Antibodies, Hybridomas*. Kennett,R.H., McKearn,T.J. and Bechtol,K.B. (eds), Plenum Press, New York, London, p. 391.
26. Renart,J., Reiser,J. and Stark,G.R. (1979) *Proc. Natl. Acad. Sci. USA*, **76**, 3116.
27. Towbin,H., Staehelin,T. and Gordon,J. (1979) *Proc. Natl. Acad. Sci. USA*, **76**, 4350.
28. Norrild,B., Bjerrum,O.J. and Vestergaard,P.F. (1977) *Anal. Biochem.*, **81**. 432.
29. Burnette,W.N. (1981) *Anal. Biochem.*, **112**, 195.
30. Gershoni,J.M. and Palade,G.E. (1983) *Anal. Biochem.*, **131**, 1.
31. Towbin,H. and Gordon,J. (1984) *J. Immunol. Methods*, **72**, 313.
32. Gershoni,J.M. (1985) *Trends Biochem. Sci.*, **10**, 103.
33. Bjerrum,O.J. and Heegaard,N.H.N. (1987) *Handbook of Immunoblotting*. CRC Press, Boca Raton, Florida, in press.
34. Jackson,P. and Thompson,R.J. (1984) *Electrophoresis*, **5**, 32.
35. Olmsted,J.B. (1981) *J. Biol. Chem.*, **256**, 11955.
36. Stanley,K.K., Kocher,H.-P., Luzio,J.P., Jackson,P. and Tschopp,J. (1985) *EMBO J.*, **4**, 375.
37. Panyim,S. and Chalkley,R. (1969) *Arch. Biochem. Biophys.*, **130**, 337.
38. Bradbury,J.M. and Thompson,R.J. (1984) *Biochem. J.*, **221**, 361.
39. Oblas,B., Boyd,N.D. and Singer,R.H. (1983) *Anal. Biochem.*, **130**, 1.
40. Holmberg,D., Ivars,F. and Edlund,T. (1983) *J. Immunol. Methods*, **61**, 9.
41. Maruyama,K., Mikawa,T. and Ebashi,S. (1984) *J. Biochem. (Tokyo)*, **95**, 511.
42. Jackson,P.J. and Luzio,J.P. (1987) In *Handbook of Immunoblotting.*, Bjerrum,O.J. and Heegaard,N.H.N. (eds), CRC Press, Boca Raton, Florida, in press.
43. Stanley,K.K., Page,M., Campbell,A.K. and Luzio,J.P. (1986) *Mol. Immunol.*, **23**, 451.
44. Towbin,H., Schoenenberger,C., Ball,R., Braun,D.G. and Rosenfelder,G. (1984) *J. Immunol. Methods*, **72**, 471.
45. Magnani,J.L., Nilsson,B., Brockhaus,M., Zopf,D., Steplewski,Z., Koprowski,H. and Ginsburg,V. (1982) *J. Biol. Chem.*, **257**, 14365.
46. Harpin,M.L., Coulon-Morelec,M.J., Yeni,P., Danon,F. and Baumann,N. (1985) *J. Immunol. Methods*, **78**, 135.
47. Hales,C.N. (1972) *Diabetologia*, **8**, 229.
48. Luzio,J.P., Newby,A.C. and Hales,C.N. (1974) *Biochem. Soc. Trans.*, **2**, 1385.
49. Luzio,J.P., Newby,A.C. and Hales,C.N. (1976) *Biochem. J.*, **154**, 11.
50. Kawajiri,K., Ito,A. and Omura,T. (1977) *J. Biochem. (Tokyo)* **81**, 779.
51. Howell,K.E., Ansorge,W. and Gruenberg,J. (1986) In *Microspheres: Medical and Biological Applica-*

tions. Rembawn,A. and Tokes,Z. (eds), CRC Press, Boca Raton, Florida, in press.

52. Richardson,P.J. and Luzio,J.P. (1986) *Appl. Biochem. Biotechnol.*, in press.
53. Westwood,S.A., Luzio,J.P., Flockhart,D.A. and Siddle,K. (1979) *Biochim. Biophys. Acta*, **583**, 454.
54. Devaney,E. and Howell,K.E. (1985) *EMBO J.*, **4**, 3123.
55. Roman,L.M. and Hubbard,A.L. (1984) *J. Cell Biol.*, **98**, 1497.
56. Gruenberg,J. and Howell,K.E. (1985) *Eur. J. Cell Biol.*, **38**, 312.
57. Richardson,P.J., Siddle,K. and Luzio,J.P. (1984) *Biochem. J.*, **219**, 647.
58. Ito,A. and Palade,G.E. (1978) *J. Cell Biol.*, **79**, 590.
59. Pontremoli,S., Melloni,E., Diamani,G., Michetti,M., Salamino,F., Sparatone,B. and Horecker,B.L. (1984) *Arch. Biochem. Biophys.*, **233**, 267.
60. Merisko,E.M., Farquhar,M.G. and Palade,G.E. (1982) *J. Cell Biol.*, **92**, 846.
61. Pfeffer,S.R. and Kelly,R.B. (1985) *Cell*, **40**, 949.
62. Matthew,W.D., Tsavaler,L. and Reichardt,L.F. (1981) *J. Cell Biol.*, **91**, 257.
63. Kronvall,G., Seal,V.S., Finstad,J. and Williams,R.C. (1970) *J. Immunol.*, **104**, 140.
64. Luzio,J.P. and Stanley,K.K. (1983) *Biochem. J.*, **216**, 27.
65. Schneider,C., Newman,R.A., Sutherland,D.R., Asser,V. and Greaves,M.F. (1982) *J. Biol. Chem.*, **257**, 10766.
66. Gurvich,A.E., Kapner,R.B. and Nezlin,R.S. (1959) *Biokhimiya*, **24**, 129.
67. Gurvich,A.E., Kuzloveva,O.B. and Turmanova,A.E. (1961) *Biokhimiya*, **26**, 803.
68. Hales,C.N. and Woodhead,J.S. (1980) *Methods Enzymol.*, **70**, 334.
69. Singer,S.J. (1974) *Annu. Rev. Biochem.*, **43**, 805.
70. Cresswell,P., Turner,M.J. and Strominger,J.L. (1973) *Proc. Natl. Acad. Sci. USA*, **70**, 1603.
71. Miki,A., Kominani,T. and Ikehara,Y. (1985) *Biochem. Biophys. Res. Commun.*, **126**, 89.
72. Helenius,A. and Simons,K. (1975) *Biochim. Biophys. Acta*, **415**, 29.
73. Newby,A.C. (1984) *Brain Receptor Methodologies, Part A.* Marangos,P.J., Campbell,I.C. and Cohen,R.M. (eds) Academic Press, NY, p. 75.
74. Helenius,A. and Simons,K. (1977) *Proc. Natl. Acad. Sci. USA*, **77**, 6568.
75. Kagawa,Y. (1972) *Biochim. Biophys. Acta*, **265**, 297.
76. Newby,A.C., Chrambach,A. and Bailyes,E.M. (1982) *Tech. Life Sci. Biochem.* **B4**, 1.
77. Pober,J.S., Guild,B.C., Strominger,J.L. and Veatch,W.R. (1981) *Biochemistry*, **20**, 5625.
78. Bailyes,E.M., Newby,A.C., Siddle,K. and Luzio,J.P. (1982) *Biochem J.*, **203**, 245.
79. Crumpton,M.J. and Parkhouse,R.M.E. (1972) *FEBS Lett.*, **22**, 210.
80. Herrman,S.H. and Mescher,M.F. (1979) *J. Biol. Chem.*, **254**, 8713.
81. Driessen,M., Weitz,G., Brouwer-Kelder,E.M., Donker-Koopman,W.E., Bastiaannet,J., Sandhoff,K., Barranger,J.A., Tager,J.M. and Schram,A.W. (1985) *Biochim. Biophys. Acta*, **841**, 97.
82. von Weicker,H., Kropp,J. and Roecke,D. (1971) *Z. Klin. Chem. u. Klin. Biochem.*, **5**, 375.
83. Hansen,R.S. and Beavo,J.A. (1982) *Proc. Natl. Acad. Sci. USA*, **79**, 2788.
84. Yarden,Y., Harari,I. and Schlessinger,J. (1985) *J. Biol. Chem.*, **260**, 315.
85. Weber,W., Bertics,P.J. and Gill,G.N. (1984) *J. Biol. Chem.*, **259**, 14631.
86. Fraser,C.M. and Venter,J.C. (1982) *Biochem. Biophys. Res. Commun.*, **109**, 21.
87. Parker,P.J., Young,S., Gullick,W.J., Mayes,E.L., Bennett,P. and Waterfield,M.D. (1984) *J. Biol. Chem.*, **259**, 9906.
88. Campbell,D.G., Gagnon,J., Reid,K.B.M. and Williams,A.F. (1981) *Biochem. J.*, **195**, 15.
89. Kennedy,S.M. and Burchell,B. (1983) *Biochem. Pharmacol.*, **32**, 2029.
90. Acuto,O., Fabbi,M., Smart,J., Poole,C.B., Protentis,J., Royer,H.D., Schlossman,S.F. and Reinherz,E.L. (1984) *Proc. Natl. Acad. Sci. USA*, **81**, 3851.
91. Brown,W.R.A., Barclay,A.N., Sunderland,C.A. and Williams,A.F. (1981) *Nature*, **289**, 456.
92. Barclay,A.N. and Ward,H.A. (1982) *Eur. J. Biochem.*, **129**, 447.
93. Randle,B.J., Morgan,A.J., Stripp,S.A. and Epstein,M.A. (1985) *J. Immunol. Methods*, **77**, 25.
94. Hsuing,L., Barclay,A.N., Brandon,M.R., Sim,E. and Porter,R.R. (1982) *Biochem. J.*, **203**, 293.
95. Hughes,E.N. and August,J.T. (1982) *J. Biol. Chem.*, **257**, 3970.
96. Kasper,L.H., Crabb,J.H. and Pfefferkorn,E.R. (1983) *J. Immunol.*, **130**, 2407.
97. Hsu,H.H., Munoz,P.A., Barr,J., Oppliger,I., Morris,D.C., Vaananen,H.K., Tarkenton,N. and Anderson,H.C. (1985) *J. Biol. Chem.*, **260**, 1826.
98. Lennon,V.A., Thompson,M. and Chen,J. (1980) *J. Biol. Chem.*, **255**, 4395.
99. Orlow,S.J., Rosenstreich,D.L., Pifco-Hirst,S. and Rosen,O.M. (1985) *J. Immunol.*, **134**, 449.
100. Mellman,I.S. and Unkeless,J.C. (1980) *J. Exp. Med.*, **152**, 1048.
101. Sakai,M., Saisu,H., Koshigoe,N. and Abe,T. (1985) *Eur. J. Biochem.*, **148**, 197.
102. Nielsen,L.S., Hansen,J.G., Andreasen,P.A., Skriver,L., Dan,K. and Zeuthen,J. (1983) *EMBO J.*, **2**, 115.

103. Inagaki,T., Ohtsuki,K. and Inagami,T. (1983) *J. Biol. Chem.*, **258**, 7476.
104. Herion,P. and Bollen,A. (1983) *BioSci. Rep.*, **3**, 373.
105. Staehelin,T., Hobbs,D.S., Kung,H. and Pestka,S. (1981) *Methods Enzymol.*, **78**, 505.
106. Cheng,K.C., Gelboin,H.V., Song,B.J., Park,S.S. and Friedman,F.K. (1984) *J. Biol. Chem.*, **259**, 12279.
107. Chernausek,S.D., Chatelain,P.G., Svoboda,M.E., Underwood,L.E. and Van-Wyk,J.J. (1985) *Biochem. Biophys. Res. Commun.*, **126**, 282.
108. Secher,D.S. and Burke,D.C. (1980) *Nature*, **285**, 446.
109. Mole,J.E., Hunter,F., Paslay,J.W., Bhown,A.S. and Bennett,J.C. (1982) *Mol. Immunol.*, **19**, 1.
110. Momoi,M.Y. and Lennon,V.A. (1984) *J. Neurochem.*, **42**, 59.
111. Meyer,L.J., Lafferty,M.A., Raducha,M.G., Foster,C.J., Gogolin,K.J. and Harris,H. (1982) *Clin. Chim. Acta*, **125**, 109.
112. Morgan,B.P., Daw,R.A., Siddle,K., Luzio,J.P. and Campbell,A.K. (1983) *J. Immunol. Methods*, **64**, 269.
113. Schroder,H.C., Wehland,J. and Weber,K. (1985) *J. Cell Biol.*, **100**, 276.
114. Heikinheimo,M., Stenman,V.H., Bang,B., Hurme,M., Makela,O. and Bohn,H. (1983) *J. Immunol. Methods*, **60**, 25.
115. Urdal,D.L., March,C.J., Gillis,S., Larsen,A. and Dower,S.K. (1984) *Proc. Natl. Acad. Sci. USA*, **81**, 6481.
116. Wang,T.S., Hu,S.Z. and Korn,D. (1984) *J. Biol. Chem.*, **259**, 1854.
117. Lathe,R. (1985) *J. Mol. Biol.*, **183**, 1.
118. Harpold,M.M., Dobner,P.R., Evans,R.M. and Bancroft,F.C. (1978) *Nucleic Acids Res.*, **5**, 2039.
119. Gray,M.R., Colot,H.V., Guarente,L. and Rosbash,M. (1982) *Proc. Natl. Acad. Sci. USA*, **79**, 6598.
120. Helfman,D.M., Feramisco,J.R., Fiddes,J.C., Thomas,G.P. and Hughes, S.H. (1983) *Proc. Natl. Acad. Sci. USA*, **80**, 31.
121. Ruther,U. and Muller-Hill,B. (1983) *EMBO J.*, **2**, 1791.
122. Weinstock,G.M., ap Rhys,C., Berman,M.L., Hampar,B., Jackson,D., Silhavy,T.J., Weisemann,J. and Zweig,M. (1983) *Proc. Natl. Acad. Sci. USA*, **80**, 4432.
123. Young,R.A. and Davis,R.W. (1983) *Proc. Natl. Acad. Sci. USA*, **80**, 1194.
124. Stanley,K.K. and Luzio,J.P. (1984) *EMBO J.*, **3**, 1429.
125. Huynh,T.V., Young,R.A. and Davis,R.W. (1985) In *DNA Cloning I—A Practical Approach.* Glover,D. (ed.), IRL Press, Oxford and Washington, DC, Vol. 1, p. 47.
126. Mueckler,M., Caruso,C., Baldwin,S.A., Panico,M., Blench,I., Morris,H.R., Allard,W.J., Lienhard, G.E. and Lodish,H.F. (1985) *Science,* **229**, 941.
127. Stanley,K.K. (1983) *Nucleic Acid Res.*, **11**, 4077.
128. Jenne,D. and Stanley,K.K. (1985) *EMBO J.*, **4**, 3153.
129. Siddle,K., Pauza,D., Bellatin,J., Lennox,E.S. and Stanley,K.K. (1986) *Biochem. Soc. Trans.*, **14**, 318.

CHAPTER 4

Separation and analysis of membrane lipid components

JOAN A.HIGGINS

1. INTRODUCTION

Lipids are a major component of membranes, accounting for between 20 and 80% of the membrane mass. Current ideas of membrane structure postulate that the lipids are arranged predominantly in the form of a bilayer into which proteins may be inserted or with which proteins are associated peripherally. One major role of membranes is to form continuous barriers as plasma membrane and as cytoplasmic compartments. The physical characteristics of lipids provide the basis of these barriers. However, lipids do not merely form an inert framework in membranes. Recent investigations have indicated that lipids have more fundamental roles. There is evidence that some responses of cells to external stimuli including hormones, neurotransmitters and growth factors are mediated by changes in membrane lipids. Hydrolysis of polyphosphoinositides produces two second messengers, inositol trisphosphate and diglyceride. The former functions to increase intracellular calcium ion concentration and the latter activates protein kinase C (1). It has been postulated that methylation of phosphatidylethanolamine to produce phosphatidylcholine is increased in response to hormones and that this change is involved in the cellular events which follow stimulation (2). There is also evidence that alterations in the membrane bilayer such as shifts of phospholipids between the leaflets, selective hydrolysis of phospholipids at focal points, or formation of non-bilayer lipid configurations may be involved in cell fusion, and intracellular vesiculation and refusion. Related to this, membrane lipids have an asymmetric transverse distribution across the bilayer. The exact role of this asymmetry has yet to be elucidated.

   The aim of the present chapter is to consider practical aspects of investigations of membrane lipids. The material considered will cover two areas. In the first section the basic techniques of membrane lipid extraction, separation and analysis will be considered with emphasis on aspects of current interest. In the second section methods of applying these techniques to specific investigations of membrane characteristics including the determination of the transverse distribution of phospholipids in membranes and investigations of the synthesis and intracellular transport of lipids will be described. This chapter will, of necessity, be selective. It is intended that it will provide the reader with the necessary technical information to approach an investigation of membrane lipids with confidence and with sufficient background to adapt the techniques to fit his or her experimental protocol. Biophysical approaches for the study of lipids and lipid−protein interactions are described in Chapter 8.

2. LIPID EXTRACTION

The aim of lipid extraction procedures is to extract membrane lipids quantitatively and without contamination with non-lipid components. Because a variety of forces, including hydrogen bonds, hydrophobic associations and electrostatic forces are involved in the binding of the membrane lipids, it is necessary to use polar solvents such as methanol or ethanol which disrupt the membrane structure and denature the membrane proteins. The solvent systems most commonly used to extract membrane lipids are based on the procedure originally described by Folch, Lees and Sloane-Stanley and include chloroform and methanol (3). The initial extract produced may be contaminated with a number of non-lipid compounds and is washed to remove these. A variety of protocols have been used for extraction of tissue lipids. However, those most commonly used for extraction of membrane lipids are described below. The choice of procedure depends mainly on whether the membrane to be extracted is in the form of a pellet, a concentrated suspension or a dilute suspension.

2.1 **Solvents and apparatus**

All lipid extraction procedures should be performed in glass using analar or re-distilled solvents. To measure solvents or to take aliquots of lipid extracts, glass dispensors or glass pipettes with automatic dispensors should be used. Positive displacement automatic pipettes can also be used for measuring lipid extracts; however, care should be taken to ensure that these are correctly calibrated for solvents and that material is not extracted from the plastic tips.

   As many lipids have double bonds, these are susceptible to peroxidation. To avoid this, re-distilled solvents should be used, and, if highly unsaturated lipids are to be extracted, dissolved oxygen should be removed by bubbling nitrogen through the solvent. Procedures should be carried out at, or below, room temperature. Lipids should not be left for any period of time in a dry form but dissolved in an organic solvent. If the extracts are to be stored, these should be kept in a dark-glass bottle or a container wrapped in aluminium foil in a freezer. If storage is for any length of time, or highly unsaturated lipids are present, 0.05% butylated hydroxytoluene should be added as an anti-oxidant.

2.2 **Extraction of membrane pellets or small volumes of concentrated suspensions**

(i)   Re-suspend the membrane pellet or disperse the suspension in glass centrifuge tubes containing at least 20 volumes of chloroform/methanol (2/1 v/v). As membranes are isolated from homogenized tissues and are readily re-suspended it is not usually necessary to homogenize the tissue extract. However, occasionally with pellets or with some preparations, membrane clumps form in chloroform/methanol. Such preparations are best extracted by addition of the necessary volume of methanol alone, dispersion of the pellet using a vortex mixer, or if this is insufficient a Polytron or Ultraturax homogenizer (see Chapter 1), followed by addition of chloroform to give a final composition of chloroform/methanol of 2/1.

   Twenty volumes of chloroform/methanol per volume of membrane pellet or suspension should yield a homogeneous single phase suspension. If two phases form add chloroform/methanol (2/1) and mix well until a single phase is produced.

(ii) Cover the tubes to prevent evaporation and allow the extract to stand for at least 15 min (this may be extended if necessary), pellet the denatured protein by centrifugation and decant the supernatants into glass centrifuge tubes.

(iii) Add one-fifth the volume of 0.05 M $CaCl_2$ to the extract and mix thoroughly using a vortex mixer or a glass rod flattened at one end. A uniform emulsion should be produced. Allow this to stand, or centrifuge the tubes in a bench centrifuge to separate the two phases.

(iv) Using a Pasteur pipette carefully remove the upper phase which consists of chloroform/methanol/water, 3/42/47. The lower phase consists mainly of chloroform containing the lipids, with the exception of gangliosides, which remain in the upper phase from which they may be isolated by dialysis against water and lyophilization.

(v) The lower phase may be washed by addition of an equal volume of methanol/water (1/1), mixing and centrifuging to separate the phases.

Whether the lipid extract is washed and the number of washes used depends on the investigation in hand. For example, if incorporation of radiolabelled precursors of lipids such as palmitoyl-CoA, S-adenosylmethionine, choline or ethanolamine into lipid is determined, there may be a significant carry-over of the labelled substrate into the lipid extract. Washing would reduce this. Alternatively, if a good recovery of total lipid is required and the presence of small amounts of water-soluble material is not detrimental, it is not necessary to wash the extract after the initial phase separation and removal of the upper phase. After removal of most of the upper phase the small amount which remains can be dispersed into the lower phase by addition of a few drops of methanol.

2.3 Acidification of the lipid extraction solvent

Acidic phosphatides, free fatty acids and polyphosphoinositides are not extracted completely by chloroform/methanol. It is necessary to add 0.25% concentrated HCl to the chloroform/methanol for complete extraction of these lipids. This can be adopted as a general method for most lipid extraction procedures. However, acid hydrolysis of plasmalogens occurs, and if these lipids are under investigation acid should not be added to the solvent used for their extraction.

2.4 Extraction of large volumes of membrane suspensions

Large volumes of membrane in aqueous media are best extracted using the procedure of Bligh and Dyer (4).

(i) To each 8 ml of membrane suspension add 30 volumes of chloroform/methanol (1/2) and mix to produce a single phase. If a large precipitate forms, this should be removed by filtration or centrifugation. If the precipitate is small this can be removed later with the upper phase.

(ii) Add 10 ml of chloroform and 10 ml of water, mix, and either allow the two phases to separate on standing or centrifuge the mixture to accelerate the separation.

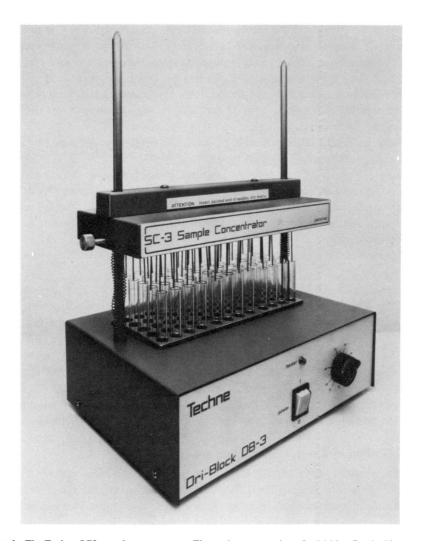

Figure 1. The Techne SC3 sample concentrator. The equipment consists of a dri-bloc fitted with removable blocks which are available for a variety of test tube or vial sizes. The solvent is evaporated by a stream of nitrogen directed through needles, the number and position of which may be adjusted according to the samples to be concentrated.

(iii)　Remove the upper phase together with any denatured protein at the interface using a Pasteur pipette. The lower phase contains the lipids with the exception of gangliosides.

2.5 Removal of solvent from the lipid extracts

Small volumes of chloroform are best removed from lipid extracts by evaporation under a stream of nitrogen in a fume cupboard. The Techne SC3 sample concentrator is excellent for this purpose (*Figure 1*). The lipid extracts are placed in a suitable sized glass container (test tubes or vials) and fitted into a dri-bloc the temperature of which

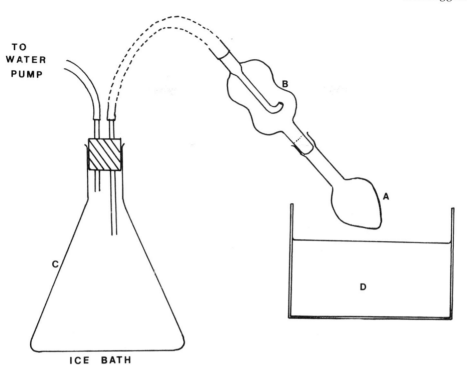

TO
WATER
PUMP

B

A

C

D

ICE BATH

Figure 2. Apparatus for the evaporation of solvents from lipid extracts. The apparatus consists of a pear-shaped flask (**A**) connected by a ground glass joint to a trap (**B**). This is connected by pressure tubing to a cooled flask (**C**) which is evacuated by a water pump. The lipid extract is placed in the pear-shaped flask, which is warmed in a water bath (**D**) and gently swirled. The solvent evaporates and condenses in the flask leaving the lipids in the pear-shaped flask.

is regulated thermostatically. A stream of nitrogen or compressed air is directed through adjustably spaced needles onto the surface of each extract solution. Using this equipment 48 samples containing up to 10 ml of chloroform each can be concentrated or dried simultaneously. Similar equipment can be constructed using either a hot plate, a heated block or a water bath to warm the extract, while nitrogen is directed into each test tube or vial through a system of Pasteur pipettes attached by a common tubing to a cylinder. Rapid evaporation of solvents can be achieved using this equipment; however, the samples should not be heated above 40−50°C if the lipids are to be analysed further.

Large volumes of chloroform are best removed from lipid extracts by rotary evaporation using a commercial apparatus. These procedures tend to be slow, especially if many samples are to be dried. An alternative method for volumes of approximately 100 ml of extract uses the system illustrated in *Figure 2*. The apparatus is simple and can be constructed in a glass workshop.

(i)   Transfer the lipid extract into a pear-shaped flask with a ground glass outlet.

(ii)   Connect this to the first trap, which is connected to a conical flask attached to a water pump. The connection should be made while the pump is running.

(iii)   Swirl the flask gently while warming it in a water bath to allow the solvent to evaporate. Any of the lipid extract which boils into the first trap is directed back into the pear-shaped flask, while evaporated solvent moves into the cooled flask and condenses.

(iv) Disconnect the pear-shaped flask while the vacuum is applied to avoid sucking back of water from the pump into the lipid extract.

A safety guard should be placed around the apparatus to protect the operator against implosion of the flask. Using this equipment a large number of lipid extracts can be rapidly evaporated to dryness in pear-shaped flasks. The extracts are then dissolved in a small volume of chloroform or chloroform/methanol (1/1) and stored in the flask for further investigation or transferred to another container.

3. SEPARATION OF LIPID CLASSES

Many methods have been developed for the separation and analysis of lipids. These include solvent fractionation, silicic acid or alumina column chromatography, thin layer chromatography (t.l.c.) or high performance liquid chromatography (h.p.l.c.). It is not possible to consider these techniques in detail here and the reader is referred to reference text-books (5−7). In general, investigations of membranes involve relatively small amounts of lipids. For this reason the technique most frequently used for separation of membrane lipids is t.l.c. on silica gel. This is relatively rapid, extremely versatile, completely separates different lipid classes and can be used qualitatively and quantitatively. Practical aspects of t.l.c. are therefore described in detail below and a number of systems for the analysis of membrane lipids are given. In addition, methods for the separation of polyphosphoinositides and inositol phosphates and for the separation of the methylated products of phosphatidylethanolamine are described in view of the importance of these lipids in membrane research.

3.1 General aspects of t.l.c.

3.1.1 *Apparatus*

(i) *Thin layer plates.* T.l.c. is usually carried out on thin layers of silica gel on a support such as glass, plastic or aluminium sheet. However, it is also possible to use layers of other adsorbents such as alumina, cellulose or Sepharose. Silica gel is available with calcium binder (silica gel G) or without (silica gel H). A variety of ready prepared plates in a range of sizes are available from commercial sources (e.g. Merck, Analabs). It is also possible to spread thin layer plates in the laboratory using apparatus such as the Desaga applicator (Brinkman Instruments). However, unless unusual adsorbents are required or many plates are needed it is probably of no economic advantage to prepare plates rather than to buy prepared ones. Unless otherwise stated, the separations described below were performed using 20 × 20 cm plates having silica gel H at a thickness of 0.25 mm sometimes with the fluorescent indicator F_{254} (Merck silica gel t.l.c. plates 60 F_{254} or Analab silica gel H t.l.c. plates).

(ii) *Tanks.* Rectangular tanks are used for running t.l.c. plates. These should be lined with filter paper cut to fit the larger sides of the tank to saturate the atmosphere. If solvent evaporates from the adsorbent layer during ascending chromatography this produces anomalous results. The tank should be prepared at least an hour ahead of the run by pouring sufficient solvent (see below) into the tank to a depth of about 1.5 cm.

The filter paper lining should be saturated with solvent and a lid fitted and sealed with petroleum jelly. The lid should be removed for the minimum time necessary to place the plate in the tank and replaced and re-sealed. A variety of tanks are available. Some of these have ridges designed to accommodate a large number of plates. However, in our experience it is not possible to separate lipids properly and reproducibly if the adsorbent layers are close together. A maximum of two plates should be run simultaneously in one tank. These should be placed so that the support sheet is towards the tank lining paper and the adsorbent layers are at a maximum distance apart. The adsorbent layer should not touch the supporting side ridges of multiplate tanks as this causes the ascending solvent to move more quickly at the outer sides of the plate producing a curved solvent front and lipid spots which run towards the centre and overlap rather than migrating evenly up the plate.

3.1.2 *Application of samples to thin layers*

For precise separation of lipid mixtures and especially under warm humid conditions, it is advisable to activate the prepared thin layer plates by heating these to 110°C for 30 min. After cooling the plates, apply the lipid samples to be separated approximately 2.5 cm from the edge of the plate. The lipids should be dissolved in a polar solvent such as chloroform/methanol (1/1), for this minimizes solvent spread and reduces the spot size thus maximizing resolution. Depending on the number of samples to be separated, thin layer plates can be lightly marked into columns using a soft pencil while taking care not to damage the adsorbent layer. With the aid of a plastic template, which covers the plate and provides a hand rest, apply lipid solutions in a line across the base of each column using a Hamilton syringe, a micropipette with a fine tip or a capillary pipette (*Figure 3*). If necessary, the lipid solution can be applied repeatedly over the same line allowing the solvent to evaporate between applications. Great care should be taken to avoid damaging the adsorbent layer as this causes irregularity in the movement of solvent up the plate and prevents proper separation of the lipids. Some investigators apply the lipid extract in a series of overlapping small spots rather than a single line. However, with practice, it is possible to streak the lipid solution in a fine line using a syringe or a pipette. The amount of lipid applied to thin layer plates of standard thickness (0.25 mm) will vary between experiments. However, as a guide, phospholipids of microsomal lipids are separated satisfactorily when up to 2 mg of lipid ($\sim$ 2 μmol) is applied in a column approximately 5 cm wide. Phosphatidylcholine comprising 50% of the microsomal lipid produces an intense band under these conditions but is clearly separated from the other phospholipids in the sample.

(i) *Use of standards*. Although the relative positions of lipids separated on thin layers are consistent with one plate type and solvent system the R_f values show some variation between plates run on different days. Pure samples of all lipids under investigation should therefore be applied to each plate at the same time as the lipid extract. About 10 μg of lipid in 10 μl of solvent should be applied in a single spot at the edge of the t.l.c. plate as indicated in *Figure 3*.

(ii) *Running (developing) thin layer plates*. Before developing, allow solvents to evaporate completely from the lipid spots and bands. Because dried lipids are susceptible to oxidation and are unstable, develop the plates immediately.

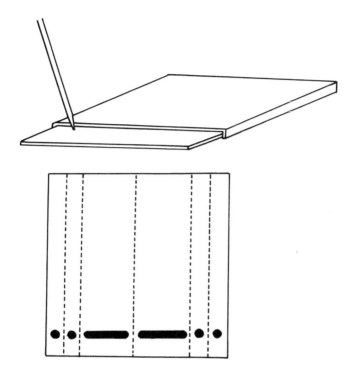

Figure 3. Application of samples to t.l.c. plates. In the upper part of the figure the thin layer plate, template and micropipette used for application of the sample are illustrated. In the lower part of the figure a typical t.l.c. plate before development is illustrated. The standards are spotted in the outer lanes which are marked lightly in soft pencil and the lipid extract is streaked in the inner lanes. A variety of patterns of application of the samples is possible depending on the number of standards and samples to be separated.

Depending on the size of the tank, use 80−150 ml of the solvent system selected (see below), allow the tank to equilibrate and insert the plates. The solvent depth should be approximately 1 cm below the lipid spots. Allow the solvent to ascend to within 1−2 cm of the top of the plate, and remove the plate, mark the solvent front lightly in pencil and allow the plate to dry in a fume cupboard.

3.1.3 *Two-dimensional t.l.c.*

To separate complex mixtures of lipids, for example lysophospholipids from their parent phospholipids, it may be necessary to use two solvent systems run at right angles to each other.

(i)   Place the lipid sample at one corner of the plate and develop the plate in the first solvent system so that the spots separated are along one edge of the plate.

(ii)   Dry the plate, turn it through 90° so that the lipid spots are along the bottom, and then develop the plate in the second solvent (*Figure 4*). Standard lipid samples cannot be run in two dimensions with the lipid mixture. However, the relative positions of the different lipids can be determined by running standard samples at the edges of the plate in each solvent system (*Figure 4*).

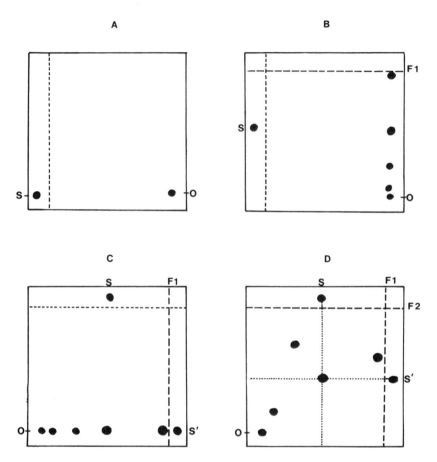

Figure 4. Two-dimensional t.l.c. **A.** The lipids to be separated are spotted at the lower right hand corner of the plate. At the left hand side a lane is lightly marked in pencil for a standard sample S. **B.** Illustrates the position of the lipids after development in the first solvent. **C.** The plate is turned through 90° so that the right-hand edge becomes the bottom edge. A second lane is marked and the same standard applied S'. **D.** After development in the second solvent the lipids move up the plate. The position of the standard is indicated by dotted lines. F1 and F2 and dashed lines indicate the solvent fronts in solvent systems 1 and 2, respectively. O indicates the origin at which the sample is applied.

3.2 Solvent systems for separation of membrane lipids

Membrane lipids include non-polar lipids (cholesterol, cholesterol esters, triglycerides, diglycerides, monoglycerides and free fatty acids) and polar lipids (phospholipids and glycolipids). With the exception of cholesterol, the non-polar lipids are minor components of membranes as are the glycolipids, Non-polar lipids associated with membranes are often not part of the structure but have become trapped within the cisternal space of the membrane vesicles during preparation; for example triglycerides destined for secretion by the liver are sequestered within the cisternae of both endoplasmic reticulum and Golgi preparations (see Chapter 1). No single system will separate all of the lipid classes. It is usually necessary to select a solvent system which

111

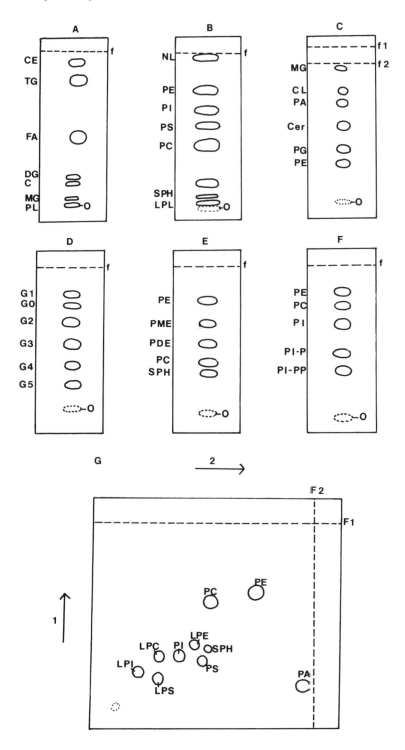

fulfills the requirements of the investigation or, if a complete analysis is necessary, to use several solvent systems.

The characteristics of the solvent systems which are most important in the separation of lipids are the overall polarity of the mixture and its acid/base nature. A number of solvent systems for specific uses are detailed below and in *Figure 5* and *Table 1*. These should provide a range of conditions covering most investigations of membrane lipids and illustrate the versatility of the technique. The list is selective, however, and many alternative systems may be found in the literature. The reader is referred, for further details of a variety of methods, to Renkonen and Kunkkenen (7) as well as to those detailed in (5,6).

3.2.1 *Separation of non-polar lipids*

Non-polar lipids are separated on thin layers of silica gel using a solvent system consisting of heptane (or light petroleum 40−60) diethyl ether/glacial acetic acid (60/40/2, by vol). Phospholipids remain at the origin and other lipids separate as indicated in *Figure 5A*.

3.2.2 *Separation of major classes of phospholipids*

Phospholipids are separated on thin layers of silica gel using the solvent system chloroform/methanol/glacial acetic acid/water (60/50/1/4, by vol). This system resolves phosphatidylethanolamine, phosphatidylserine, phosphatidylinositol, phosphatidylcholine and sphingomyelin (*Figure 5B*). Lysophospholipids remain close to the origin and are not completely resolved from each other. Cardiolipin (a mitochondrial phospholipid) and other acidic phospholipids move ahead of phosphatidylethanolamine. Non-polar lipids move to the solvent front. If the non-polar lipids interfere with the investigation, they may be completely removed to the top of the plate, after allowing the plate to dry, by redevelopment in the same direction in ether to 2−3 cm past the first solvent front.

3.2.3 *Separation of acidic phospholipids*

Skipski *et al.* (8) have described a number of systems for separation of acidic phospholipids using two development steps in the same dimension. After application of the lipids the plates are developed in acetone/light petroleum (1/3, v/v). The solvent is allowed

Figure 5. Systems for separation of lipids by t.l.c. on silica gel H. **A.** Heptane/diethylether/glacial acetic acid, 60/40/2. **B.** Chloroform/methanol/glacial acetic acid/water, 60/50/1/4. **C.** Acetone/light petroleum, 1/3 to f1, followed by chloroform/methanol/glacial acetic acid/water, 80/13/8/0.3 to f2. **D.** Propanol/water, 7/3. **E.** n-propylalcohol/proprionic acid/chloroform/water, 3/2/2/1. **F.** Methanol/chloroform/ammonia/water, 48/40/5/10. **G.** Separation of phospholipids and lysophospholipids by two-dimensional t.l.c. Solvent system 1, chloroform/methanol/ammonia, 65/35/5 to F1. Solvent system 2, chloroform/methanol/glacial acetic acid/water, 10/2/4/2/1. TG, triglyceride; DG, diglyceride; MG, monoglyceride; C, cholesterol; CE, cholesterol ester; FA, fatty acid; PL, phospholipid; NL, neutral lipid; PE, phosphatidylethanolamine; PI, phosphatidylinositol; PS, phosphatidylserine; PC, phosphatidylcholine; SPH, sphingomyelin; LPL, lysophosphatidylcholine; PG, phosphatidylglycerol; Cer, ceramide monohexoses; PA, phosphatidic acid; CLP, cardiolipin; PME, phosphatidylmonomethylethanolamine; PDE, phosphatidyldimethylethanolamine; PI-P, phosphatidylinositol monophosphate; PI-PP, phosphatidylinositol diphosphate; G0−G5, gangliosides; LPC, lysophosphatidylcholine; LPE, lysophosphatidylethanolamine; LPS, lysophosphatidylserine; LPI, lysophosphatidylinositol.

Table 1. A selection of solvent systems for the separation of membrane lipids.

Neutral lipid separation

  Heptane/diethylether/glacial acetic acid, 60/40/1
  Heptane/diethylether/glacial acetic acid, 80/20/1

Phospholipid separation

  Chloroform/methanol/glacial acetic acid/water, 25/15/4/2
  Chloroform/methanol/glacial acetic acid/water, 60/50/1/4
  n-propanol/propionic acid/chloroform/water, 3/2/2/1

Acidic phospholipid separation

  Solvent system I: acetone/light petroleum, 1/3
  Solvent system II: chloroform/methanol/glacial acetic acid/water, 80/13/8/0.3
  Both solvent systems are used in the same dimension.

Polyphosphoinositide separation

  n-propanol/4 M ammonia, 2/1
  Chloroform/methanol/4 M ammonia, 9/7/2
  Chloroform/methanol/28% ammonia/water, 40/48/5/10

Ganglioside separation

  Chloroform/methanol/2.5 M ammonia, 60/40/9
  Propanol/water, 7/3

Two dimensions systems for phospholipid separation

Dimension 1	*Dimension 2*

Chloroform/methanol/ammonia
  65/35/5
	Chloroform/methanol/acetone/acetic acid/water
	  10/2/4/2/1

Chloroform/methanol/7 M ammonia
  90/54/11
	Chloroform/methanol/acetic acid/water
	  90/40/12/2

Chloroform/methanol/acetic acid/water
  50/20/7/3
	Chloroform/methanol/40% aqueous methylamine/water
	  13/7/1/1

to dry under a stream of nitrogen and developed in the same direction with chloroform/methanol/glacial acetic acid/water (80/13/8/0.3, by vol). The first solvent system removes non-polar lipids to the top of the plate so that they do not overlap the acidic phospholipids. Using this system, cardiolipin, phosphatidic acid, ceramidemonohexoses, phosphatidylglycerol and phosphatidylethanolamine are resolved (*Figure 5C*).

3.2.4 *Separation of phospholipids and lysophospholipids*

In order to separate lysophospholipids and phospholipids, for example after treatment of membranes with phospholipase A, it is necessary to use a two-dimensional system. A suitable system uses chloroform/methanol/ammonia (65/35/5, by vol) in the first dimension and chloroform/methanol/acetone/acetic acid/water (10/2/4/2/1, by vol) in the second dimension (*Figure 5G*). Standard lipids of interest should be used at least in the first separations to check the locations of these. Specific stains (see below) can also be used to identify lipid classes.

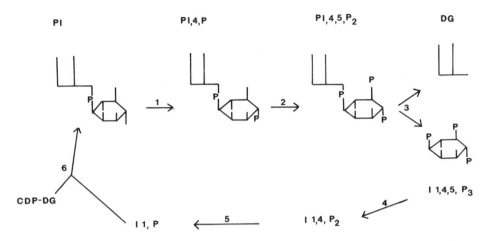

Figure 6. Interactions between the polyphosphoinositides. For details see the text. The proposed site of activation by hormones is via reaction 3.

3.2.5 *Separation of gangliosides*

Gangliosides are minor membrane components. However, they are considered to be of importance in a number of membrane functions. Gangliosides are complex cerebrosides in which the ceramide residue is linked to carbohydrate containing glucose−galactosamine−sialic acid residues. There are many types of ganglioside differing in their carbohydrate structure and a detailed consideration of these is outside the scope of this chapter. However, systems suitable for the separation of gangliosides include propanol/water (7/3, v/v) (*Figure 5D*) or chloroform/methanol/2.5 M ammonia (60/40/9, by vol). Solvent systems containing acid should be avoided because these cause degradation of the gangliosides.

3.2.6 *Separation of intermediates in the methylation of phosphatidylethanolamine to yield phosphatidylcholine*

The methylation of phosphatidylethanolamine has been implicated in receptor-mediated response of a variety of cells to hormones. Phosphatidylethanolamine, phosphatidyl-monomethylethanolamine, phosphatidyldimethylethanolamine and phosphatidylcholine are resolved on silica gel using the solvent system, n-propyl alcohol, proprionic acid/chloroform/water (3/2/2/1, by vol) (*Figure 5E*). This system is also suitable for the separation of the major classes of phospholipid, yielding sharp bands. It has the disadvantage however, that development of a 20 cm plate takes approximately 4 h compared with 1.5−2 h for the system given above.

3.3 **Separation of the polyphosphoinositides and their derivatives**

Polyphosphoinositides have been found to play a major role in signal transmission for hormones, neurotransmitters and growth factors. The initial response to the hormone or ligand is apparently increased hydrolysis of phosphatidylinositol-4,5-phosphate to yield diglyceride and inositol-1,4,5-trisphosphate (*Figure 6*). Both of these function as second messengers, the diglyceride activating protein kinase C, and the inositol tris-phosphate mobilizing calcium from membrane-bound stores such as the cisternae of the endoplasmic reticulum (1). Inositol trisphosphate is hydrolysed to yield inositol bis-phosphate, inositol monophosphate, and finally inositol. Inositol is then incorporated into phosphatidylinositol which is phosphorylated to form first phosphatidyl-inositol-4-phosphate and then phosphatidyl-inositol-4,5-bisphosphate (see Appendix VI).

Separation of the three phosphatidylinositides can be achieved by t.l.c. of the phospholipids, or by alkaline hydrolysis of the phospholipids followed by paper chromatography or by ion-exchange chromatography including h.p.l.c., to separate the inositol phosphates. The method chosen will depend on the investigation in hand. In many experiments the phospholipids are labelled with [^{32}P]- or with [^{14}C]inositol. The lipids and their derivatives can then be traced using the radiolabel as a marker and by comparison with authentic standards.

3.3.1 *Thin layer chromatography of polyphosphoinositides*

Calcium ions retard movement of phosphatidyl-4,5-inositol (9). It is therefore essential to use silica gel H plates for separation of the polyphosphoinositides. If plates are prepared in the laboratory silica gel H containing 1% potassium oxalate as a chelating agent should be used (9). Using silica gel H plates, methanol/chloroform/ammonia/water (48/40/5/10, by vol) or N propanol 4N ammonium hydroxide (2/1, v/v) will resolve phosphatidylinositol-4-phosphate and phosphatidylinositol-4,5-phosphate from other phospholipids (*Figure 5F*). Phosphatidylinositol overlaps sphingomyelin, phosphatidyl-serine, phosphatidylethanolamine and phosphatidic acid. If it is necessary for phos-phatidylinositol to be separated, the spot containing this lipid can be removed, the lipids eluted (see below), re-applied to a thin layer plate and separated using one of the systems in the earlier section for resolution of phospholipids. Alternatively a two dimensional system has been reported which separates polyphosphatidylinositols and phospholipids. Chloroform/methanol/28% ammonia (65/38/8, by vol) is used in the first dimension and butanol/water/glacial acetic acid (6/1/1, by vol) in the second dimension (10).

3.3.2 *Separation of inositol phosphates after alkaline hydrolysis*

The following procedure for the analysis of polyphosphoinositides and for the preparation of inositol phosphates is based on the work of Ballou and co-workers (11,12) modified by Ellis (13), Downes and Michell (14) and described by Irvine *et al.* (15) for the preparation of inositol phosphates from brain phosphatidylinositols.

One procedure is given but this may be scaled down or modified.

(i) *Alkaline hydrolysis*

(1)   Treat lipid equivalent to 10 mg of phosphorus with 12 ml of 1 M KOH and boil under reflux for 30 min.

(2)   Cool the mixture, adjust the pH to 3.0 with formic acid and extract the fatty acids released by alkaline hydrolysis extracted using several heptane washes.

(ii) *Chromatography of inositol phosphates on Dowex columns*

(1)   Adjust the pH of the solution to 8.0 with ammonia and apply the solution to a 1.5 × 3 cm Dowex 1 × 8−400 column (formate form).

(2) Elute the column successively with 15 ml each of

    (a)   0.1 M formic acid/0.2 M ammonium formate to elute INOSITOL MONOPHOSPHATES (predominantly 1 form).

    (b)   0.1 M formic acid/0.4 M ammonium formate to elute INOSITOL BIS-PHOSPHATES (1,5 and 4,5 forms).

    (c)   0.1 M formic acid/1.0 M ammonium formate to elute INOSITOL TRIS-PHOSPHATES (1,4,5 and 2,4,5 forms).

There are many inositol phosphates. However, the forms isolated from brain consist predominantly of the 4,5-bisphosphate and the 1,4,5-trisphosphate.

(3) Desalt the eluants by addition of Dowex 50 H^+ followed by filtration to remove the resin. The extracts may be freeze-dried and stored at $-20°C$.

(iii) *Paper chromatography of inositol phosphates*. Inositol phosphates may be separated on Whatman number 1 paper using isopropanol/saturated ammonia/water (7/5/2, by vol) as a solvent system in descending chromatography. After $5-10$ days the approximate R_f values for the three inositol phosphates are, inositol monophosphate, 0.6; inositol bisphosphate, 0.35; and inositol trisphosphate, 0.16 (11). These are detected on the paper using the following procedure.

(1) Mix 5 ml of 60% perchloric acid with 10 ml of 1 M HCl and 25 ml of 4%, w/v ammonium molybdate.

(2) Make this up to 100 ml with water.

(3) Spray the paper chromatogram with this reagent.

(4) Allow the paper to dry at room temperature and place it under a strong u.v. source.

(5) The phosphate-containing lipids appear as blue spots, or bands.

3.3.3 *Analysis of inositol phosphates produced experimentally (see also Appendix VI)*

Inositol trisphosphate is an important second messenger in many cell types. Therefore it is frequently necessary to assay the phosphorylated bases rather than the phospholipids, and this is most easily achieved by the use of Dowex ion-exchange resins essentially as described above.

(i) Stop the reaction under investigation with 10% trichloroacetic acid and remove the denatured protein by centrifugation.

(ii) Remove 1.0 ml aliquots of the supernatant; neutralize these with dilute NaOH; make each aliquot up to 5 ml with water, and apply it to columns of 1 ml of Dowex 1 × 8 200−400 ion-exchange resin (formate form). These are conveniently prepared in disposable Pasteur pipettes with a glass wool plug.

(iii) Elute the three inositol phosphates as described above using 15 ml of each eluant.

(iv) Desalt the fractions isolated and freeze-dry them for further investigation.

(v) The fractions may be dissolved in distilled water, their identity checked using paper chromatography as above, and/or assayed by counting of radioactivity or phosphate determination depending on the experiment.

3.4 **Detection of lipid on t.l.c. plates**

A number of methods for the detection of lipids on t.l.c. plates are given below. These include general lipid stains, phospholipid stains and specific stains. Many of the reagents used are sprayed finely and evenly onto the adsorbent layer. This is conveniently achieved using a spray bottle powered by an aerosol can. These may be purchased from suppliers of t.l.c. plates. If desired, part of the plate can be covered to protect the lipids and only the standards at the edge of the plate stained.

3.4.1 *Iodine vapour*

The most satisfactory rapid, non-destructive, general method for detection of lipids on thin layer plates is iodine vapour. After removal from the t.l.c. developing tank, the plate is allowed to dry in a fume cupboard and then placed in a lidded tank of suitable size containing a few crystals of iodine. This should be prepared in advance and kept in a warm place to allow the crystals to sublime. The lipids appear as brown−yellow spots or bands. This may take any time from a few minutes to hours depending on the amount of lipid on the plate. The lipid spots are outlined lightly using a soft pencil and the iodine allowed to fade. This may be accelerated by warming the plate. As iodine quenches it is essential that the stain is removed before radioactive lipids are counted using a scintillation counter. If the lipids are to be assayed by a colorimetric method a trace of residual iodine is usually not a problem.

3.4.2 *Phospholipid stain*

The procedure of Vaskovsky and Kostevsky (16) provides an extremely sensitive method for staining phosphate-containing lipids.

To prepare the stain make up the following solutions.

> (a) Dissolve 16 g of ammonium molybdate in 120 ml of water.
> (b) To 40 ml of concentrated HCl add 10 ml of mercury and 80 ml of solution a. Shake the mixture for 30 min and filter.
> (c) To 200 ml of concentrated H_2SO_4 add all of solution b and the remainder of solution a. Make up to a litre and store in a dark glass bottle.

(i) After removing the plates from the t.l.c. developing tank, allow the solvent to evaporate completely and spray the plates evenly with reagent c above. This is best achieved by placing the plate in a cardboard box from which one side has been removed. Place the box on its side with the plate at the back and the open side upwards in a fume cupboard. Spray the plate keeping the corrosive spray within the box.

(ii) After a few minutes, the phospholipid-containing spots appear as dark blue bands or spots. The colour does not interfere with subsequent colorimetric analysis of the phospholipids.

3.4.3 *Stain for amino groups*

Phosphatidylethanolamine, phosphatidylserine and other lipids with free amino groups can be detected specifically using ninhydrin.

(i) Allow the t.l.c. plates to dry after removal from the t.l.c. developing tank and spray them with 0.25% ninhydrin dissolved in acetone.

(ii)   The amino lipids appear as pink−purple spots within 1−2 h at room temperature, or within 15 min if the plates are heated to about 100°C. It is advisable to wear gloves when using the ninhydrin spray.

3.4.4 *Stain for choline-containing lipids*

The Dragendorff stain (7) specifically stains choline-containing lipids, i.e. phosphatidyl-choline and sphingomyelin. To prepare this stain, make up the following solutions.

    (a)   1.7 g of basic bismuth nitrate in 100 ml of 20% acetic acid.

    (b)   10 g of potassium iodide in 25 ml of water.

    (c)   Mix 20 ml of solution a with 5 ml of solution b and add a further 70 ml of water and use this as the spray reagent.

(i)   After removal of the plate from the t.l.c. developing tank, allow the solvent to evaporate, and spray the plate.

(ii)   After a few minutes, choline-containing lipids appear as orange coloured spots or bands.

3.4.5 *Stain for glycolipids*

α-Naphthol specifically stains glycolipids, including cerebrosides, sulphatides, glycosyl-diglycerides and gangliosides (5). To prepare this stain, make up the following solutions.

    (a)   α-Naphthol 0.5 g in 100 ml of methanol/water (1/1).

    (b)   95 ml of concentrated H_2SO_4 plus 5 ml of distilled water.

(i)   Allow the t.l.c. plate to dry after development and spray it with solution a.

(ii)   Air dry the plate and then re-spray with solution b taking the same precautions as using the Vaskovsky−Kostevsky reagent (see above).

(iii)   Heat the plate to 120°C until the colour develops. Glycolipids stain purple−blue and other polar lipids yellow.

3.4.6 *Stain for gangliosides*

Gangliosides can be specifically detected with the resorcinol stain (5).

    (a)   Dissolve 2 g of resorcinol in 100 ml of distilled water. This is stable. At least 4 h before use add 10 ml of this solution to 80 ml of concentrated HCl containing 0.5 ml of copper sulphate 0.1 M (2.5 g of $CuSO_4.5H_2O$ in 100 ml of water) and dilute the mixture with distilled water to 100 ml.

(i)   After development of the t.l.c. plate allow this to dry and spray the plate with reagent a.

(ii)   Cover the plate with a second glass plate of the same dimensions and heat the sandwiched plates to 120°C. Gangliosides appear as blue−violet spots.

4. DETERMINATION OF LIPIDS SEPARATED BY T.L.C.

4.1 **Elution of lipids**

For further analysis, lipids can be eluted from the t.l.c. plate using the following procedure.

(i)   Draw around the lipid spot using a needle or the sharp edge of a microscope slide or spatula. Using the same instrument, divide the area of the spot into sections using horizontal and vertical lines.

(ii)   Place the thin layer plate on a sheet of smooth surfaced paper and, using the flat edge of a microscope slide or a spatula, scrape off each area of silica gel completely.

(iii)   Tap the powder onto the sheet of paper and using this as a funnel transfer the gel into a conical 15 ml centrifuge tube.

(iv)   To elute the lipids, a solvent should be used which is more polar than that required to move the lipid on the silica gel plate. For most lipids a mixture of chloroform/methanol (1/1, v/v) is appropriate.

(v)   Resuspend the silica gel in approximately 15 ml of solvent, mix thoroughly using a vortex mixer and pellet the silica gel using a bench centrifuge.

(vi)   Carefully decant off the solvent or remove it with a Pasteur pipette into a suitable container.

(vii)   Add a further 15 ml of solvent and repeat the extraction several times.

(viii) Pool the extracts and remove the solvent as described above.

4.2 Determination of phospholipids

Phospholipid phosphorus can be determined on aliquots of the eluted lipids or directly on the scraped silica gel spot without elution using the following procedure, which is based on the method of Fiske and Subbarrow (17) modified by Bartlett (18).

Reagents

(a)   Perchloric acid, 7%.

(b)   Ammonium molybdate, 2.5%.

(c)   Standard phosphate solution, 3.58 g of $Na_2HPO_4.12H_2O$ dissolved in 1 litre. Add a few drops of chloroform to prevent bacterial growth. Just before use dilute this stock solution 1 to 10 to give a concentration of 1 μmol/ml.

(d)   Aminonaphtholsulphonic acid reagent. This can be purchased in a ready prepared mixture or can be made up from the reagents: 30 g of sodium bisulphate; 5 g of sodium sulphite; 0.5 g of 1,2,4-aminonaphtholsulphonic acid. Mix these reagents by grinding them together in a pestle and mortar. Dissolve the reagents in 250 ml of distilled water, leave to stand for 3 h in the dark and filter into a dark glass bottle. This reagent is stable for about 8 weeks in the refrigerator.

Procedure

(i)   Scrape the phospholipid-containing spots as above into Pyrex test tubes. As the phosphate assay is extremely sensitive it is essential to use clean test tubes. These should be washed only with phosphate-free detergents or with distilled water containing a few millilitres per litre of concentrated HCl followed by many rinses with glass distilled water. The same procedures should be used for all glassware. Only glass distilled water should be used for preparing solutions.

(ii)   Pipette 1.5 ml of perchloric acid into each test tube in the fume cupboard using a glass pipette with an automatic dispenser.

(iii)   Top the tube with a marble of appropriate diameter and heat the mixture to 230°C for twice the time required for the solution to change from charred brown to colourless or pale yellow (30 min is usually adequate). The digestion can be performed in a heated block or a micro Kheldahl rack in a fume cupboard and extreme care should be taken as hot perchloric acid may be explosive with organic material.

(iv)   Allow the tubes to cool. Add 2.5 ml of distilled water. Mix well and pellet the silica gel by centrifugation.

(v)   Remove the supernatant carefully using a Pasteur pipette and transfer it to a second test tube.

(vi)   Add 1.5 ml of ammonium molybdate and mix well. Add 0.2 ml of amino-naphtholsulphonic acid reagent and mix well.

(vii)   Cap the tubes with marbles and heat them for 9 min in a boiling water bath, cool the tubes to room temperature and read the blue colour at 830 μM against a reagent blank prepared in exactly the same way but without phosphate.

(viii)   Prepare a standard curve using $0-2.0$ μmol of phosphate, and omit the digestion and centrifugation steps. The μmoles of phospholipid in the digested lipid can be determined directly by reading the absorbance from the standard curve. This method is linear to 2.0 μmol of phosphate and sensitive to less than 0.05 μmol.

To determine phospholipid phosphorus in total lipid extract or fractions eluted from t.l.c. plates, pipette an aliquot of the lipid solution into a test tube. Evaporate the solvent to dryness under a stream of nitrogen. Add perchloric acid and digest and assay the phospholipid as above omitting the centrifugation step.

It is advisable to remove a sample of silica gel from the t.l.c. plates used and assay its phosphate content before determining phospholipids by the method above. Note that Merck silica gel 60 F254 plates have no detectable phosphate, whereas Merck silica gel/Kieselguhr F254 plates have a high phosphate content and cannot be used for this procedure.

4.3 Determination of esters

Ester bonds in triglycerides, diglycerides, monoglycerides, cholesterol esters or phospholipids are determined by the method of Snyder and Stephens (19).

Reagents

(a)   Dissolve 5 g of $Fe(ClO_4)_3.6H_2O$ in 10 ml of 70% perchloric acid and 10 ml of distilled water. Dilute the solution with cold absolute alcohol to 100 ml. Store the solution at 4°C. This solution is stable for several months.

(b)   Immediately before use mix 4 ml of reagent a with 3 ml of 70% perchloric acid and dilute to 100 ml with cold absolute alcohol.

(c)   Dissolve 2 g of hydroxylamine hydrochloride in a small volume of water and dilute to 50 ml with absolute alcohol.

(d)   Dissolve 4 g of NaOH in a small volume of distilled water and dilute to 50 ml with absolute alcohol.

(e)   Mix equal volumes of reagent c and reagent d. Filter the mixture and use the supernatant as reagent e. Reagents c, d and e should be prepared just before use.

(f)   Standard ester solution in chloroform. Triolein 1 μmol/ml or methylstearate

1 μmol/ml can be used. Take into account that the triglyceride has three ester groups when using this as a standard.

Procedure

(i)   Dissolve eluted lipids in chloroform and pipette aliquots into tubes. Evaporate the solvent to dryness under a stream of nitrogen.

(ii)   Add 1.0 ml of reagent e. Top the tube with a marble and heat the mixture to 65°C for 2 min in a block heater or a water bath.

(iii)   Allow the tubes to cool and add 3.0 ml of reagent b. Mix and allow the mauve colour to develop for 30 min.

(iv)   Read the absorbance against a reagent blank at 530 μm.

(v)   Prepare a standard curve in exactly the same way using the ester standard solution f, 0−10 μmol. μmoles of triglyceride, diglyceride, or monoglyceride can be read directly from the standard curve. Divide the μmoles of ester by 3, 2 or 1, respectively, to determine the amount of each lipid.

This assay is sensitive to above 0.05 μmol of ester bond, and because it measures ester bond and not a specific compound it can only be used after separation of lipids.

4.4 Determination of cholesterol and cholesterol esters

Cholesterol and cholesterol esters can be determined on total lipid extracts or after separation and elution of lipids. On unseparated lipid extracts the total cholesterol (i.e. free and esterified) is determined followed by precipitation of the free cholesterol with digitonin. The esterified cholesterol is thus total cholesterol minus free cholesterol.

Reagents

(a)   Dissolve 2.5 g of $FeCl_3.6H_2O$ in 85% orthophosphoric acid to give a final volume of 100 ml. Store this in a brown bottle at room temperature. Discard when a precipitate forms.

(b)   Dilute 4 ml of reagent a to 50 ml with concentrated H_2SO_4. Cool the mixture and store this at room temperature. Discard if the reagent becomes cloudy.

(c)   Dissolve 1 g of digitonin in 50 ml of 95% ethanol. Dilute the solution to 100 ml with water. Store in a dark bottle for up to 6 months.

(d)   Standard cholesterol solution is made by dissolving 100 mg of cholesterol in 100 ml of glacial acetic acid. Dilute this 1 in 10 for assay (0.01% or 0.258 mM).

Procedure

(i)   Pipette aliquots of the lipid extract in chloroform into stoppered test tubes. Remove the solvent under nitrogen.

(ii)   Add 6.0 ml of glacial acetic acid, mix and add 4.0 ml of reagent b. Mix and read the purple colour after 10 min at 550 μm against a reagent blank.

(iii)   Prepare a standard curve in the same way replacing part of the 6.0 ml of glacial acetic acid with the cholesterol standard (0−1.5 μmol of cholesterol).

Precipitation of free cholesterol

(i)   Pipette an aliquot of the lipid extract into 15 ml conical centrifuge tubes. Evaporate the solvent under a stream of nitrogen and add 1.0 ml of acetone/95% ethanol (1/1 v/v).

122

(ii)   Add 1.0 ml of digitonin reagent c. Allow the tubes to stand for 10 min.
(iii) Centrifuge the precipitate down at 300 r.p.m. for 5 min, discard the superna-
       tant and drain the tubes for 5 min.
(iv)   Resuspend the pellets in 4.0 ml of acetone using a vortex mixer and recentrifuge.
       Dry the pellet under a gentle stream of nitrogen.
(v)   Dissolve the pellet in 6.0 ml of glacial acetic acid and proceed as for the deter-
       mination of total cholesterol.

The assay is linear to 1.5 μmol of cholesterol and sensitive to 0.05 μmol.

Because of its clinical importance a number of kits are available for the determination of cholesterol by an enzymic assay based on cholesterol oxidase. Although these
are more sensitive than colourimetric assays, they are intended for use on serum samples
and are less successful when applied to lipids isolated from membranes. Boehringer,
Sigma and BDH are examples of suppliers of such kits.

4.5 Determination of free fatty acids

Free fatty acids can be estimated by a colorimetric assay based on production of the
copper salt of the fatty acid. This is then estimated by reaction with diethyldithiocarbamate (20).

Reagents

(a)   Dissolve 6.45 g of $Cu(NO_3)_2.3H_2O$ in 100 ml of water. Mix 10 volumes of this
      with 1 volume of 1 M acetic acid and 9 volumes of 1 M triethanolamine in water.
      Prepare the mixture immediately before use from its three constituents, which
      may be kept as stock solutions.
(b)   Dissolve 0.1 g of sodium diethyldithiocarbamate in 100 ml of butanol. Store the
      solution at 4°C and use within a week.

Procedure

(i)   Pipette the fatty acid in chloroform into a 15 ml stoppered test tube.
(ii)   Add 2.5 ml of copper reagent a and mix by shaking the stoppered tube for 2 min.
(iii) Separate the two phases by centrifugation. Remove the aqueous upper layer with
       a syringe or a Pasteur piette and remove 3.0 ml of the lower phase taking care
       not to contaminate this with upper phase.
(iv)   Add 0.5 ml of reagent b, mix and read the yellow colour at 440 μm against a
       reagent blank.

This method is linear to 100 μg of fatty acid and sensitive to 10 μg.

5. INVESTIGATIONS OF THE TRANSVERSE DISTRIBUTION OF MEMBRANE LIPIDS

There have been a number of critical reviews of investigations of the transverse distribution of membrane lipids (21−23). These provide information of the range of membrane preparations which have been investigated and the variety of methods used but
do not consider practical details of experimental procedures. In this section detailed
methods will be described for determining the distribution of membrane phospholipids
using rat liver microsomal membranes as a model system. These methods have a general

(vii) Cap the tubes with marbles and heat these in a boiling water bath for 9 min. Cool the tubes and read the absorbance of the solution at 830 μm. Determine μmoles of phosphate released from a standard curve prepared as described earlier.

(viii) The activity of mannose-6-phosphatase is linear with time dependent on the amount of protein present. Calculate the μmoles of phosphate released per min during the linear part of the reaction.

The latency of mannose-6-phosphatase is:

$$\frac{\text{activity with taurocholate} - \text{activity without taurocholate}}{\text{activity with taurocholate}} \times 100\%$$

Microsomal vesicles have a latency above 90%. This should be retained during treatments with probes of the phospholipids.

5.2.3 *Latency of ethanol acyltransferase*

Ethanol acyltransferase catalyses the transfer of palmitate from palmityl CoA to ethanol. This enzyme is latent in microsomal vesicles and has a more ubiquitous tissue distribution than glucose-6-phosphatase, which is found only at high levels in liver endoplasmic reticulum (26). Ethanol acyltransferase thus provides an additional method of determining the integrity of microsomal membranes and can be used for microsomes lacking mannose-6-phosphatase.

(i) *Assay of ethanol acyltransferase*

Reagents

(a) 0.175 M Tris-HCl, pH 7.0, containing 1 mg/ml albumin and 8 mM MgCl$_2$.
(b) 400 μM [^{14}C]- or [^{3}H]palmityl CoA, 100 μCi/μmol in above buffer.
(c) Absolute alcohol.
(d) 1.0% deoxycholate in above buffer.

Procedure

(i) Set up a series of test tubes containing 0.14 ml of buffer and 10 μl of absolute ethanol. To half the tubes add 20 μl of deoxycholate and to the other half add 20 μl of buffer.

(ii) After treatment of the microsomes with the probe remove aliquots containing approximately 2 μg of protein in 10 μl and transfer to the test tubes.

(iii) Start the reaction by addition of 20 μl of labelled palmityl CoA. Allow the reaction to proceed at room temperature for 2 min and stop the reaction by addition of 1.5 ml of isopropyl alcohol/heptane/water (80/20/20, by vol). Allow the mixture to stand for 5 min.

(iv) Add 1.0 ml of heptane and 0.5 ml of water and mix. Allow the two phases to separate. Remove the upper phase and wash this with 0.05 M NaOH in 50% ethanol to remove palmitic acid.

(v) Pipette an aliquot of each heptane phase into scintillation vials, remove the solvent under a stream of nitrogen add scintillation fluid and determine the radioactivity in the samples.

The latency of ethanol acyltransferase is

$$\frac{\text{d.p.m. in heptane phase with deoxycholate} - \text{d.p.m. in the heptane phase without deoxycholate}}{\text{d.p.m. in the heptane phase with deoxycholate}} \times 100\%$$

As with mannose-6-phosphatase the latency of ethanol acyltransferase is greater than 90% in impermeable microsomal vesicles.

5.2.4 *Permeability of other membrane vesicles*

The integrity of membrane vesicles prepared from different organelles can be determined using similar approaches to those described above.

Leakage of lactic dehydrogenase, a soluble cytosolic enzyme, into the incubation medium is a satisfactory indicator of the integrity of isolated cells. Similarly haemoglobin is a marker of the permeability of red blood cells. Lactic dehydrogenase, trapped during homogenization in plasma membrane vesicles, has been used to demonstrate the impermeability of these preparations (27). Golgi preparations contain secretory proteins which can be labelled as described for the microsomal membranes and used as an indicator of loss of impermeability (28). Golgi membranes also contain latent galactosyltransferase which can be used to determine vesicle integrity in a similar way to mannose-6-phosphatase in microsomes (28).

5.3 **Retention of microsomal membrane structure**

It is difficult to demonstrate that the structure of the phospholipid bilayer is not perturbed during experimental procedures to investigate the transferase asymmetry of membranes. Using the electron microscope to examine sections and freeze-fractured preparations of membranes before and after treatment with the probe should indicate whether the membrane bilayer structure is retained and whether any major structural alteration is produced. However, this would not detect transmembrane rearrangements of the phospholipids. Ideally, both inside-out and outside-out membrane vesicles should be examined using the same probe. If complementary results are obtained this indicates that the membrane structure is not rearranged by the probe. However, as noted above preparations of inside-out impermeable vesicles from membranes other than red blood cells are not available.

As a compromise, in the absence of an ideal control, studies of the transverse distribution of phospholipids of microsomal and other membranes should use more than one probe, for it is unlikely that different techniques would produce the same structural perturbation. In control experiments it is also necessary to demonstrate that results obtained with the closed vesicle are not a consequence of the specificity of the probe used. Membrane vesicles should be opened experimentally and treated with the probe to demonstrate that under these conditions the phospholipids are freely available.

5.3.1 *Methods for preparing open microsomal vesicles*

A number of methods have been developed for making microsomal membrane vesicles permeable. The efficiency of these methods can be assessed by determining loss of vesicle contents or latency of mannose-6-phosphatase or ethanol acyltransferase.

(i) *Sodium carbonate treatment*. This causes microsomal vesicles to open and the membranes to fuse forming flat sheets (29).

(1) Resuspend microsomes at a concentration of less than 1 mg protein/ml suspended in 100 mM sodium carbonate, pH 11.0. Allow the suspension to stand for at least 30 min on ice.

(2) Pellet the microsomes by centrifugation at 105 000 *g* for 50 min. If it is necessary to completely separate the membrane and supernatants, which contain the contents of the vesicle, increase the centrifugation to 120 min and resuspend the membrane pellet in sodium carbonate and re-centrifuge the suspension to isolate the membranes.

(3) Wash the pellet with the appropriate buffer to remove the sodium carbonate solution before resuspending the microsomes for further investigation.

(ii) *French press treatment*. This causes microsomal vesicles to fragment and to collapse (30).

(1) Resuspend the microsomes in the buffer to be used for subsequent studies at a concentration of 5 mg protein/ml.

(2) Place the suspension in a French pressure cell and increase the pressure to 20 000 p.s.i.

(3) Allow the cell to equilibrate for 5 min and *carefully* open the valve to allow the membrane suspension to escape at the approximate rate of a drop every second.

The suspension, which becomes noticeably clearer after this treatment, may be used immediately without pelleting as the membrane concentration is appropriate for further investigation.

(iii) *Detergent treatment*. Microsomal vesicles may also be made permeable by addition of detergents to the membrane suspension. Taurocholate, 0.4%, deoxycholate, 0.05% and lysolecithin, 0.005%, are all effective. These can be used immediately for further investigation provided the probe used is active in the presence of the detergent.

Thus, microsomal membranes may be made permeable by alkaline pH and low osmotic pressure, mechanical sheering or by chemical methods. The use of these methods has also been established for Golgi preparations, and potentially they may be applied to other subcellular fractions.

6. DETERMINATION OF THE TRANSVERSE DISTRIBUTION OF MEMBRANE PHOSPHOLIPIDS USING HYDROLYTIC ENZYMES

6.1 Principle

Membrane vesicles are incubated with and without phospholipase under conditions in which they remain impermeable. The lipids are extracted, the phospholipids separated, and the extent of hydrolysis determined by comparing membranes treated with phospholipase and those untreated.

6.2 Phospholipases

Phospholipases hydrolyse different bonds in phospholipids (*Figure 7*). Phospholipases A_1 and A_2 hydrolyse the acyl groups at positions 1 and 2 of phospholipids, respective-

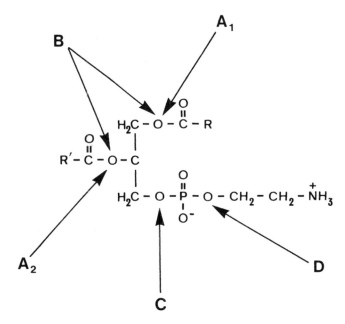

Figure 7. Sites of hydrolysis of glycerophospholipids by phospholipases.

ly producing lysolecithin. Phospholipase C hydrolyses the phosphoryl base producing diglyceride, and phospholipase D hydrolyses the base leaving phosphatidic acid.

Potentially all of these enzymes may be used as probes. To determine their suitability for a particular study it is necessary to demonstrate that the phospholipase used does not open the membrane vesicle under investigation. In the case of the microsomal membranes, phospholipase C, from *Clostridium perfringens*, and phospholipase D have been used successfully as probes of phospholipids, while phospholipase C, from *Bacillus cereus*, and phospholipase A from snake venoms or bee venoms, cause the membrane vesicles to open.

A variety of phospholipases are available from different suppliers (Sigma, Boehringer, Worthington), and the most purified form of the enzyme available should be used. van den Bosch (31) has reviewed the phospholipases and the methods for their purification.

6.3 Procedures

(i)   Resuspend the membrane vesicles at a concentration of 5 mg protein/ml in a buffer which does not cause them to be opened. For microsomal membranes 0.14 M NaCl buffered to pH 7.4 with 0.05 M Tris-HCl or 0.05 M phosphate buffer, or bicarbonate has been used.

(ii)   Calcium ions are necessary for the action of phospholipase C. These should be added to give a concentration of at least 0.25 mM. If the membranes have been isolated or incubated in the presence of a chelating agent it is necessary to increase the calcium ion concentration to 1 mM.

(iii) Add phospholipase to half the samples to give a concentration in the range indicated below.

(a)   Phospholipase C − 1 unit/ml to 10 unit/ml.
(b)   Phospholipase A − 0.5 unit/ml to 5 unit/ml.
(c)   Phospholipase D − 1 unit/ml to 10 unit/ml.

The rate of hydrolysis will vary with the phospholipase concentration. If rapid equilibration is necessary, as in radiolabelling experiments, a high concentration of phospholipase should be used with a short incubation time. The volume of the incubation may be varied depending on the amount of membrane proteins available. Investigations of rat liver microsomes have generally used 5 mg of membrane protein in 1.0 ml, while investigations of Golgi membranes or plasma membranes have used 0.5 mg of membrane protein in a volume of 0.1 ml.

(iv) Incubate the membranes with and without phospholipase for a suitable time. Stop the reaction by addition of chloroform/methanol, 2/1, together with EDTA, 10 mM, which chelates calcium ions and prevents further action of phospholipase C during extraction.

(v) Extract the lipids as described earlier and assay the total phospholipid phosphorus on aliquots of the lipid extract. If phospholipase C is used as a probe, percentage hydrolysis of the phospholipid is calculated using the formula:

$$100 - \left(\frac{\text{phospholipid phosphorus in membranes treated with phospholipase}}{\text{phospholipid phosphorus in membrane incubated without phospholipase}} \times 100 \right)$$

The hydrolysis of individual phospholipids is then determined by separation of the major classes of phospholipids by t.l.c. The phospholipid spots are scraped and assayed as described above and the percentage hydrolysis of each phospholipid determined in the same way as the total hydrolysis.

If phospholipase A is used as a probe, the total phospholipid phosphorus is unchanged, as the products of the enzymes action are lysophospholipids. For full analysis of the extent of hydrolysis of phospholipids by phospholipase A, it is necessary to separate the lipids by two-dimensional t.l.c. as described above. Alternatively a one-dimensional system may be used to separate the major classes of phospholipids and the percentage hydrolysis of these determined by comparison with controls incubated without phospholipase A.

In preliminary experiments, the time course of hydrolysis of membrane phospholipids should be determined. Hydrolysis should plateau at approximately 50%, but this may vary with the phospholipase used, because not all the membrane phospholipids are necessarily substrates. In addition, the phospholipids may not be evenly distributed in the membrane bilayer. Hydrolysis of microsomal phospholipids plateaus at 50% within 5 min when incubated with phospholipase C at a concentration of 10 unit/ml, and in 15 min with phospholipase C at a concentration of 1 unit/ml.

On the basis of results from these experiments the percentage of each membrane phospholipid hydrolysed at equilibrium, i.e. in the outer leaflet of the membrane bilayer, may be calculated.

To demonstrate that the results obtained are not due to the specificity of the phospholipase used, the membrane vesicles should be opened using the methods described above and incubated with phospholipase. In order to assign a particular phospholipid to the membrane outer leaflet it is necessary to show that this is completely hydrolysed in open vesicles. If a phospholipid is not hydrolysed in either open or closed vesicles it cannot be assigned to either side of the bilayer and a different probe must be used to determine its distribution.

7. DETERMINATION OF THE TRANSVERSE DISTRIBUTION OF MEMBRANE PHOSPHOLIPIDS USING CHEMICAL LABELS

7.1 Principle

Membrane vesicles are incubated with a chemical labelling reagent and the percentage of the phospholipase which reacts is determined. Chemical labels for phospholipids are not as well developed as those used for proteins (see Chapter 5) and are restricted to the aminophospholipids, phosphatidylethanolamine and phosphatidylserine. Labelling reagents which have been used include formylmethionine methyl phosphate, fluorodinitrobenzene and trinitrobenzene. For any of these probes to be valid the criteria indicated above must be fulfilled that the reagent does not penetrate the membrane, nor cause the membrane vesicles to become leaky. The procedures for the labelling of microsomes with trinitrobenzene sulphate (TNBS) are described below. This reagent has been used most extensively for determining the distribution of aminophospholipids of plasma membranes of whole cells and of a variety of subcellular membrane preparations.

7.2 Procedures

(i)   Resuspend microsomes (1 − 12 mg of protein) in 5 ml of 280 mM mannitol buffer pH 7.4, containing 70 mM sucrose, 40 mM sodium bicarbonate, 1 mM $MgCl_2$ and 2 mM succinate. The pH of the buffer is important as microsomal vesicles become leaky and TNBS penetrates the vesicles above pH 8.5.

(ii)   Add TNBS to give a final concentration of 3.0 mM. Between 1 and 12 mM TNBS has been used with 8 mg of membrane protein with similar results.

(iii)   Incubate the mixture for the range of times at 7°C, room temperature, or 37°C. The rates of labelling of the membrane phospholipids increase with increased temperature but the final level of reaction is the same. In some experiments, for example with whole cells, it is necessary to use low temperatures to prevent penetration of TNBS.

(iv)   Stop the reaction by addition of 1 ml of 30% trichloroacetic acid. Precipitate the denatured membranes using a bench centrifuge and resuspend the pellet in 1.0 ml of methanol. Add 8 ml of chloroform/methanol/concentrated HCl (2/1/0.01, by vol). Allow the extracts to stand for 30 min and add 2.0 ml of 0.02 M KCl in 0.1 M HCl to form two phases. Discard the upper phase and wash the lower phase with 2.0 ml of 0.2 M KCl in 0.1 M HCl before drying this under vacuum or nitrogen. Separate the phospholipids on t.l.c. plates using chloroform/methanol/glacial acetic acid/water, 60/50/1/4. The derivatized phospholipids are yellow and can be marked on the t.l.c. plate before detection of the unreacted aminophospholipids using ninhydrin spray. The approximate R_fs of the lipids are: trinitrophenol phosphatidylethanolamine, 0.86; phosphatidylethanolamine, 0.65; trinitrophenyl phosphatidylserine, 0.6; and phosphatidylserine, 0.55.

(v) Scrape the phospholipid-containing spots and determine the phospholipid-phosphorus. Calculate the percentage of phospholipid available for reaction using the formula:

$$\frac{\text{phospholipid phosphorus in reacted phospholipid}}{\text{phospholipid phosphorus in unreacted phospholipid plus reacted phospholipid}} \times 100$$

The phosphorus in the reacted and unreacted phospholipid should be equal to the phospholipid in the control microsomes incubated without TNBS. The percentage of phosphatidylethanolamine or phosphatidylserine which reacts with TNBS in unopened microsomes is the percentage of that phospholipid in the outer leaflet of the membrane bilayer.

The microsomal vesicles in control experiments should be opened to demonstrate complete accessibility of the aminophospholipids to TNBS in permeable vesicles. In microsomes approximately 30% of the aminophospholipids react with TNBS. This increases to more than 90% when the microsomal vesicles are opened.

8. DETERMINATION OF THE TRANSVERSE DISTRIBUTION OF MICROSOMAL PHOSPHOLIPIDS USING PHOSPHOLIPID EXCHANGE PROTEINS

8.1 Principle

Phospholipid exchange proteins (PL-TP) catalyse the exchange of phospholipid molecules between membranes, lipoproteins, or liposomes. These proteins have been isolated from a number of tissues including liver, heart and brain. Although the functions of PL-TP have not been completely clarified, there is evidence that *in vivo* they are involved in transfer rather than exchange of phospholipids between membranes in cells and may therefore play important roles in membrane biogenesis and turnover. *In vitro*, PL-TP catalyse a molecule for molecule exchange of phospholipids between the outer leaflets of two membrane species. As these proteins apparently function *in vivo* and do not alter the structure of membranes, they provide a unique tool for investigations of the transverse distribution of phospholipids.

The membranes under investigation are prepared with the phospholipid radiolabelled (donor) and are incubated with an excess of a second membrane or phospholipid-containing species (acceptor) in the presence of a PL-TP. The two membranes are separated and the percentage transfer of radioactivity from donor to acceptor and the specific activities of the phospholipids determined. From this data and the relative amounts of phospholipid present in the two membrane preparations, it is possible to calculate the percentage of each phospholipid available for exchange and hence in the outer leaflet.

8.2 Phospholipid exchange proteins

PL-TP specific for phosphatidylcholine, for phosphatidylinositol, and a non-specific PL-TP which will transfer phosphatidylethanolamine have all been purified from bovine liver cytosol (32). The choice of PL-TP used will depend on the experimental system.

Table 2. Procedures for labelling liver microsomal phospholipids.

Phospholipid labelled	Isotope and dose (μCi/100 g)[a]		Time after injection before preparation of microsomes
All phospholipid	Sodium ^{32}P	500 – 1000	16 h
Glycero phospholipids	[^{14}C]- or [^{3}H]Glycerol	5 – 20	1 – 16 h
Phosphatidylcholine	[^{14}C]- or [^{3}H]Choline[b]	5 – 20	16 h
	[^{14}C]- or [^{3}H](Methyl)methionine	5 – 20	1 – 16 h
Phosphatidylethanolamine	[^{14}C]- or [^{3}H]Ethanolamine[b]	5 – 20	16 h
Phosphatidylinositol	[^{14}C]- or [^{3}H]Myo-inositol[c]	5 – 20	1 – 16 h

[a]Injected intraperitoneally or intravenously into the animal.
[b]As the phospholipid bases are incorporated initially into the outer leaflet of the membrane bilayer it is necessary to allow some time to elapse to allow uniform labelling.
[c]Myo-inositol is subject to bacterial degradation. This should be made up in sterilized water and samples removed for injection under sterile conditions.

8.3 Acceptor membranes

Acceptors which have been used in investigations of membrane phospholipids include liposomes, prepared from pure phospholipids, plasma lipoproteins and membranes including red blood cells and mitochondria. The only restrictions on the choice of acceptor are that it can be prepared at a concentration sufficient to give an excess of phospholipid over the donor membrane of up to 10-fold and that the donor and acceptor membranes can be easily separated and recovered quantitatively.

8.4 Radiolabelling the donor membrane

Membrane phospholipids are labelled by intraperitoneal or intravenous injection of a radiolabelled precursor prior to sacrifice of the animal and preparation of the membranes. Some procedures for labelling liver microsomes are given in *Table 2*. To label other membranes similar methods can be used and, in preliminary experiments, the extent of incorporation of the precursor determined by extraction and analysis of the phospholipids. One procedure for determining the availability of microsomal membrane phosphatidylcholine for exchange with liposomal phospholipid is described below. Potentially this can be used for other membranes and phospholipids provided the appropriate controls are performed.

8.5 Liposome preparation

(i) Pipette phosphatidylcholine containing 2% phosphatidic acid in organic solvent into a thick walled tube or vial. Allow the solvent to evaporate while rotating the tube to spread the lipid in a film over the inner surface.
(ii) Add 0.25 M sucrose, 0.01 M Tris, 0.001 M EDTA, pH 7.4, and sonicate the mixture using an M.S.E. sonicator at setting 5 for 10 min. The tube should be cooled by immersion in ice during the sonication. The quantity of phospholipid used and the volume of buffer will vary depending on the experiment (see below).
(iii) Centrifuge the sonicated mixture for 90 min at 105 000 *g* to pellet any undispersed lipid and remove the upper three quarters of the supernatant preparation. Ex-

tract an aliquot of this with chloroform/methanol, 2/1, and determine the phospholipid content as described above.

(iv)   Incubate radiolabelled microsomes equivalent to $0.5-1.0$ mg phosphatidylcholine ($\sim 3-5$ mg microsomal protein) with liposomes containing $5-10$ mg of phosphatidylcholine in a final volume of 1.0 ml, 0.25 M sucrose, 0.01 M Tris buffer, 0.001 M EDTA, pH 7.4, with and without PC-TP for a range of times at 25°C. At the end of the incubation period carefully layer the microsomal suspension onto 9.0 ml of 0.75 M sucrose in centrifuge tubes and centrifuge at 105 000 g for 90 min. The microsomes form a pellet and the liposomes remain in the load layer.

(v)   Extract the lipids of both the liposomes and microsomes, separate these by t.l.c., and elute the phosphatidylcholine. Take aliquots of this to determine the phospholipid phosphorus and radioactivity. From this data determine the percentage of the labelled phospholipid transferred from microsomes to liposomes and the specific activity of the phosphatidylcholine in d.p.m./μmol in each fraction.

Exchange between the phosphatidylcholine of microsomes and liposomes is indicated by a transfer of radioactivity from one fraction to another and a fall in the specific activity of the microsomal lipid and a rise in that of the liposomes. Under the conditions described above exchange is usually complete within an hour. This will vary with the activity of the PC-TP preparation. However, microsomes should not be incubated for longer than two h, as this results in the vesicles becoming leaky. The PC-TP concentration in the incubation should be increased to avoid long incubation times.

At equilibrium, i.e. when the exchange process is complete, provided all of the phosphatidylcholine in the microsomes is available for exchange with 2/3 of the liposomal phosphatidylcholine (that fraction of the phosphatidylcholine in the outer leaflet of liposomes) then:

$$\text{d.p.m. transferred} = \frac{\text{PC-lipo}}{\text{PC-lipo} + \text{PC-micro}} \times \text{c.p.m.-total}$$

where: PC-lipo is the exchangeable pool of liposomal PC (2/3 total); PC-micro is the total microsomal PC and d.p.m.-total is the total radioactivity in microsomal sample initially.

If the transfer of radioactivity is less than the theoretially complete exchange, this indicates that not all of the microsomal PC is available and hence that a proportion is in the inner leaflet of the membrane. This proportion can be calculated:

$$\text{d.p.m.transferred} = \frac{\text{PC-lipo}}{\text{PC-O micro} + \text{PC-lipo}} \times \frac{\text{PC-O micro}}{\text{PC-micro}} \times \text{d.p.m.-total}$$

where: PC-O micro is the amount of PC in the outer leaflet of the microsomes.

The specific activity of the liposomal PC is 2/3 that of the outer leaflet PC at equilibrium and the specific activity of the inner leaflet PC is the same as that of the microsomes before exchange. The measured specific activities should be consistent with those calculated from the percentage PC in the outer leaflet of the microsomal membrane.

Experiments using PC-TP have indicated that all of the PC of rat liver microsomes is exchangeable. There is apparently a transmembrane movement of PC across microsomal membranes, which invalidates the use of PL-TP for investigations of these membranes. However, PC-TP has been used successfully in investigations of red blood cell membranes and potentially may be used for other membranes using procedures similar to those described.

In all investigations of the transverse distributon of phospholipids, it is necessary to demonstrate that membrane vesicles remain closed during experimental manipulation. An additional important control is to ensure that the acceptor and donor membranes are separated after the exchange process. Liposomes tend to stick to membrane vesicles so it is especially important to check that this does not occur by performing control experiments in which both species are labelled with a non-exchangeable marker. A trace of [^{14}C]cholesterol oleate is a suitable marker for liposomes while microsomes and other cell membranes are labelled by [^{3}H]leucine injected either a short time before sacrifice of the animal (30 min) to label the secretory protein contents or a longer time (24 h) to label the membrane proteins. Using these labels as tracers the degree of cross-contamination between the liposomes and membranes during the experimental procedure can be determined.

9. INVESTIGATIONS OF THE SYNTHESIS, TURNOVER AND INTRACELLULAR MOVEMENT OF MEMBRANE PHOSPHOLIPIDS

The above sections describe the experimental techniques which can be applied to a variety of investigations of membrane lipid synthesis and intracellular transport. In such studies radiolabelled precursors of membrane phospholipids are incorporated either *in vivo* or *in vitro* pulse or pulse−chase experiments. The kinetics of appearance of newly synthesized phospholipids within the luminal compartments (for secretion) or the cisternal or cytoplasmic leaflets of the membrane bilayers are then determined.

9.1 Incorporation of labelled precursors into phospholipids

In vivo precursors are incorporated into membrane lipids as shown in *Table 2*. In chase experiments it is usually necessary to dilute the radioactive precursor by following the first injection by a second injection of non-labelled precursor. This is essential if the labelled precursor forms a slowly turning over pool within the cells under investigation.

In vitro the endoplasmic reticulum (recovered as a microsomal fraction − see Chapter 1) is the major site of synthesis of membrane phospholipid in hepatocytes, although recent studies have indicated that other membranous organelles also have some biosynthetic activity. Radiolabelled precursors of phospholipids can be incorporated *in vitro* into membranes which have the appropriate enzyme activity.

9.1.1 *Incorporation of cytidine di-phosphocholine or cytidine di-phosphoethanolamine into phosphatidylcholine or phosphatidylethanolamine*

(i)   Incubate membranes (up to 11.0 mg protein) with cytidine di-phospho[^{14}C]choline or cytidine di-phospho[^{14}C]ethanolamine, 0.8 M (specific activity ∼10^6 d.p.m./μmol), in 0.05 M phosphate buffer, pH 7.4, containing 0.87% NaCl, 1.0 mM EGTA and 15 mM MgCl$_2$.

9.1.2 *Incorporation of methyl groups from S-adenosylmethionine into phosphatidylcholine*

Methyl groups are transferred from S-adenosyl[^{14}C]methylmethionine to phosphatidylethanolamine. Whether the three methylation steps are catalysed by one or two enzymes is a point of controversy. However, for optimum synthesis of phosphatidylcholine by this pathway, incubate membranes (up to 5 mg protein) with 1 mM S-adenosyl[^{14}C]-methyl methionine (specific activity ~ 10^6 d.p.m./μmol) in 50 mM Tris-HCl, pH 8.5, in 0.87% NaCl containing 10 mM $MgCl_2$.

9.1.3 *Incorporation of bases into phospholipids by base exchange*

Endoplasmic reticulum membranes have enzymes which catalyse the exchange of bases in the incubation medium for those in the membrane phospholipid. The role of these enzymes *in vivo* is not clear. However, the pathway may be exploited to label membrane phospholipids and to follow their subsequent movement within the bilayer.

(i)   Incubate microsomes (2−3 mg protein) with 0.75 mM [^{14}C]base (ethanolamine, choline, serine), specific activity approximately 10^6 d.p.m./μmol, and 1.0 mM $CaCl_2$ in Hepes buffer, pH 7.4, containing 0.25 M sucrose.

(ii)   After incorporation of labelled precursors into membranes, the reaction may be stopped by addition of an excess of unlabelled precursor to the incubation medium or by isolation of the membrane by centrifugation.

9.2 Identification of the site of newly synthesized phospholipid

After labelling phospholipids *in vivo* isolate the subcellular organelles under investigation (see Chapter 1). Membranes labelled *in vitro* are investigated without further purification.

Incorporation of precursors into phospholipids is determined by extraction of the membrane lipids, separation of these by t.l.c, elution of the phospholipid and determination of the specific activity. The rate of incorporation of precursors into the different phospholipids may thus be determined.

The percentage of the labelled phospholipid in the outer leaflet of the membrane bilayer and the specific activities of the outer and inner leaflet pools may be determined using phospholipases, chemical labelling reagents or phospholipid transfer proteins as indicated above. As with investigations of the transverse distribution of phospholipids, it is necessary to perform control experiments on opened membrane vesicles. The newly synthesized, i.e. labelled phospholipids which are not accessible to the probe in closed vesicles should become available when the vesicles are opened.

Phospholipids are sequestered into the cisternal space of cell fractions, especially the Golgi membranes and to a lesser extent into the endoplasmic reticulum. To determine the extent of sequestration of labelled phospholipids the membrane vesicles should be opened using sodium carbonate or the French pressure cell as described above. The membrane fractions are isolated by a prolonged centrifugation step (e.g. 2 h at 105 000 g), followed by a further centrifugation of the supernatant (1 h at 105 000 g) to ensure a more complete separation of the membranes and vesicle contents. The supernatant lipids are extracted by the method of Bligh and Dyer (4) and the membrane lipids with chloroform/methanol, 2/1. Using these experimental procedures, the appearance of labelled phospholipids in subcellular organelles, in the two leaflets of the membrane bilayer and in lipids destined for secretion may be determined. The interrelationships between membrane phospholipids and the possible product—precursor relationships in a variety of membranes and tissues may thus be investigated.

10. REFERENCES

1. Berridge,M.J. and Irvine,R.F. (1984) *Nature*, **312**, 315.
2. Hirata,F. (1985) In *Phospholipids of the Nervous System*. Horrocks,L.A., Kanfer,J.N. and Porcellatti,G. (eds), Raven Press, New York, Vol. **II**, p. 99.
3. Folch,J., Lees,M. and Sloane-Stanley,G.H. (1957) *J. Biol. Chem.*, **226**, 497.
4. Bligh,E.G. and Dyer,W.J. (1959) *Can. J. Biochem. Physiol.*, **37**, 911.
5. Kates,M. (1972) *Techniques of Lipidology*. Elsevier, North-Holland.
6. Christie,W.W. (1982) *Lipid Analysis*. Second Edition, Pergamon Press.
7. Marinetti,G.V. (1976) *Lipid Chromatographic Analysis*. Marcel Dekker.
8. Skipski,V.P., Barclay,M., Reichman,E.S. and Good,J.J. (1967) *Biochim. Biophys. Acta*, **137**, 80.
9. Gonzalez-Sastre,F. and Folch-Pi,J. (1968) *J. Lipid Res.*, **9**, 532.
10. Tolbert,M.E.H., White,A.C., Aspry,K., Cutts,J. and Fain,J.N. (1980) *J. Biol. Chem.*, **255**, 1938.
11. Grado,C. and Ballou,C.E. (1961) *J. Biol. Chem.*, **236**, 54.
12. Tomlinson,R.V. and Ballou,C.E. (1961) *J. Biol. Chem.*, **236**, 54.
13. Ellis,R.B., Galliard,T. and Hawthorne,J.N. (1963) *Biochem. J.*, **88**, 125.
14. Downes,C.P. and Michell,R.H. (1981) *Biochem. J.*, **198**, 133.
15. Irvine,R.F., Brown,K.D. and Berridge,M.J. (1984) *Biochem. J.*, **221**, 269.
16. Vaskovsky,V.E. and Kostevsky,E.Y. (1968) *J. Lipid Res.*, **9**, 396.
17. Fiske,C.H. and Subbarrow,Y. (1925) *J. Biol. Chem.*, **66**, 325.
18. Bartlett,G.R. (1959) *J. Biol. Chem.*, **234**, 466.
19. Snyder,F. and Stephens,N. (1959) *Biochim. Biophys. Acta*, **34**, 244.
20. Duncombe,W.G. (1965) *Biochem. J.*, **88**, 7.
21. Etamadi,A.-H. (1980) *Biochim. Biophys. Acta*, **604**, 423.
22. Op den Kamp,J.A.F. (1979) *Annu. Rev. Biochem.*, **48**, 47.
23. Van Deenen,L.L.M. (1981) *FEBS Lett.*, **123**, 3.
24. Kreibich,G., Debey,P. and Sabatini,D.D. (1973) *J. Cell Biol.*, **58**, 436.
25. Arion,W.J., Ballas,L.M., Lange,A.J. and Wallin,B.K. (1976) *J. Biol. Chem.*, **251**, 4901.
26. Polokoff,M. and Bell,R.M. (1978) *J. Biol. Chem.*, **253**, 7173.
27. Higgins,J.A. and Evans,W.H. (1978) *Biochem. J.*, **174**, 563.
28. Higgins,J.A. (1984) *Biochem. J.*, **219**, 261.
29. Fujiki,Y., Hubbard,A.L., Fowler,S. and Lazarow,P.B. (1982) *J. Cell Biol.*, **93**, 97.
30. Higgins,J.A. and Hutson,J.L. (1984) *J. Lipid Res.*, **25**, 1295.
31. van den Bosch,H. (1982) In *Phospholipids*. Hawthorne,J.N. and Ansell,G.B. (eds), Elsevier Biomedical, Amsterdam/New York/London, p. 313.
32. Westerman,J., Kamp,H.H. and Wirtz,K.W.H. (1983) *Methods Enzymol.*, **98**, 581.

CHAPTER 5

Solubilization and reconstitution of membrane proteins

OWEN T.JONES, JULIE P.EARNEST and MARK G.McNAMEE

1. INTRODUCTION

The purification and characterization of membrane proteins presents several unique problems not normally encountered with soluble proteins. Proteins that are intimately associated with the lipid bilayer of membranes are essentially insoluble in water and require the use of detergents or other disruptive agents for both solubilization and purification. Such proteins are classified as 'integral' membrane proteins according to the Fluid−Mosaic Model of membrane structure (1), and these proteins will be the primary focus of this chapter.

Since the techniques used to liberate integral membrane proteins from the membrane also destroy the native membrane structure, a challenge in many membrane studies is to ensure that the membrane-associated function of the protein is not altered or destroyed by the solubilization and purification procedures. One approach is to re-incorporate the protein back into a membrane structure after purification, a process generally referred to as reconstitution. For some membrane proteins, such as ion channels or transport proteins, function can only be meaningfully defined and assayed in a re-constituted membrane system. Other membrane proteins may act primarily as enzymes or binding proteins (receptors) and useful data can often be obtained using the solubilized protein.

In addition to the valuable functional information derived from solubilization and reconstitution techniques, the same procedures can also be exploited to manipulate membrane proteins into structures convenient for detailed structural analysis. It is clear that detergents will play a key role in the development of techniques for crystallizing membrane proteins in a state suitable for diffraction analysis (2).

Although the need to use detergents and reconstitution techniques usually represents a complication in the analysis of membrane proteins, there are several useful aspects of membrane systems that can greatly enhance the opportunity to characterize a complex biological process. Good procedures for separating different types of membranes on the basis of size, density, surface charge and other physical characteristics have been developed (see Chapter 1). Thus, it may be possible to partially purify a membrane-associated function 10- to 100-fold while it is still associated with a particular membrane fraction. The tissue homogenization, cell disruption and membrane fractionation schemes provide opportunities to identify the subcellular localization of important cell functions. In some systems, such as higher plants, the difficulty of purifying and identifying different membrane fractions has been a major problem in purifying plasma membrane proteins, although progress is now being made (3,4; see Chapters 1 and 2).

The goal of this chapter is to outline some of the experimental strategies that have proved useful in solubilizing, purifying and reconstituting membrane proteins. Unfortunately, no single detergent or reconstitution scheme has yet emerged that can be applied with confidence to all membrane proteins. The state-of-the-art in membrane protein research is still very much an art. An effort is made to provide relevant information about the most widely-used detergents and to offer practical suggestions for their use. Whenever possible, unifying principles are presented and discussed.

The chapter also outlines the criteria that should be used in evaluating the success of a reconstitution experiment. Factors such as membrane homogeneity, size, protein distribution and orientation, leakiness, etc., must all be considered when designing assays to test the function of the reconstituted membrane protein. Not surprisingly, all of the important factors are affected (sometimes in insidious ways) by the detergents, lipids and procedures used during solubilization and reconstitution.

Throughout the chapter examples from a wide range of different membrane systems will be presented. However, the authors have the most direct experience with the nicotinic acetylcholine receptor from the electric ray *Torpedo californica*, one of the best characterized membrane proteins, and the acetylcholine receptor will be featured as a model membrane protein. Several other articles have appeared over the years which attempt to provide insights into the often confusing world of detergents and reconstitution (4,5), and the readers will be referred when appropriate to other sources of information for specific theoretical or practical information that cannot be accommodated here.

2. SOLUBILIZATION OF MEMBRANE PROTEINS

As indicated in Section 1, the first steps in the purification of a particular membrane protein will usually involve the selection of an appropriate biological source and the isolation of a membrane fraction enriched in the protein. The next stage involves solubilization with detergents and it is the aim of this section to provide an overview of the theoretical and practical strategies for the solubilization of integral membrane proteins.

In physico-chemical terms solubilization is defined as 'preparation of a thermodynamically stable isotropic solution of a substance normally insoluble in a given solvent by the introduction of an additional amphiphilic component or components' (6). Such a definition is equally applicable when studying solubilization and reconstitution of biological membranes consisting of lipid and protein components insoluble in their aqueous environment. Suitable amphiphiles are detergents. When added to a biological membrane, the detergent usually partitions into the membrane until a saturation point is reached. At higher detergent-to-lipid ratios, detergent−lipid−protein mixed micelles start to form (see *Figure 1* and also refs. 7 and 8).

2.1 Criteria for solubilization

The most widely used criterion for solubilization is retention of a protein, or protein activity, in the supernatant after high-speed centrifugation, usually 105 000 *g* for 1 h (4). Most membranes will sediment under these centrifugation conditions and a dis-

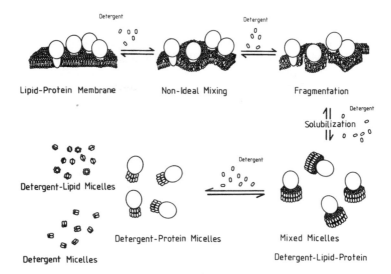

Figure 1. Scheme for the interaction of detergents with biological membranes.

tinction between solubilized and insoluble material can be made. However, it is important to realize that such a definition of solubilization is operational and may depend on factors such as the temperature and density of the medium, particularly if additives such as sucrose and glycerol are present.

Other ways to assess solubilization have been used. Chromatography on gel filtration columns with large pore sizes can separate a void volume and a retained fraction corresponding to insoluble (or aggregated) material and solubilized material, respectively (4,9). Again the distinction between soluble and insoluble may be hazy and will depend on the size fractionation range of the gel matrix.

In principle, electron microscopy (EM) is an excellent method for distinguishing between soluble and insoluble material. However, during sample preparation using conventional staining techniques, detergent dilution may occur giving rise to partial reconstitution of membranes (10).

Solubilization is usually accompanied by a marked decrease in the solution turbidity and this may be detected by a decrease in either light scattering or an increase in light transmission by the solution. However, even fully solubilized proteins may not give optically-clear solutions due to the presence of large micelles which contain both lipid and protein.

The recent application of ^{31}P n.m.r. to monitor detergent solubilization (11,12) should prove to be a particularly powerful tool, but may not be accessible to many laboratories. Using this technique it is possible to monitor the conversion of the broad resonance peak characteristic of phospholipids in the membrane to the narrow signal characterizing phospholipids in the micelle.

An important consideration in solubilization studies is whether or not a detergent is selectively extracting particular (possibly essential) components from the membrane. Although selective extraction of lipid classes from sarcoplasmic reticulum membranes has not been observed with the non-ionic detergent $C_{12}E_8$ (12; see *Figure 2*), it is well established that digitonin preferentially solubilizes cholesterol-containing membranes

(13,14). Whenever a new detergent is used it is probably advisable to check for selective extraction of membrane components. For lipids this can be done by either two-dimensional thin-layer chromatography (t.l.c.) (15; see Chapter 3) or gas−liquid chromatography (g.l.c.) of the fatty acid methyl esters (16). Extraction of proteins can be monitored by total protein assays or by measuring the activity of specific enzymes or ligand-binding proteins (see Section 4).

2.2 Selection of detergents

The success of a solubilization, purification and reconstitution protocol is often determined by the choice of a suitable detergent. Much of the confusion in choosing a detergent arises from the large number of compounds available, many of which are identical but sold under different names. The structures, alternate names and some of the properties of the detergents most commonly used in biochemical studies are given in *Tables 1−4*. In general, no single detergent has emerged as the best choice in all cases. There are probably three reasons for this:

(i)   genuine differences in the effects of detergents on specific membrane proteins;
(ii)   lack of systematic solubilization and reconstitution strategies; and
(iii) the complexity of the interaction between molecules of such distinct chemical structures as proteins, lipids and detergents.

Consequently, the selection of a particular detergent for solubilization and reconstitution is based more on precedent and empirical factors than on scientific principles.

There are, of course, certain requirements for a particular detergent to be useful. Many of the advantages and disadvantages of the common detergents are summarized in *Table 2*. In general, the detergent should solubilize but not denature the protein,

Table 1. Chemical and physical properties of commonly-used detergents.

Detergent[a]	m.p. °C	Mol. wt monomer	Mol. wt[e] micelle	CMC[e] % (w/v)	M	References
SDS	206	288	18 000	0.23	8.0×10^{-3}	6,7,27
Cholate	201[b]	430	4300	0.60	1.4×10^{-2}	4,7,17,27
Deoxycholate	175[b]	432	4200	0.21	5.0×10^{-3}	7,17,27
C_{16}TAB	230[c]	365	62 000	0.04	1.0×10^{-3}	6,7,27
Lyso PC (C_{16})	−	495	92 000	0.0004	7.0×10^{-6}	6,7,23
CHAPS	157[c]	615	6150	0.49	1.4×10^{-3}	4
Zwittergent 3-14	−	364	30 000	0.011	3.0×10^{-4}	4
Octyl glucoside	105[c]	292	8000	0.73	2.3×10^{-2}	4
Digitonin	235[c]	1229	70 000	−	−	4,7
$C_{12}E_8$	−[d]	542	65 000	0.005	8.7×10^{-5}	117
Lubrol PX	−[d]	582	64 000	0.006	1.0×10^{-4}	4
Triton X-100	−[d]	650	90 000	0.021	3.0×10^{-4}	4,6,7
Tween 80	−[d]	1310	76 000	0.002	1.2×10^{-5}	5

[a]See *Figure 2* for structures.
[b]Based on free acid.
[c]Decomposes.
[d]Viscous liquid at room temperature.
[e]Determined at 20−25°C.

142

Table 2. Properties of commonly-used detergents[a].

Property	SDS	C16	CHO	DOC	LYS	CHA	ZWI	OGL	DIG	C12	T80	LUB	TNX
Strongly denaturing[b]	+	+	−	−	+/−	−	+/−	−	−	−	−	−	−
Dialysable	+	+	+	+	−	+	+/−	+	−	−	−	−	−
Ion-exchangeable[c]	+	+	+	+	−	−	−	−	−	−	−	−	−
Complexes ions[d]	+	−	+	+	−	−	−	−	−	+/−	+/−	+/−	+/−
Strong A_{280}[d]	−	−	−	−	−	−	−	−	−	−	−	−	+
Assay interference[d]	−	−	−	−	−	−	−	−	−	+/−	+/−		+/−
Cold precipitates	+	+	−	+	−	−	−	−	−	−	−	−	−
High cost[e]	−	−	−	−	+	+	+	+	+	+	−	−	−
Availability[e]	+	+	+	+	+	+	+/−	+	+	+/−	+	+	+
Toxicity[f]	−	−	−	−	−	−	−	−	+	−	−	−	−
Ease of purification	+	+	+	+	+/−	+	+	−	+	−	−	−	−
Radiolabelled[g]	+	−	+	+	+	−	−	+	−	+	+	+	+
Defined composition	+	+	+	+	+	+	+	+	−	−	−	−	−
Autooxidation	−	−	−	−	−	−	−	−	−	+	+	+	+

[a]SDS, sodium dodecyl sulfate; C16, hexadecyl trimethylammonium bromide; CHO, cholate; DOC, deoxycholate; LYS, lysophosphatidylcholine; CHA, CHAPS; ZWI, Zwittergent 3-14; OGL, octyl glucoside; DIG, digitonin; C12, C12E8; T80, Tween 80; LUB, Lubrol PX; TNX, Triton X-100. See *Table 1* for structures.
[b]Strongly denaturing refers to disruption of secondary and tertiary protein structure.
[c]Ionic detergents are unsuitable for ion exchange chromatography.
[d]See text.
[e]See *Table 3*.
[f]All detergents are harmful notably as skin and lung irritants. Digitonin is a cardiac glycoside.
[g]See *Table 4*.

it must be readily available in a pure form and it should preferably be inexpensive. A list of common sources of the detergents shown in *Figure 2* is given in *Table 3*. In many studies it may also be necessary to quantify detergent addition and removal. This is best achieved using radiolabelled detergents and a list of the available labelled compounds and common suppliers is given in *Table 4*. An important consideration is that the requirements for solubilization may be entirely different from those ideal for reconstitution. For example, a detergent like Triton X-100, which is widely used to solubilize biological membranes, can cause difficulties in reconstitution since it is not easily removable (see Section 3). Several of the factors important in detergent performance are discussed below.

2.2.1 Detergent structure and charge

Detergents usually contain spatially distinct hydrophobic and hydrophilic regions and they can be conveniently classified as being either ionic or non-ionic (see *Figure 2*). Ionic detergents may be anionic (e.g. the bile salts cholate and deoxycholate), cationic (e.g. alkyltrimethylammonium salts) or zwitterionic (e.g. the recently developed compound CHAPS). Non-ionic detergents include compounds such as octyl-β-D-glucopyranoside (octyl glucoside) and polyoxyethylene derivatives such as Triton X-100, Lubrol PX and the Tween series.

The most noticeable difference among the detergent structures is that the polar portion of the non-ionic detergents (often an uncharged polyoxyethylene chain) is usually much larger than that of the ionic detergents. An interesting feature of the bile salts, revealed from the stereochemical structures, is that the cyclopentaphenanthrene ring possesses both a hydrophobic and a hydrophilic face (17).

An important consideration with the polyoxyethylene derivatives is that the commercial compounds may be heterogeneous due to the broad distribution of polyoxyethylene chains arising from the polymerization processes used in their synthesis. Additionally, the polyoxyethylene derivatives are prone to autooxidation particularly in the presence of heavy metal ions (18). Autooxidation may be prevented by the addition of 0.2 mol% butylated hydroxytoluene, which acts as a free radical scavenger. The charge on the detergent can be important particularly if ion-exchange steps are necessary during protein purification. Additionally, some charged detergents, particularly the bile salts, may form insoluble complexes with divalent metal ions or precipitate at low pH (pH 6.8) as a result of their carboxyl groups (17,19).

Another consideration is that the detergent may interfere with important assays for characterizing the solubilized material. For example, many non-ionic detergents contain high levels of phosphates which seriously affect lipid analyses. Other detergents, notably

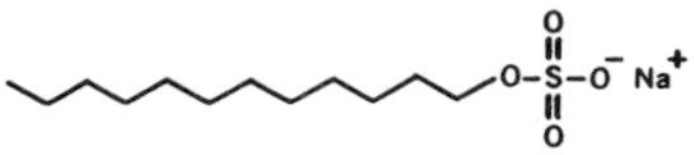

SDS
Sodium Dodecyl Sulfate
Sodium Lauroyl Sulfate

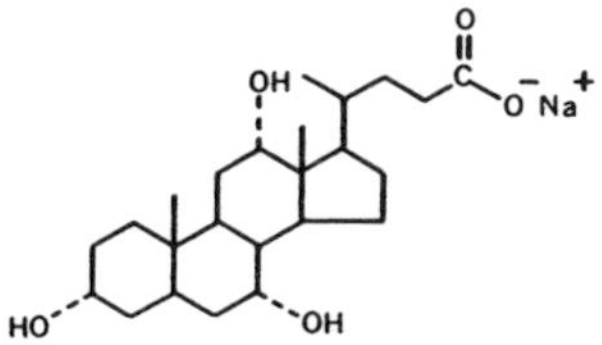

Cholate
3α,7α,12α-Trihydroxy-5β-cholan-24-oate

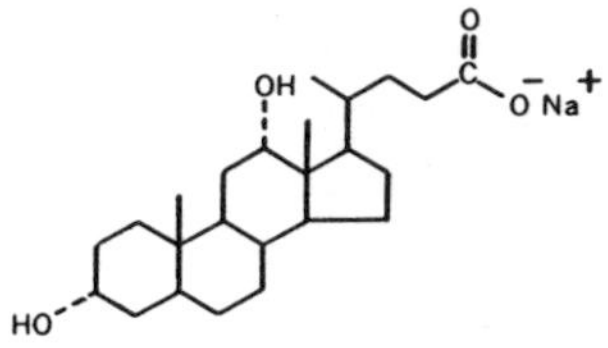

Deoxycholate
3α,12α-Dihydroxy-5β-cholan-24-oate

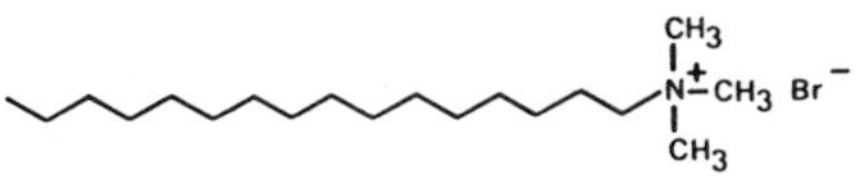

C₁₆TAB
Hexadecyl trimethylammonium Bromide
Cetrimide

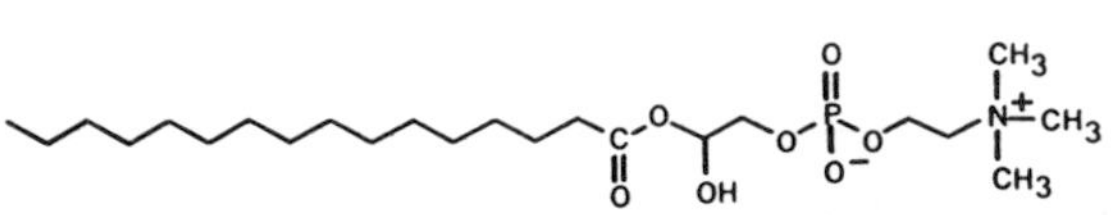

Lyso PC
Lysophosphatidylcholine
Lysolecithin

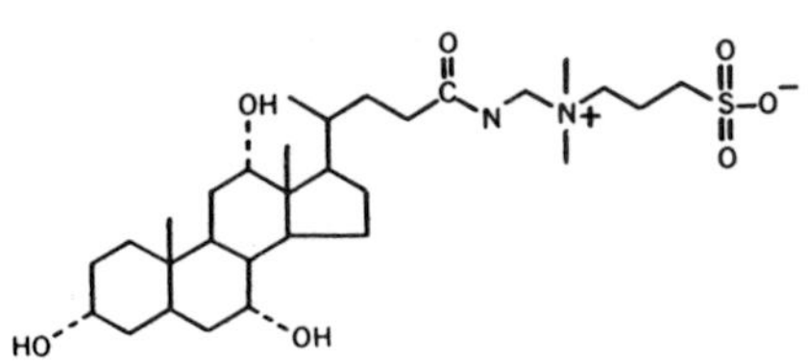

CHAPS
3-[(3-Cholamidopropyl)-Dimethylammonio]-
-1-propane sulfonate

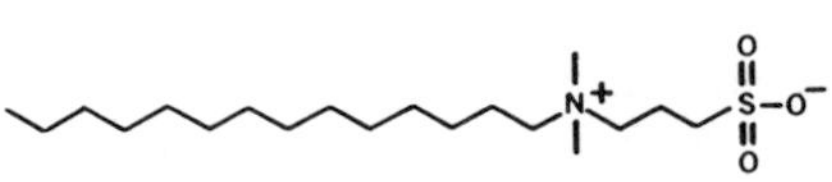

Zwittergent 3-14
N-tetradecyl-N,N-dimethyl-3-ammonio-
-1-propane sulfonate

144

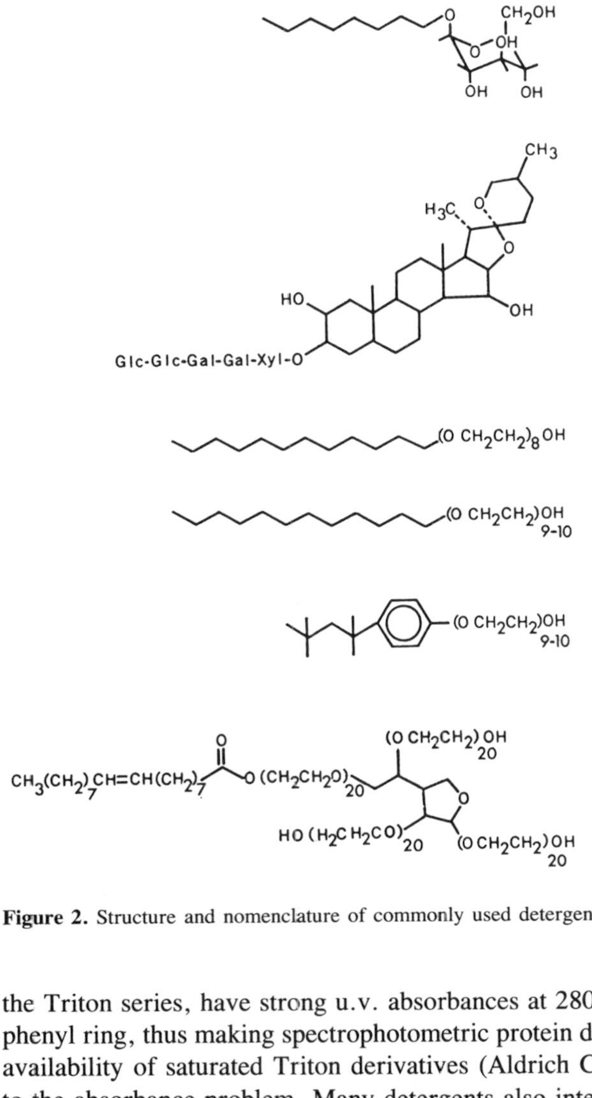

Figure 2. Structure and nomenclature of commonly used detergents.

the Triton series, have strong u.v. absorbances at 280 nm due to the presence of the phenyl ring, thus making spectrophotometric protein determination difficult (20). The availability of saturated Triton derivatives (Aldrich Chemical Co.) offers a solution to the absorbance problem. Many detergents also interfere with colorimetric protein determinations. In some instances, the detergent can interfere with one or more components of the assay mixtures, for example by sequestering ligands and enzyme substrates. Impurities in polyoxyethylene derivatives have been shown to inhibit purified dopamine β-monooxygenase (21) and adenylate cyclases (22).

2.2.2 Critical micelle concentration

All amphiphiles possess a capacity to form structures known as micelles. Micelles may be defined as thermodynamically stable colloidal aggregates, spontaneously formed by amphiphiles above a certain concentration range known as the critical micelle concentration (CMC) at temperatures above the critical micelle temperature (7). Below the CMC the amphiphile is dispersed in its monomeric form. Excellent reviews of the

Table 3. Detergent availability.

Detergent[a]	Unit cost ($/g)[b]	Purity %	Source[c]
SDS	0.14−0.64	98−100	A,C,F,S
$C_{16}TAB$	0.08−0.11	95−100	A,F,S
Cholate	0.17−1.2	96−99	A,C,F,S
Deoxycholate	0.20−0.49	96−98	A,C,F,S
Lyso PC (C_{16})	322−350	99−100	C,F,S
CHAPS	8−21	98−100	A,C,F,S
Zwittergent 3-14	5.32	99−100	C
Digitonin	9−22	75−99	A,C,F,S
Octyl glucoside	26−56	99−100	A,C,F,S
$C_{12}E_8$	52	97	C
Lubrol PX	0.1	N.S.[d]	S
Tween 80	0.01−0.38	75−100	A,F,S
Triton X-100	0.01−0.04	N.S.	A,C,F,S

[a]See *Figure 2* for structures.
[b]Prices quoted are those for 1985−86 and are approximate.
[c]A, Aldrich Chemical Co.; C, Calbiochem; F, Fluka Chemicals; S, Sigma Chemical Co. N.B. Only major suppliers in USA and Britain quoted.
[d]Probably greater than 95%.

Table 4. Sources and availability of radiolabelled detergents.

Detergent	Label	Unit price ($/μCi)[a]	Source[b]
SDS	^{35}S	0.8	Am
	^{14}C	4.5	RPI
Cholate	^{3}H	0.5	NEN
	^{14}C	4.0−6.3	Am,NEN,RPI
Deoxycholate	^{14}C	3.4−5.8	Am,ICN
Lyso PC (C_{16})	^{14}C	16.7−33.5	Am,NEN
Octyl glucoside	^{14}C	1.0[c]	ARC[c]
$C_{12}E_8$	^{14}C	5.0	RPI
Triton X-100	^{3}H	1.0	NEN
Tween 80	^{14}C	2.80	RPI

[a]Prices quoted are those for 1985−86 and are approximate.
[b]Key: Am, Amersham Radiochemicals; ICN, ICN: NEN, New England Nuclear; RPI, Research Products International; ARC, American Radiolabeled Chemicals.
[c]Octyl glucoside is available by custom synthesis. A recent order for 4 mCi was obtained for $4000. American Radiolabeled Chemicals offers custom synthesis at much lower rates than other suppliers.

thermodynamics, structures and properties of detergent and detergent−membrane micelles, which are outside the scope of this chapter, are available (6,17,23−25). The concept of micelle formation is relevant to solubilization and reconstitution studies since it appears there is some correlation between the ability to form micelles and the concentration of detergent required for solubilization (26).

The CMC has practical consequences affecting reconstitution directly and indirectly. For example, the solubilizing concentration of a detergent is increased by factors which raise the CMC (see below). An indirect consequence of the CMC is that detergents with high CMCs are more easily removed from a solution by dialysis against detergent-free buffer than those with low CMCs. Detergent removal is a key step in reconstitution as discussed in Section 3. The CMCs of several detergents most widely used in reconstitution studies are reported in *Table 1* (see ref. 27 for a more comprehensive list).

Many factors affect the CMC. Most significant are modifications in both the polar and non-polar regions of the detergent (25). In general, the CMC is lowered as the size of the non-polar region is increased. Counteracting this tendency are structural features which interfere with the packing properties of the detergents. For example, double bonds and chain branches act to increase the CMC. Disruption of packing also occurs through modifications in the polar regions. The electrostatic repulsion between the headgroups of the charged detergents raises their CMCs by around 100-fold compared with their non-ionic analogues (25). Increasing the ionic strength lowers the CMC particularly with the ionic detergents (28). The effect of increasing temperature on ionic detergents is only slight (above the critical micelle temperature), but leads to a marked reduction in the CMCs of non-ionic detergents, presumably by an increased hydrophobic effect (7,21,24).

When the detergent concentration in solution is significantly higher than the CMC, micellar size increases until the micelles aggregate and come out of solution. The temperature at which this occurs is called the 'cloud point'. This behaviour can be exploited to solubilize and purify some membrane proteins. Triton X-114, for example, has a cloud point around 20°C. Membrane proteins solubilized in Triton X-114 at 0°C will precipitate out of solution at 20°C accomplishing preferential solubilization and purification in one step (29). For method see Section 2.2.1, Chapter 6.

2.3 Factors affecting stabilization of solubilized membrane proteins

2.3.1 *Lipid−detergent−protein ratios*

Because of difficulties in measuring detergent−membrane binding constants (non-ideal mixing), it is not easy to directly relate the CMC of a detergent to its potential for interacting with and solubilizing lipid−protein bilayers (8). The absence of a rigorous description of solubilization has not compromised reconstitution studies and a semi-quantitative understanding has been gained from practical experience.

The approach used in most studies is to maximize solubilization and at the same time preserve protein activity. The most effective way to do this is to examine solubilization over a wide range of detergent-to-lipid ratios, and such experiments should be considered an essential feature of any solubilization/reconstitution strategy. Since it is easier to measure protein than lipid concentrations, the detergent-to-protein ratio is often expressed in the literature. As the levels of solubilization depend on both the lipid and micellar detergent concentrations, some measure of the effective concentration of detergent would be useful. Only recently have attempts been made to derive such a parameter. The most useful of these is the ϱ (rho) parameter, introduced by Rivnay and Metzger (26), where:

$$\varrho = \frac{[\text{Detergent}]\text{-CMC}_{\text{eff}}}{[\text{Phospholipid}]}$$

147

and CMC$_{eff}$ describes the CMC determined under specific experimental conditions. The CMC$_{eff}$ is preferred to the CMC as the latter is often lowered in the presence of lipids and proteins (23), but in practice the literature values of CMCs are used (see *Table 1*). The ϱ factor attempts to describe the molar ratio of micellar detergent to phospholipid employed. Unfortunately, due to the number of variables in a reconstitution protocol, it is difficult to calculate and compare ϱ values for various proteins reported in the literature but some generalizations can be made. For example, an increase in solubilization is predicted to occur as the ϱ parameter is raised. For the IgE receptor 90% solubilization is only achieved when ϱ is greater than 2. However, as ϱ is increased the chance of denaturation also increases. For the acetylcholine receptor and the IgE receptor, solubilization at ϱ values between 2 and 10 are optimal (with most reconstitution performed at ϱ values of around 2). When acetylcholine receptor membranes (2.5 mg/ml protein, 1.2 mM native phospholipid) are solubilized in 1% (w/v) cholate, conditions which give a value for ϱ of 9, ion channel activity is preserved on reconstitution. If however the cholate concentration is raised to 2% (w/v), where ϱ now is around 29, under otherwise identical conditions there is a marked inactivation of the ion channels (30). For carnitine acylcarnitine transferase, values of $1.5-2.0$ are required in both solubilization and reconstitution steps (31). Unfortunately, the situation is probably more complex than implied by ϱ values alone, as no account is taken of differences in lipid− detergent interactions.

2.3.2 *Role of lipids in protein stability*

Although the role of phospholipids in stabilizing membrane proteins is only beginning to be understood (32), it is now well established that complete removal of lipids from biological membranes inactivates most membrane proteins $(33-35)$ even in the presence of detergents. Under some solubilization (or reconstitution) conditions such delipidation is detected by precipitation of aggregated protein.

The central role which lipids play in preserving protein stability suggests several practical steps which can be taken to minimize denaturation. The simplest of these relate to the detergent itself. For example, the activity of a particular detergent can be changed by altering factors such as the temperature, ionic strength and pH. Alternatively, simply lowering the detergent concentration is possible, but may compromise the level of solubilization. If high levels of detergent can be employed, it may then be beneficial to minimize the time the protein spends in detergent solution, for example by using rapid reconstitution methods.

Delipidation of proteins during purification on affinity columns can present a problem and it may be necessary to supplement all wash solutions with lipids (36,37). In many instances the concentration of native lipids is enough to offer sufficient protection from most solubilizing concentrations of detergent. However, it has become apparent that extraction can be performed at detergent concentrations that would otherwise cause denaturation, providing that additional 'protective' lipids are added (38,39). The protective lipids presumably lower the ϱ factor described above. It is not clear whether some lipids offer better protection than others. It is certain, however, that the sensitivity of some proteins to denaturation can depend on the combination of detergent and lipid employed. For example, the IgE receptor is more susceptible to denaturation using CHAPS as a detergent in conjunction with asolectin, than with native (tumour-derived) lipids (26). Additionally, successful reconstitution of the acetylcholine receptor has been achieved using octyl glucoside in conjunction with native *Torpedo* lipids, but protein inactivation occurs when octyl glucoside is used with asolectin (40). An intriguing idea is that the success of CHAPS, cholate and digitonin is related to the ability of these steroid derivatives to be able to substitute for cholesterol — a lipid which may stabilize the structures of certain membrane proteins (41).

2.3.3 *Role of ligands in stabilizing function*

A feature of many membrane proteins is that they possess more than one conformational state characterized by different affinities for particular ligands. Examples include receptor molecules (42,43), enzymes (44,45) and ion channels (46). Switching between conformational states is often found to be sensitive to alterations in the lipid environment (47−49), and solubilization may favour one state compared with another. Often detergents promote irreversible affinity transitions to some degenerate state associated with denaturation possibly by the delipidation mechanisms discussed above. Some protection against denaturation is often seen on inclusion of ligands or substrates in the solubilization mixtures. The ligands presumably lock the protein into a stable conformational state. For example, experiments on the β-adrenergic receptor show that agonists, but interestingly not antagonists, protect the protein during solubilization by deoxycholate (50). Similar ideas have been exploited in reconstitution experiments with ATPases by including ATP in the incubation media (33,51), and with succinate dehydrogenase by adding succinate (52).

2.4 **Solubilization and purification of nicotinic acetylcholine receptor from** *Torpedo californica*

To illustrate some of the general guidelines outlined above a detailed protocol for purification of *Torpedo* acetylcholine receptor (AChR) is provided in *Tables 5, 6* and *7*. From 200 g of electroplax approximately 20 mg of purified AChR can routinely be obtained. Note that additional lipids are added to the detergent solution during the washing and elution steps of the affinity chromatography. If the lipids are omitted, the purification proceeds normally, but the final AChR preparation does not give functional ion channels following reconstitution.

3. GENERAL STRATEGIES FOR RECONSTITUTION OF MEMBRANE PROTEINS

The ideal end product in the reconstitution of a membrane protein is usually a suspension of uniformly sized unilamellar vesicles with a uniform lipid-to-protein ratio. The vesicle bilayer should have the permeability properties of the native membrane: it should be able to trap impermeable ions or other substances and should have sufficient internal volume to allow measurement of trapped substances. There should be no residual detergent present. The membrane proteins should be uniformly oriented within the membrane, and should be intercalated into the bilayer in the same fashion as in the native membrane. The bilayer must have a composition which supports native protein function.

Even though all of these criteria are rarely met by current reconstitution methods, much useful information about the structure and function of membranes and proteins

Table 5. Preparation of *Torpedo californica* electroplax membranes (see also Chapter 3, Table 6).

All steps are carried out at 0−4°C as quickly as possible.

1. Partially thaw 200 g of frozen *T. californica* electroplax tissue, cut it into small pieces (1 cm³) with scissors and/or a scalpel and then place tissue in a 600 ml beaker.
2. Add 200 ml of homogenization buffer (HB-4, see below).
3. With stirring, add 0.4 g of iodoacetate and 0.2 ml of 200 mM PMSF (in ethanol) to give 1 mg/ml iodoacetate and 0.1 mM PMSF (protease inhibitors).
4. Homogenize for 4 × 30 sec with a Brinkmann Polytron PCU-2 homogenizer at setting 7.
5. Centrifuge at 5500 r.p.m. (5000 *g*) for 10 min in a Sorvall GSA rotor [or 6500 r.p.m. (5000 *g*) in an SS-34 rotor].
6. Filter through four layers of cheesecloth and keep supernatant.
7. Optional: resuspend pellets in 80 ml of HB-4 and repeat steps 3−6, adding 0.2 g of iodoacetate and 0.1 ml of PMSF.
8. Centrifuge the combined supernatants in a Beckman Ty 35 rotor at 30 000 r.p.m. for 60 min or in a Ty 19 rotor at 19 000 r.p.m. for 125 min. (Multiple runs of the Ty 35 may be necessary for large preparations.)
9. Suspend pellets in 5−10 ml of dialysis buffer (DB-1). Use a hand homogenizer to get a smooth suspension. Freeze in cryotubes in liquid nitrogen.

Homogenization buffer (HB-4)

    10 mM sodium phosphate (1.38 g NaH₂PO₄/l)
    5 mM EDTA (1.85 g/l)
    5 mM EGTA (1.90 g/l)
    0.02% NaN₃ (0.2 g/l)
    pH 7.5

Dialysis buffer (DB-1)

    100 mM NaCl (5.84 g/l)
    10 mM MOPS (2.09 g/l)
    0.1 mM EDTA (0.037 g/l)
    0.02% NaN₃ (0.2 g/l)
    pH 7.4

can nevertheless be derived from reconstituted systems. However, there is considerable room for improvement.

3.1 General methods of reconstitution

Reconstitution of a membrane protein typically begins with a mixed micellar solution of purified protein, detergent and lipid (either endogenous or exogenous). In some cases the mixed micelle is composed of only protein and detergent: whether the protein under investigation requires lipids for stability must be determined as discussed above. If lipid was not added during solubilization or purification (or if more lipid is required), it is typically added at this stage to give a final lipid-to-protein molar ratio, defined here as (ϕ), of 2000−10 000. This excess of lipid is usually required if the reconstituted vesicles are to be used for flux or transport studies, although a value of 100−2000 may be desirable for special purposes. Lipid can be added as detergent−lipid micelles or can be dried in a thin film and the detergent−protein solution stirred over it until the lipid is incorporated. It is desirable to maintain all lipid suspensions under an atmosphere of inert gas.

Table 6. Solubilization and purification of *Torpedo* acetylcholine receptor.

All steps are carried out at $0-4°C$.

1. Thaw crude *Torpedo* membrane fractions corresponding to 200 g of electric tissue.
2. Dilute membranes to 2 mg protein/ml using DB-1 (see *Table 5*). (A volume of 200 ml is about right for a 200 g membrane preparation.)
3. With stirring, add 20% sodium cholate (in DB-1) to give 1% cholate (10 ml).
4. Stir gently for 20 min.
5. Centrifuge at 34 000 r.p.m. in a Ty 35 rotor for 45 min. While centrifugation is in progress, wash the affinity column with buffer in preparation for purification of AChR (see step 7).
6. Remove the supernatant and save it for affinity column purification.
7. Pre-equilibrate the Affi-Gel 401/bromoacetylcholine affinity column with DB-2 (DB-1 + 1% cholate). (See *Table 7* for preparation of the affinity column.)
8. Apply the supernatant to the 25 ml affinity column at a flow-rate of 1.5 ml/min.
9. Wash the column with at least three column volumes of DB-3 (DB-1 + 1% cholate + 1 mg/ml DOPC)[a].
10. Collect fractions during the wash and monitor A_{280}. Absorbance should drop to baseline levels after extensive washing.
11. Elute AChR by washing the column with 50 ml of DB-4 (DB-3 + 10 mM Carb, prepared by adding 91 mg solid carbamylcholine chloride to 50 ml of DB-3). Collect fractions of 2 ml and monitor the A_{280}.
12. Pool the peak fractions (typically fractions with $A_{280}>0.5$) and measure the volume and A_{280}. Protein concentration in mg/ml is $0.6 \times A_{280}$.
13. Keep the purified AChR at $0-4°C$ and use purified fractions for reconstitution as soon as possible. It is much better to remove the cholate and then freeze the *membranes* in liquid nitrogen.
14. Wash the column with several column volumes of DB-1 and then with 0.01 M acetate (pH 4).

[a]Dioleoylphosphatidylcholine (DOPC) from Avanti is dried *in vacuo* and suspended in DB-1 + 1% cholate. For a typical preparation, 10 ml of DOPC (20 mg/ml in chloroform) is dried on a rotatory evaporator, dried for an additional $1-3$ h in a vacuum desiccator, and then dissolved in 200 ml of DB-3. 50 ml of DB-3 is set aside for step 11. It is at this stage that exhaustive lipid exchanges or adjustments in lipid:protein ratios can be made for certain kinds of experiments.

3.1.1 *Detergent removal*

Vesicle formation requires removal of the detergent from the mixed micelles so that a phospholipid bilayer can form spontaneously. The most common method of detergent removal is dialysis. The micellar solution is enclosed within a dialysis bag and dialysed against a large volume of detergent-free buffer, usually in the cold. Typically, a 200-fold excess of buffer with three or four changes over several days, is used. An advantage of this method is its simplicity: several different samples (composed of different lipids, for example) can be dialysed simultaneously with little attention or special equipment. The disadvantage is the length of time required to remove detergent. Not only can this be an inconvenience, but also prolonged exposure to detergent during dialysis has been shown to damage certain membrane proteins. This phenomenon is probably related to the inevitable existence of detergent−protein micelles, even in the presence of excess lipid (see *Figure 1*). If protease inhibitors were not added during solubilization or purification they might be helpful if added at this stage.

Many factors can affect the rate of detergent removal, and care should be taken to ensure that the level of residual detergent is less than 0.02% (w/v) or one hundreth of its original level. Cholate seems to be the most frequently used detergent for dialysis.

Table 7. Affinity column preparation for AChR purification.

For a 200 g equivalent membrane preparation, a 25 ml column works well. The procedure below is for a 50 ml column, since Bio-Rad sells the Affi-Gel 401 resin in 50 ml aliquots. Since the columns are not stable indefinitely, the procedure should be scaled down for smaller preparations.

1. To 50 ml of packed Affi-Gel 401 (Bio-Rad) in a 1.5 × 20 ml Econo-Column (Bio-Rad) add 50 ml of 20 mM dithiothreitol (DTT) in 0.1 M Tris, pH 8.0. Suspend well and allow to incubate for 20 min[a].
2. Wash the column with water to remove the DTT. Assay both the gel and the final wash solution for the concentration of sulphhydryl groups (see below). The wash should be at background levels.
3. Add an equal volume of 100 mM NaCl–50 mM sodium phosphate, pH 7 to the packed gel in the column. Suspend well, add 300 mg of bromoacetylcholine, and immediately mix the gel suspension by shaking. Allow the reaction to proceed for 30 min with occasional shaking. (The high phosphate concentration is necessary to maintain the pH at 7.) See below for preparation of bromoacetylcholine.
4. Wash the column with water and test the gel for free sulphhydryls. There should be very few sulph-hydryls left. Repeat the addition of bromoacetylcholine if necessary.
5. Add 50 mg of iodoacetamide to serve as an alkylating agent for any residual sulphhydryls and to serve as an anti-microbial agent. Store the column until needed.
6. The column may be used more than once (typically three times during a 3–4 week period). It should always be kept in low ionic strength buffer at pH 4 to minimize ligand hydrolysis. If the yield of purified AChR falls below 50% of the optimal yield, it is time to make a new column.

Assay for sulphhydryls

(i)   Prepare a stock solution of 10 mM dithionitrobenzoate (DTNB) and keep frozen (39.6 mg/10 ml carefully titrated with NaOH to pH 7).
(ii)   Add 0.33 ml of 10 mM DTNB to 10 ml of 0.1 M Tris, pH 8.0.
(iii) To 50 μl of gel or to 50 μl of eluate add 1 ml of dilute DTNB. (Cut the end off an Eppendorf tip to avoid clogging by gel samples.)
Qualitatively, the reduced gel will give a bright yellow colour. After reaction with bromoacetylcholine there will be no colour. The reaction can be quantitated by measuring A_{412} and by weighing the amount of gel used after aspirating away excess buffer. The qualitative assay is sufficient.

Preparation of bromoacetylcholine

To 18.4 g (0.1 mol) of solid choline bromide, add 24.2 g (0.1 mol) of bromoacetylbromide, dropwise with stirring. The addition requires about 40 min. The viscous mixture is cooled in an ice-bath and stirred for an additional 75 min. To the reaction mixture 75 ml of absolute ethanol is slowly added. The white crystalline solid is filtered by suction. The solid is re-crystallized from 400 ml of isopropanol to give crystals with a m.p. of 136–138°C. Typical yield is 24 g.

---

[a]This step ensures that the gel is in a completely reduced state. It may not be necessary. It is also possible to carry out this and subsequent steps in a test tube or beaker with batch washes. We have found it convenient to do everything in the same column.

Triton X-100 is normally quite difficult to remove by this method due to its low CMC. It was therefore not often used for reconstitution until it was found that BioBeads SM-2 or Amberlite XAD-2, hydrophobic resins which absorb detergents, can accelerate Triton X-100 removal (53,33). In both instances, the washed resins are simply added to the dialysing buffer. This technique has been used in the reconstitution of the sodium channel (37).

A second method of detergent removal is to dilute the mixed micellar solution below the detergent's CMC, which allows the lipids and proteins to form vesicles. Residual detergent can be removed by gel filtration or short dialysis. This technique was first described for reconstitution of cholate-solubilized cytochrome oxidase (54) and has been recently updated for reconstitution of rhodopsin (55).

(i)   Add lipid-free octyl glucoside-solubilized rhodopsin to a thin film of egg phos-
        phatidylcholine (PC) ($\phi=80-300$), vortex and equilibrate at 4°C for 4 h.
(ii)   Add this solution dropwise to rapidly stirred buffer so that the final octyl glucoside
        concentration is brought to 10 mM ($\varrho < -9$).
(iii)   Remove detergent by dialysis against an 18-fold excess of buffer, and concen-
        trate the vesicles using an Amicon PM-10 membrane.

An additional method for detergent removal is to pass the lipid−detergent−protein
solution through a gel filtration column. As the mixture passes through the column,
the detergent concentration is lowered thus allowing lipid−protein vesicles to form.
Gel filtration has been found to be highly effective in reconstituting several membrane
proteins, including AChR (28), ($Ca^{2+}+Mg^{2+}$)-ATPase (9) and the lactose transporter
(56). Although this method is rapid, it is not convenient for large numbers of samples.
Additionally, detergent removal by gel filtration is reported to give a broader size
distribution of vesicles than by dialysis (57). For a more complete review of methods
for removal of unbound detergents see ref. 58.

3.1.2 Reconstitution by insertion into pre-formed vesicles

Reconstitution of membrane proteins by incorporation into pre-formed vesicles has been
demonstrated in several cases. This technique (*Figure 4b*) has been shown to result
in more uniformly oriented protein (59). Proteins with small hydrophobic 'tails', such
as colicin Ia or cytochrome b_5, insert into liposomes spontaneously, although specific
lipids are often required. This technique has been used to reconstitute the hepatic
asialoglycoprotein receptor (60).

In some cases detergent-solubilized proteins are mixed with pre-formed liposomes,
as in the reconstitution of guanylate cyclase (61).

(i)   Add buffer to a thin film of lipids so that the final lipid concentration is in the
        millimolar range.
(ii)   Sonicate the suspension until clarified using a bath sonicator; this may require
        up to several hours, and should be done under nitrogen or argon. (For a more
        complete description of liposome techniques, see ref. 62.)
(iii)   Incubate the membrane-bound form of guanylate cyclase, purified and solubilized
        in 0.1% (w/v) Lubrol PX, with liposomes for 1−5 h and collect reconstituted
        vesicles by centrifugation at 300 000 *g* for 30 min (for a variation on this tech-
        nique, see the freeze−thaw method below).

The enzymic activity associated with these vesicles demonstrated that, unlike several
of the membrane proteins discussed in this chapter, guanylate cyclase in liposomes had
a 10-fold lower V_{max} than in soluble form. Enzymic activity was greatly affected by
both the phospholipid composition and the nature of the fatty acyl chain, although the
effects were not a function of chain length or saturation.

3.1.3 Reconstitution by sonication/freeze−thawing

A final method of reconstitution which involves no detergent at all is the technique
of sonication and freeze−thawing. An aliquot of lipid-free, and detergent-free protein,
is combined with an aqueous dispersion of lipids. The mixture is then sonicated (63),
or quick-frozen in liquid nitrogen, thawed and sonicated (64). The advantages of this

technique are that it is quick and avoids exposure to detergents. The disadvantages are that most membrane proteins cannot be stripped of both lipid and detergent without irreversible denaturation. They can also become irreversibly damaged by more than a few seconds of exposure to sonic oscillation. An additional disadvantage is that variations in the power and design of sonicators make it very difficult to standardize the sonication conditions.

The chlorophyll *a/b* light-harvesting complex has been reconstituted using a hybrid technique (65). Chloroplast membranes were solubilized in 1.5% (v/v) Triton X-100 at a final chlorophyll concentration of 3 mM. This suspension was added to an equal volume of sonicated lipids (10 mg/ml), stirred, then freeze−thawed three times. The sample was then stirred over BioBeads SM-2 for 45 min to removal residual Triton X-100, centrifuged to remove BioBeads and then freeze−thawed three times. Mostly single-walled vesicles were formed, with some larger multilamellar sheets.

3.2 Characterization of vesicle morphology

A complete study of any reconstitution protocol requires some knowledge of the morphology of the products. Ideally, the reconstituted vesicles should be homogeneous in shape, size and protein distribution. However, a problem unique to reconstituted membranes is the presence of both lipid and lipid−protein vesicles, a heterogeneity which can seriously complicate the analysis of proteins whose function is solute translocation (40). In this section some of the techniques widely used to characterize vesicle morphology are described.

3.2.1 *Electron microscopy*

Visualizing the reconstituted membrane using EM provides the best means of characterizing the distribution of sizes and shapes present in the sample. The two techniques which are used most often are negative-staining and freeze-fracture. It is beyond the scope of this chapter to provide the detailed methods for these techniques (see ref. 66), but advantages and disadvantages of each technique are summarized. Techniques used on lipid vesicles (66) are usually applicable to reconstituted membranes, although the choice of negative-stain may vary.

Negative-stain EM involves blotting a diluted sample of reconstituted membrane, typically 0.1 mg lipid/ml, onto a copper mesh grid which has been coated with a polymerized film such as Parlodian or Formvar. Carbon is shadowed onto the film for support and the sample on the grid is then exposed to a solution of heavy metal stain, such as uranyl acetate or sodium phosphotungstate, and blotted dry. When staining reconstituted vesicles, both the tonicity and pH of the washing solution and the stain must be appropriate, so that the vesicle morphology will remain as constant as possible (67). Negative-stain microscopy is very quick, (perhaps 30 min from sample to electron microscope), and a large number of samples can be scanned in an afternoon. It has the added advantage of allowing fairly high resolution examination of the image, provided precautions are taken to avoid electron doses which vapourize the sample (see *Figure 3* for examples). Three-dimensional image reconstructions of membrane protein structure are typically done on negative-stained samples (68).

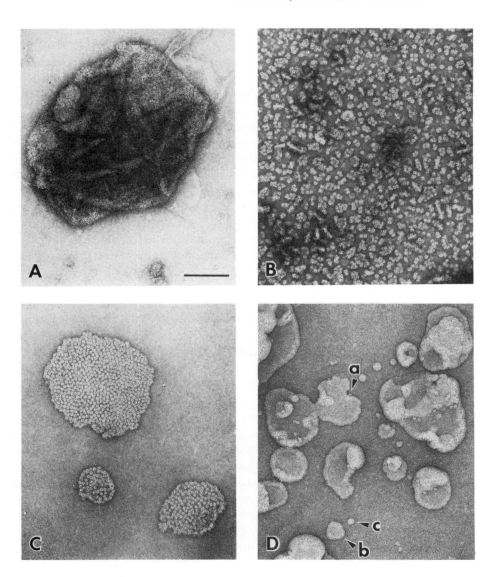

Figure 3. Electron micrographs of acetylcholine receptor. **(A)** Native membrane vesicle from crude homogenate of *Torpedo californica* electroplax. Bar = 0.1 μm. Rosettes of AChR are clearly visible. **(B)** Preparation from **A**, solubilised with sodium cholate in the presence of lipid (dioleoylphosphatidylcholine) and purified on a BAC-affinity column. AChR rosettes are visible as dimers, monomers and oligomers. Magnification as **A**. **(C)** Preparation from **B**, dialysed against cholate-free buffer. Bilayer sheets are formed (ϕ=218). Magnification as **A**. **(D)** Preparation from **B**, equilibrated with 2/3 volume of lipid/cholate suspension (60 mg/ml asolectin, 4% cholate), then dialyzed against cholate-free buffer. Vesicles exhibit good ion flux properties (ϕ=10 000). Dispersed dimers and monomers are visible, with an edge view shot at (a). Vesicle at (b) is a protein-free liposome, while structure at (c) is a micelle, as its diameter is smaller than the 24 nm minimum for bilayer vesicles. Magnification as **A**. All EM samples were prepared by applying 25–250 ng of protein to a parlodian-coated, carbon-shadowed copper grid, washed extensively with buffer or 1% uranyl acetate, then stained with 1% uranyl acetate. Samples were examined on a Phillips EM 400 electron microscope with an anti-contamination trap, at 100 kV.

155

Freeze-fracture EM involves quick-freezing a pelleted sample of reconstituted membranes and fracturing the frozen pellet so that the interior of the bilayer is exposed. Fracture faces are then shadowed with platinum at a 45° angle and vertically shadowed with carbon for support. The platinum−carbon film then provides a relief or replica of the membrane surface. Integral membrane proteins are seen as either bumps or pits, although it is not always the case that one protein corresponds to one bump [or intramembraneous particle (IMP)]: several proteins may coalesce to form one IMP. An advantage of this technique is that protein distribution in the bilayer is more accurately assessed and it is possible to unequivocally demonstrate whether a vesicle is uni- or multi-lamellar, and whether there are proteins trapped inside. The disadvantage is that the resolution is limited by the size of the platinum particles which are coating the surface of the protein, so quantitative measurement of particle size is difficult, although not impossible. Additionally, the method is of limited use in determining size distributions of vesicles as the apparent vesicle diameter depends on the depth of the fracture plane in relation to the vesicle — only the widest dimensions observed represent the true vesicle diameters.

Recent applications of cryoelectron microscopy suggest that this technique will prove invaluable in visualizing reconstituted membranes (69). In this case the aqueous sample is applied to an uncoated grid, blotted, then plunged into freezing ethane. Vitreous ice is formed by the rapid freezing, and direct visualization of biological materials is possible by maintaining the grid and sample at liquid nitrogen temperatures using a specially adapted cold stage in the EM. This technique avoids the artifacts induced by stain or dehydration and the sample is observed in its appropriate ionic environment.

3.2.2 *Gel filtration and other techniques*

The sizes and distributions of reconstituted vesicles can often be determined by gel chromatography using resins with large pore sizes. It is possible to calibrate the filtration columns using latex beads of defined size and good agreement in size distribution of liposomes determined by EM and calibrated sizing columns has been reported (8). An important advantage of the sizing columns is the ease with which protein incorporation into vesicles of defined sizes can be assessed. A major disadvantage is their long elution times (several days). Furthermore it is possible that the gels alter the vesicle distribution or adsorb material onto the column.

Characterization of reconstituted vesicles can also be achieved using centrifugation methods (70). Fong and McNamee (71) used sucrose density centrifugation of reconstituted AChRs to show multiple populations of membranes with varying amounts of protein. When the samples were freeze−thawed before centrifugation, the populations appeared to coalesce into one major peak, although electron microscopic evidence indicated a population of vesicles that was still more heterogeneous than the gradient profile suggested (J.Earnest, unpublished observations).

3.3 **Factors which affect the morphology of the reconstituted membrane**

The size and shape of the reconstituted membrane vesicle is often critical when assaying the function of integral membrane proteins. The function of ion channels, for example, is normally assayed by measuring the amount of radioactively labelled ions which are taken up into the sealed reconstituted vesicles. It is useful to remember that a vesicle of 25 nm diameter (such as those frequently formed by cholate dialysis), which has an internal volume of 2×10^{-18} ml per vesicle, will contain only one molecule of solute at a solute concentration of 1 mM (72). Even when there is a channel present to allow ions into the vesicle the internal volume can be an undesirable limiting factor in the assay of protein function. A second important feature of the reconstituted membrane is the orientation of the protein within the bilayer. Quantitative assay of protein function is often difficult in a reconstituted system, unless the degree of orientation of the protein can be accurately determined, and unless the impermeability of the vesicle bilayer can be trusted. These properties of the reconstituted membrane should always be carefully characterized (see below).

3.3.1 *Lipid-to-protein ratio*

Although several membranes are known to contain low lipid-to-protein ratios *in vivo*, and still exist as sealed vesicles, it has not been possible to achieve such ratios in reconstituted membranes. The AChR-containing membranes from *Torpedo californica* electroplaques, for example, can be purified as large (400 nm diameter), tightly sealed vesicles of lipid-to-protein molar ratio (ϕ) of around 300. Reconstituted AChR membranes at $\phi = 300$, however, are predominantly bilayer sheets and show no measurable trapped volume (J.Earnest, unpublished observations). The molecular basis for this discrepancy is unclear, but may involve cytoskeletal stabilizing elements in native membranes which are missing in the reconstituted systems. Most studies on reconstituted membrane proteins whose functions involve ion trapping have shown $\phi = 8000 - 12\,000$ to be the most successful range, although it is unclear whether this excess of lipid is necessary for vesicle impermeability or to maximize vesicle size.

3.3.2 *Type of detergent*

The detergent present in the mixed protein−lipid−detergent micelle can affect vesicle size. Micelles containing cholate, egg PC and glycophorin, for example, form vesicles of 25 nm diameter with an internal volume of 2×10^{-18} ml/vesicle, while otherwise identical micelles containing octyl glucoside instead of cholate, form vesicles of 230 nm diameter with an internal volume of 5×10^{-15} ml/vesicle (72). These studies also demonstrated that the initial detergent-to-lipid ratio affects vesicle morphology; larger more uniform vesicles were formed using an initial octyl glucoside-to-egg PC molar ratio of 15:1 than when the initial molar ratio was 5:1. On the other hand, it has been shown that the initial detergent-to-lipid ratio can critically affect the function of a reconstituted protein. For example, it appears that when the detergent-to-lipid ratio is sufficiently high such that displacement of annular lipids occurs many membrane proteins become irreversibly inactivated (33−35; see also Section 2.3.2).

Other aspects of reconstitution can be affected by the type of detergent present in the mixed micelle. Reconstitution by cholate dialysis, using dioleoylphosphatidylcholine (DOPC) and AChR at $\phi = 200$, results in lipid bilayer sheets (200 nm diameter) with densely packed intercalated receptor. Substitution of octyl glucoside for cholate results in aggregated precipitates of AChR, and protein-free liposomes (J.Earnest, unpublished observations).

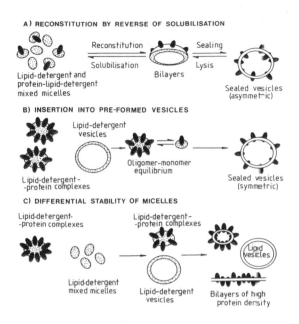

Figure 4. Schemes for reconstitution following removal of detergent from solubilized membrane components (adapted from ref. 59).

3.3.3 *Type of lipid*

The function of many membrane proteins depends upon the composition of the surrounding lipid (32). Reconstitution of the protein in question, in a range of defined lipid bilayers, allows examination of the specific effects of lipids on protein function. However, careful characterization of membrane morphology must accompany such comparative studies, as subtle changes in lipid composition can greatly change the structure of the reconstituted membrane. Early studies on reconstitution of AChR for example showed that AChR was readily incorporated into asolectin vesicles after cholate dialysis, but when egg PC was used the AChR formed chains or rings and was excluded from the liposomes (73). Even more dramatically, AChR incorporates into purified native lipid vesicles, also after cholate dialysis, but precipitates and is not incorporated into vesicles formed from native lipids which have been depleted in cholesterol (71). Substitution of dielaidoylphosphatidylcholine (DEPC, C18:1 PC *trans* double bond) for DOPC (C18:1 PC *cis* double bond) changes the reconstituted AChR membranes from membrane bilayer sheets with evenly dispersed AChR to umbrella-shaped bilayers with AChR present only along the margin. Unless the reconstituted membranes under investigation are examined microscopically, what is in fact an effect of a lipid on membrane structure may be erroneously interpreted as a specific effect of a lipid on protein function.

Reconstitution of rhodopsin by octyl glucoside dialysis was found to depend on lipid composition as well as lipid-to-protein ratio (55). Delipidated rhodopsin (26 μM) in octyl glucoside (200 mM) was equilibrated with phospholipids at $\phi = 300$. Test lipids included egg PC, DOPC, dimyristoylphosphatidylcholine (DMPC), palmitoyloleoylphosphatidylcholine (POPC) and disk lipids. In almost every case, the final lipid-to-protein molar ratio after removal of detergent was 30–50, even though the initial ratio was 300. Only with DMPC did the final value reflect the initial conditions. DMPC differs from the other lipids in that its acyl chains are saturated and short. The authors suggest that except in the case of DMPC, the octyl glucoside−phospholipid micelles have lower stability than the octyl glucoside−phospholipid−rhodopsin micelles. The excess lipids would then form protein-free vesicles early in dialysis. The micelles whch contain protein do not vary much in lipid-to-protein ratio and form vesicles at a later stage of dialysis (see *Figure 4c*).

3.3.4 *Method of detergent removal*

The speed of detergent removal can be important in the reconstitution of function (74,75), perhaps due to the sensitivity of some membrane proteins to prolonged exposure to detergents. Speed of detergent removal can also affect membrane structure. Reconstitution of AChR into asolectin, DOPC and DEPC ($\phi = 10\ 000$) results in strikingly different membrane structures when cholate is removed by dialysing for 4 days. In this case asolectin forms unilamellar vesicles with intercalated, dispersed AChR; DOPC forms protein-free vesicles with aggregated AChR; and DEPC forms large 'umbrellas' of lipid with AChR adhered to the margin. In contrast, when the cholate is removed quickly, by either cholate dilution (76) or gel filtration (73) all three types of lipid form unilamellar vesicles with AChR incorporated into the bilayer. Although the primary motivation for systematic study of variables in reconstitution is to successfully reconstitute *function* of the membrane protein, protein function cannot be tested if the overall morphology of the reconstituted membrane is inappropriate.

3.3.5 *Effects of freeze−thawing*

Lipid vesicles have been shown to spontaneously fuse below their phase transition temperature (66). This characteristic of lipid bilayers has been exploited in the reconstitution of AChRs to form larger vesicles which appear to show a greater flux response than unfrozen vesicles. Anholt *et al.* (39) demonstrated that vesicle size was increased upon freeze−thawing only when the asolectin−AChR vesicles were supplemented with cholesterol. Before freeze−thaw 3% of the vesicles had a diameter greater than 100 nm, while after freeze−thaw 24% were greater than 100 nm in diameter. Another study of several lipid compositions, however, found that the internal volume increase observed in frozen vesicles was not correlated with cholesterol content (77). A common problem encountered in reconstitution is vesicle heterogeneity. Reconstitution of the AChR into several lipid mixtures by cholate dialysis resulted in two distinct vesicle populations, as separated on a sucrose density gradient, one of which contained only lipid (71). After several freeze−thaw cycles the two peaks corresponding to the two vesicle populations merged into a single peak which migrated at intermediate sucrose density. This type of preparation is typically used for assay of ion flux, as its ability to trap radioactive cations is increased, although it is unclear whether increased trapped volume, 'sealing' of membrane vesicle or other factors are responsible for the apparent increase in activity. Negative-stained electron micrographs of freeze−thawed AChR membranes show a surprising degree of heterogeneity and a marked increase in multilamellar vesicles (10). There is also evidence from both negative-stain and freeze-fracture EM that freeze−thawing of certain membranes may cause some proteins to be trapped in the intervesicular space, and not intercalated into the bilayer (10). If freezing is desirable for long-term vesicle storage, sucrose can be added to prevent the formation of multilamellar vesicles upon thawing (78).

3.4 **Characterization of the reconstituted vesicles**

3.4.1 *Measurement of internal volume*

A simple method for estimating the volume of aqueous medium trapped within a reconstituted vesicle, or liposome, is to prepare the vesicles in the presence of 0.1 M sodium or potassium chromate. External chromate is removed by gel filtration; trapped chromate is released by solubilization of an aliquot in 10 mM Triton X-100, then assayed by measuring absorbance at 380 nm. To derive a value for the internal volume expressed in litres per mole of lipid, the chromate concentration of the aliquot is standardized to the total chromate and lipid concentrations in the sample (62).

Other markers of internal volume include 0.1 mM 6-carboxyfluorescein, which is measured by its fluorescence (62) or radioactive sugars such as [^{14}C]lactose (79) or [^{3}H]dextran (80), which are measured by liquid scintillation counting. Care must be taken to ensure that an internal marker does not alter either the function of the reconstituted protein or the reconstitution procedure itself.

Ion channels like the AChR and the sodium channel facilitate internal volume measurement. Vesicles are incubated overnight with radioactive permeant ions, external ions are removed on ion-exchange columns (37,81) and the radioactivity of ions trapped within the vesicles is measured.

3.4.2 *Measurement of vesicle permeability*

It is especially important to assay bilayer permeability when protein orientation is to be characterized, or when vectorial translocation is to be measured. A thorough study of the effects of glycophorin on the permeability of DOPC vesicles was made by Van der Steen *et al.* (80) in which trapped [^{3}H]dextran and K$^+$ were monitored simultaneously.

(i)   Prepare vesicles in the presence of 0.2% (w/v) [^{3}H]dextran.

(ii)   Remove external dextran by chromatography on Sepharose 4B-CL.

(iii) Pellet vesicles from the void volume at 100 000 g for 1 h, then apply them to a Sephadex G-50 column equilibrated with choline chloride, to remove external K$^+$.

(iv)   Follow the efflux of K$^+$ using a potassium electrode, and assay the trapped volume by liquid scintillation counting.

Assuming first order kinetics for the efflux of K$^+$, the time course for the ratio of external K$^+$ to the internal volume allows the half-time for potassium efflux to be determined.

3.4.3 *Measurement of protein orientation*

The most common method of determining the orientation of a membrane protein is to measure some aspect of its function in the presence and absence of a solubilizing detergent. Several membrane proteins bind radioactive ligands or toxins with a high affinity and specificity to sites localized on the outside of the vesicles. Vesicles with bound ligand/toxin may then be separated from unbound toxin by column chroma-tography or filtration through ion-exchange filters (e.g. 81,82). Often the specificity and the insensitivity of the binding reaction at equilibrium to the presence of detergent enable the fraction of outward-facing proteins to be measured by binding assays before and after vesicle solubilization.

Enzymic function can be measured with and without detergents but the detergent must be carefully chosen. For example, for cytochrome c oxidase only Tween 80 [3% (v/v)] did not affect enzyme activity (83).

Many membrane proteins are glycosylated on the extracellular domain. The accessi-bility of sialic acid to hydrolysis by neuraminidase, with and without detergent, can be used to assay protein orientation (78,80,84). Protein-containing vesicles are treated with neuraminidase, with or without Triton X-100, and the released sialic acid assayed. The disadvantage of this technique is its lack of specificity: if the reconstituted protein is slightly impure, contamination by other glycosylated proteins will interfere with the results.

Monoclonal antibodies, specific for external or internal domains of the membrane protein, can be used to measure sidedness. Anholt et al. (85) used monoclonal anti-bodies to immunoprecipitate AChRs which were previously labelled with radioactive toxin. The extent of binding of monoclonal antibodies to AChRs, with or without 0.3% (w/v) saponin (a permeabilizing agent) was used to characterize sidedness (85).

One disadvantage of all the above methods is that if the reconstituted membrane is permeable to the vectorial label in the *absence* of detergent (or if the membrane is not a vesicle at all, but is in some other conformation), the assay will exaggerate the ran-dom orientation. Also, unintercalated protein trapped within a sealed vesicle (80) will lead to erroneous interpretations of sidedness assays. These pitfalls can be avoided by careful characterization of internal volumes and permeability, in conjunction with EM.

There are at least two methods of determining protein sidedness using only EM, although these assays may be less reliable than the biochemical assays described above. Firstly, fixation of reconstituted vesicles using tannic acid−osmium, followed by em-bedding and thin-sectioning was developed by Saito et al. (87,88). In each of these studies only the native extracellular protein domain stained heavily, and the appearance of a dense layer on the inside surface of a cross-sectioned vesicle was interpreted as non-vectorial protein orientation. To use this technique, permeability properties of the bilayer during staining should be carefully characterized, and asymmetric staining of the protein demonstrated.

Secondly, freeze-fracture EM has been used to determine sidedness of reconstituted proteins. It was first demonstrated by Chen and Hubbell (89) that rhodopsin always partitioned to the extracellular face of the fractured native bilayer, while rhodopsin partitioned equally to both faces when reconstituted into PCs. These workers suggested the vectorial orientation seen *in vivo* is eliminated during reconstitution. The equal distribution of intramembraneous particles on concave and convex surfaces was used to show that rhodopsin in egg PC was randomly oriented (55). Caution should be used, however, before this technique is applied to other proteins, or even to rhodopsin under varying conditions of reconstitution. Many of the vectorially oriented integral mem-brane proteins of the erythrocyte appear to partition fairly evenly between fracture faces. Also, changes in bilayer composition, glycosylation or protein−protein interactions may alter the normal distribution between fracture faces (K.Fisher, personal communication).

Finally, it should be noted that for many functional studies of reconstituted protein it is not necessary to have all the proteins outward facing. The ion channel of the AChR, for example, is opened by the binding of agonists to the extracellular domain of the protein, so ion flux is seen only through properly oriented channels (47).

3.4.4 *Factors which affect orientation*

Although membrane proteins *in vivo* almost always have a uniform vectorial orientation, reconstitution frequently results in randomization. Several variables in the reconstitution procedure have been shown to influence orientation, but few consistent patterns emerge from these observations.

The M13 coat protein was shown to insert into DMPC vesicles with the N terminus consistently on the external surface. However, if reconstitution was performed at the DMPC transition temperature, both C termini and N termini were present on the external surface (90).

Orientation can be affected by lipid−protein ratio: at $\phi > 88$, sarcoplasmic reticulum Ca^{2+}-ATPase reconstituted into sarcoplasmic reticulum lipids, showed uniform vectorial orientation while at $\phi < 88$ the protein oriented randomly (88).

Glycophorin reconstituted into PC vesicles has been shown to orient $50-75\%$ outside-out (72,80). However, at low lipid-to-protein ratios ($\phi = 300$) about 30% of the glycophorin was entrapped within the vesicle and was not in the bilayer at all (80). Internally trapped protein (seen also with reconstituted AChR after freeze−thawing, J.Earnest, unpublished data) would be incorrectly perceived as non-vectorially inserted protein. This result argues for using several methods to determine protein sidedness, and emphasizes the value of EM in reconstitution.

The aggregation state of the protein and lipids during recombination can also affect orientation. The spike protein from Semliki Forest virus membranes was shown to be either monomeric or oligomeric, depending on the octyl glucoside concentration (59). When reconstitution into egg PC was controlled so that the oligomeric form was present, the protein was oriented 95% outward, while reconstitution of the monomer resulted in only 70% outward orientation. A hypothesis for the molecular basis of these observations is pictured in *Figure 4a* and *b*. These experiments, along with those of Carroll and Racker (83), suggest that if vectorial orientation is required, reconstitution methods which involve incorporation of proteins into pre-formed vesicles may offer the best prospects.

At lipid-to-protein ratios in which sealed vesicles are formed (around $\phi = 1000$), the reconstituted AChR has been shown to orient 75% (85) to 95% (77) outside-out under a variety of conditions. In contrast, the sodium channel, a protein of similar size as the AChR, appears to orient randomly in all studies thus far (91). Bacteriorhodopsin is another integral membrane protein which seems to orient randomly in a reconstituted membrane (63,89). Protein shape may influence protein orientation. For instance, the highly asymmetric funnel structure of the AChR (92) compared with the largely hydrophobic barrel shape of bacteriorhodopsin (68) may explain the greater tendency of the receptor to be outward facing.

Table 8. Reconstitution of AChR in lipid vesicles.

1.	Suspend 600 mg of asolectin[a] in 10 ml of DB-1. Sonicate at room temperature in a bath sonicator to give a reasonably homogeneous suspension (10−30 min). Add solid sodium cholate (0.4 g) to give a 4% cholate solution and sonicate again until solution is almost clear (10 min).
2.	Adjust the purified AChR solution to a protein concentration of 1.5 mg/ml by dilution with DB-3 (*Table 6*) if necessary. (Depending upon the number of fractions pooled, protein concentration may be lower but should not be below 1 mg/ml.)
3.	To two volumes of purified AChR add one volume of the asolectin−cholate mixture prepared in step 1. The final concentrations of the components should be (approximately) 2% cholate, 20 mg/ml lipid and 1 mg/ml AChR corresponding to a lipid-to-protein mole ratio of 6700 to 1.
4.	Dialyse the suspension against 4 l of DB-1 for at least 48 h with three changes of buffer using Spectra-Pore dialysis tubing that has been pre-soaked in DB-1 + 1% cholate.
5.	The dialysed membranes can be stored for at least 2 weeks at 4°C without any noticeable loss of functional integrity. They may also be stored frozen in liquid nitrogen. However, the freeze−thaw process will alter the size and degree of heterogeneity of the vesicles.

[a]Associated Concentrates, Woodside, NY.

3.5 Reconstitution of the acetylcholine receptor

Table 8 provides a step-by-step procedure that has been successfully used to prepare reconstituted membranes containing functional *Torpedo* AChR. The procedure uses dialysis to remove the detergent and gives vesicles suitable for ion flux measurements. The same procedure should work for most membrane proteins.

4. FUNCTIONAL ANALYSIS OF RECONSTITUTED MEMBRANES

One major goal of reconstitution studies is to provide an appropriate membrane environment in which to measure and characterize the functional properties of specific membrane proteins. Many of the strategies described in previous sections for effective solubilization and reconstitution are designed to preserve the functional integrity of the membrane proteins and to maximize the opportunities for convenient functional assays. In this section, some of the general approaches for assaying membrane protein function of both solubilized and reconstituted proteins are outlined.

The major types of assays commonly used are binding assays, enzyme assays and transport/permeability assays. In addition, some reconstituted systems permit demonstration of functional coupling between different proteins.

4.1 Binding and enzyme assays

These assays can generally be used at all stages of purification and can be applied to the native membrane protein, the solubilized protein and the reconstituted protein. Many of the theoretical and practical aspects of enzyme and binding assays are similar for both soluble and membrane proteins and will not be described in detail. For example, it is necessary to ensure that the binding is saturable, specific, stereospecific, reversible and inhibitable. There are several unique problems associated with membrane proteins, however, that need to be considered.

In the solubilized state, the detergent may affect the effective concentration of substrates or ligands available to the protein. For example, if the substrate is amphiphilic it may partition into detergent micelles and not interact appropriately with the active site. The detergent may also affect the conformational state of the protein thus altering the properties of the enzyme. It is also possible that the detergent will expose active sites or allosteric sites that are masked in a membrane environment. In practical terms, the complications require that some effort be made to ensure that the assay conditions accurately measure the full activity of the protein under investigation. Both the K_m and the V_{max} should be measured in the membrane and in the solubilized state for enzymes. For proteins in which only a binding process is being measured, the K_d and the total number of sites must be measured, usually by Scatchard analysis. Differences could reflect an artifactual problem with substrate availability or could reflect interesting differences in the properties of the protein.

Membrane-associated proteins offer at least one favourable advantage when designing binding assays. A key problem in most binding assays, unlike most enzyme assays, is the need to separate the bound from the free ligand by some physical technique. With membranes, centrifugation or filtration using defined-pore or glass-fibre filters is usually successful. Although the process of separation can perturb the equilibrium established between ligand and binding site, the separations are sufficiently rapid that good data can be obtained. The use of microcentrifuges, an air-driven ultracentrifuge or filtration manifolds make it possible to process large numbers of samples on an analytical scale. In some cases, filtration techniques can also be applied to solubilized proteins. A major problem in filtration is that the washing techniques necessary to reduce background levels of unbound ligand on the filters will often displace bound ligands unless they are bound with high affinity (i.e. $K_d = 10^{-7}$ M or lower). Several examples of binding and enzyme assays that have been useful in different systems will be described below to provide a practical introduction to the available strategies (see also ref. 93). *Table 9* provides a general summary of the separation techniques that are typically used for membrane-bound and solubilized proteins. An unpublished method using oil-layers is detailed in Appendix III.

Table 9. Summary of methods for ligand binding assays of solubilized proteins.

Method	Comments
Gel filtration	Slow; not suitable for multiple samples; detergent may interfere with separation.
Filtration	Fast; multiple samples; not suitable if dissociation is fast ($t < 15$ sec); wide selection of filter types.
Ultracentrifugation	Slow; limited number of samples; trapping of unbound ligands (but use of oil layer can minimize trapping — see Appendix IV).
Equilibrium dialysis	Very slow; true equilibrium measurement.
Flow dialysis	Fast; can vary ligand concentration over wide range on a single sample; non-specific binding often a problem.
Polyethyleneglycol precipitation	Fast; can use filtration or centrifugation.
Immunological methods	Requires specific antibodies.

For additional details and references, see ref. 119.

4.1.1 *Acetylcholine receptor*

The AChR specifically binds α-bungarotoxin, a positively charged snake venom protein, with high affinity ($K_d = 10^{-11}$ M) and radioiodinated derivatives of the toxin are available. If excess labelled toxin is allowed to interact with the AChR in detergent solution for $30-60$ min at room temperature (typically 50 μl containing 0.2% Triton X-100, 30 nM toxin and 5 nM AChR), the AChR−toxin complex can easily be separated from unbound toxin by passing the mixture through a DEAE-cellulose filter. The positively charged filter binds the negatively-charged complex and allows unbound toxin to be washed through. The very tight binding of the toxin to the AChR permits exhaustive washing of the filters (47,81,94). Since the assay involves ion-exchange techniques, it is essential to use a non-ionic detergent (such as Triton X-100). Membrane samples can be assayed by first incubating the membranes with appropriate ligands and then diluting them in detergent for the filtration part of the assay. A step-by-step outline of the procedure is provided in *Table 10*.

The apparent K_ds of small activating ligands can be determined using this approach by measuring their ability to compete for toxin-binding sites. The assay has also been used to detect a time-dependent change in the affinity of the AChR for activating ligands, such as carbamylcholine. Pre-incubation of AChR with carbamylcholine shifts the AChR into a very high affinity binding state (95), which can be distinguished from the resting binding state by measuring the *rate* of toxin binding. The low-to-high affinity shift is characteristic of functional AChR in a membrane environment and the shift is an essential prerequisite for ion channel activity. Thus, the binding assay is an effective method for evaluating the success of a reconstitution process. Interestingly, the shift is not observed for solubilized AChR and the assays must be carried out in the absence of detergent. Although centrifugation could be used to separate the membranes from the unbound toxin, the filtration method is simpler and better suited to large numbers of samples. The results obtained with the competition binding assay agree reasonably well with results obtained by direct measurements of radiolabelled acetylcholine and carbamylcholine binding (42).

In the case of the direct measurements of small ligand binding using tritiated acetylcholine and carbamylcholine, it was necessary to carefully control the filtration conditions to avoid variability due to rapid dissociation of the reversibly bound ligand. Glass-fibre filters (Whatman GF/F) were used and the time of filtration under vacuum was precisely controlled (42). Trapped isotope and non-specific binding was corrected for by pre-incubating samples with excess unlabelled toxin. A notable problem in the binding assays was the variability in specific activities among different batches of labelled isotope. The actual values were directly measured using electrophoresis to purify the acetylcholine and isotope dilution with unlabelled ligand to measure specific activity. In all cases, the measured values differed significantly from the values reported by the manufacturer (42).

Another approach that has been successfully used in a few cases involves labelling of the protein with spin label or fluorescent probes. If the probes are sensitive to the binding of other ligands, the probe can provide a continuous and dynamic measure of binding processes. Changeux and co-workers (96) have used a fluorescent analogue of acetylcholine to measure the binding events in both native and reconstituted AChR

Table 10. Toxin binding assay for AChR.

Principle

Incubate detergent-solubilized AChR with excess radiolabelled α-bungarotoxin at 4°C for 30−60 min in order to occupy all available binding sites. Separate toxin−AChR complexes from unbound toxin by filtering the mixture through DEAE filters. The toxin is positively charged and does not bind to the filters, whereas the complex is negatively charged and binds to the filters (at low ionic strength).

Buffers

NMT-100	100 mM NaCl	NMT-10	10 mM NaCl
	10 mM MOPS		10 mM MOPS
	0.2% Triton X-100		0.2% Triton X-100
	pH 7.4		pH 7.4

Toxins

Radiolabelled toxin — [^{125}I]α-bungarotoxin 2−4 μM in 5 mM sodium phosphate, pH 6.5 with 1 mg/ml BSA, 1−2 × 10^5 d.p.m./pmol. Available from NEN and Amersham.
Unlabelled toxin — 20 μM α-bungarotoxin in NMT-100.

Step-by-step procedure

1. Dilute AChR into NMT-100 so that 50 μl contains 2−3 pmol of binding sites (40−60 nM). Four 50 μl aliquots are needed for each sample assay.
2. Place 50 μl aliquots into each of four plastic test tubes (10−15 ml total volume per tube). Two of the tubes will be control tubes that will be pre-treated with unlabelled toxin in order to correct for non-specific binding and the other two will be used to determine total binding.
3. To the two control tubes add 15 μl of unlabelled toxin. To the other two tubes add 15 μl of NMT-100. Allow all tubes to incubate at room temperature for 20 min.
   For each complete set of assays, prepare two background tubes to monitor non-specific toxin binding to the filters in the absence of AChR. Place 65 μl of NMT-100 into each of these tubes.
4. Dilute the radiolabelled toxin into NMT-100 to a concentration of 130 nM. Each sample requires 4 × 50 μl. An additional 200 μl is required for an internal standard and for the background sample.
5. Add 50 μl of the diluted radiolabelled toxin to each tube. Incubate for 45 min at room temperature.
6. Set up a vacuum filtration unit with *two* Whatman DE-81 filters per well. Pre-wash filters with 4 ml of NMT-10 under low vacuum. (A 10-well Hoefer apparatus using 24 mm filters works well, but any filtration unit should work.)
7. Dilute each sample with 5 ml of NMT-10 and immediately apply to the filter and allow to flow through under mild suction. Fill the tube with NMT-10 and wash the filter. Repeat the wash step. Remove the nearly dry filters using a pair of tweezers and place them into disposable plastic test tubes.
8. Count the dried filters in a gamma counter optimized for ^{125}I detection.
9. For an internal standard, place two filters in a counting tube and spot 50 μl of the radiolabelled toxin directly on the filters.

The internal standard permits calculation of the specific activity of the toxin in c.p.m./pmol toxin. After subtracting the non-specific counts from the total counts, the number of pmol of bound toxin in each sample can be determined.
The specificity of binding is so high in this assay that the non-specific binding is essentially zero even for very crude preparations. The control tubes and the background tubes give essentially the same values. In practice, it is not necessary to run a pair of controls for each sample. In fact, only one set of controls is usually run for an entire set of samples assayed at the same time.

membranes with very good time resolution. They have detected the low-to-high affinity shift and have some evidence for a very low affinity state that may represent the resting AChR. Raftery and co-workers (97) have used a different probe covalently attached to the AChR as an indirect monitor of the binding of acetylcholine. They have evidence for an additional class of binding sites distinct from the two high affinity sites.

4.1.2 *Beta-adrenergic receptors*

Regulation of adenylate cyclase activity by the β-adrenergic receptor involves the interaction of three different proteins in the membrane: the receptor itself, the G_s protein and the adenylate cyclase. The G_s protein is a multisubunit protein that can bind GTP (and related analogues) and stimulates adenylate cyclase activity when complexed with GTP. Non-hydrolysable analogues of GTP lead to persistent activation of the adenylate cyclase. The β-adrenergic receptor (and other hormone receptors) regulate the activation of the G_s protein and thus indirectly regulate adenylate cyclase (98,99). Solubilization, purification and reconstitution of the proteins have played a key role in defining and characterizing the essential components of such a complex system (93). At least five functional assays can be associated with the overall process:

(i)   the binding of ligands such as [^{3}H]dihydroalprenolol (an agonist) or [^{125}I]iodo-hydroxybenzylpindolol (an antagonist) to the β-adrenergic receptor;
(ii)   the binding of GTP or non-hydrolysable GTP analogues to the G_s protein;
(iii)   the activation of the G_s protein, as measured by its ability to stimulate adenylate cyclase activity;
(iv)   GTPase enzyme activity associated with G_s; and
(v)   the adenylate cyclase activity itself.

Several different groups have examined various aspects of the regulatory steps in both crude and purified extracts (100–104). Ross and co-workers (100,101) have concentrated on the interactions between the β-adrenergic receptor from turkey erythrocytes and the G_s protein from rabbit liver that lead to activation of the G_s protein. Separate assays using lymphoma cell plasma membranes that lack the G_s protein have established that exogenously-added G_s protein in the activated state (e.g. with GTP bound) will stimulate the adenylate cyclase (105). The strategy was to purify the receptor and the G_s protein separately in detergent solutions and then to co-reconstitute them into the same lipid vesicles. G-50 Sephadex chromatography was used to remove the detergents. Since the receptor and the G_s protein were purified separately using different protocols and additional lipids were also included in the solutions, the final solubilized mixture was very complex and contained four different detergents: digitonin, Lubrol, deoxycholate and cholate.

The detailed analyses point out some of the inherent problems always associated with membrane proteins. For example, Lubrol 12A9, the detergent used to solubilize the G_s protein, inhibits the GTPase activity but not the binding of GTP or its analogues (101). Previous studies (carried out in Lubrol) had mistakenly concluded that GTPase activity was not associated with the G_s protein. The GTPase activity is recovered in the reconstituted system after removal of the detergent. Detergent solubilization also eliminates antagonist binding (but not that of [^{3}H]dihydroalprenolol) to the receptor. Full binding activity is recovered in the reconstituted system. For the small ligand binding assays, glass-fibre filters (Whatman GF/F) were used to rapidly separate protein–ligand complexes from free ligand (106).

4.1.3 *Calcium channels*

In purifying the voltage-sensitive calcium channel from skeletal muscle, Curtis and

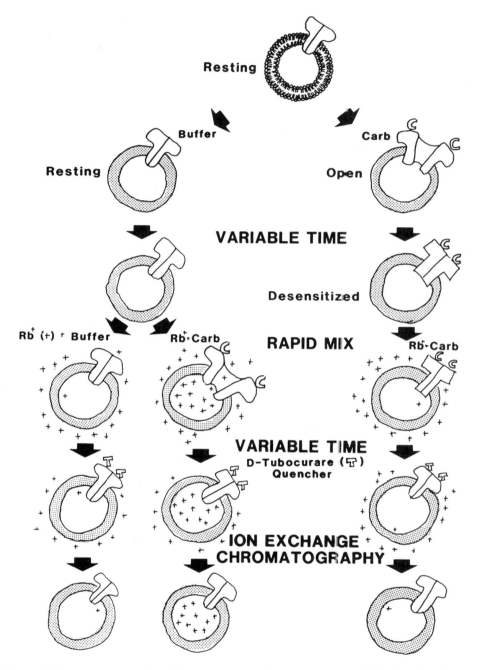

Figure 5. Schematic representation of the reaction protocols used to measure rapid kinetics of ion flux in reconstituted vesicles containing the AChR. The left side of the figure represents activation of AChR ion channels by agonists and a detailed protocol is provided in *Table 11*. The right side includes a pre-incubation step which permits analysis of the time-dependent desensitization of ion channel activation by agonists such as carbamylcholine. The influx assays can be carried out in the second to minute time range using manual mixing techniques. Measurements of *initial rates* require the use of rapid mixing devices for the pre-incubation, influx and quenching steps. A schematic illustration of the rapid quench-flow system is provided in *Figure 6*.

Catterall (107) were confronted by two significant problems in addition to the relatively low concentration of the channels in the tissue. First, since the triggering event for channel function is a voltage change, the best assay for function requires an intact membrane across which a stable membrane potential can be established. As discussed in the next section, it is difficult to design a purification procedure when such stringent reconstitution conditions are required for routine assays. Second, the binding of specific antagonists, such as [³H]nitrendipine, cannot be assayed after solubilization of the protein in detergent (digitonin) due to very high non-specific binding of the ligand to the detergent micelles. The binding problems were partially overcome by first labelling purified *membranes* with [³H]nitrendipine, washing away all non-specifically bound label, and then solubilizing and purifying the entire complex. The recovery of labelled protein was monitored at all steps of purification and the specific activities corrected for the slow dissociation of the ligand from the complex.

4.2 **Transport/permeability assays**

The major value of reconstitution techniques arises from the possibility of measuring membrane-dependent transport or permeability functions under controlled conditions. Where such transport utilizes ion gradients or membrane potentials, the presence of sealed membranes is required in order to create the compartments necessary for such vectorial processes. One of the most elegant and important reconstitution experiments was carried out by Racker and Stoeckenius (108). They re-incorporated bacteriorhodopsin, a light-driven proton pump isolated from the purple membranes of *Halobacterium halobium*, and the ATPase complex from mitochondria into a single lipid vesicle. They observed that light could drive the synthesis of ATP from ADP in the reconstituted vesicles. Since the only functional link between the two proteins was the generation of a proton gradient and its associated membrane potential, the experiment provided good evidence for the role of the protonmotive force in driving ATP synthesis.

The ability to measure the uptake or release of ions or molecules from sealed, reconstituted vesicles provides a major tool for characterizing the minimum requirements for the function of an ion channel or a transport system.

4.2.1 *Acetylcholine receptor*

In the case of the AChR, the binding of activating ligands such as acetylcholine is known to trigger the opening of a cation-specific ion channel. Prolonged exposure of the AChR to the activating ligand leads to the reversible inactivation of the channel, a process known as desensitization. Many of the key features of the activation and inactivation pathways have been inferred from electrophysiological studies on intact receptors and from ion flux studies on isolated membranes (for reviews, see 40,57). Reconstituted membranes containing purified AChR offer a way of determining if the isolated protein retains all the functional aspects attributed to the receptor. They also offer the possibility of obtaining much better data that will lead to a quantitative mechanism for AChR ion permeability control. *Figures 5, 6* and *7* provide schematic outlines of two ion permeability protocols currently used to monitor the functional properties of AChR-containing membranes and *Table 11* provides a step-by-step outline of a simple influx assay routinely used to monitor AChR function in reconstituted membranes.

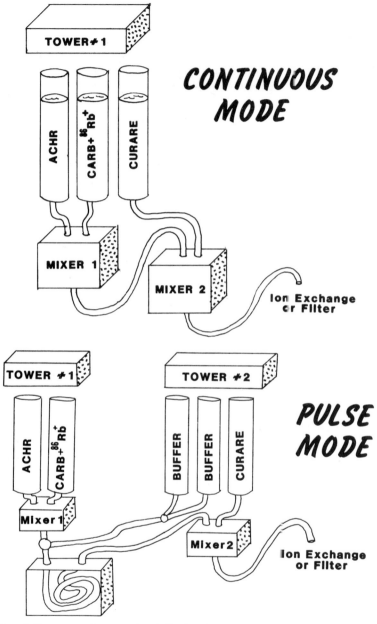

Figure 6. Quench-flow system for measuring AChR ion channel activation. Two major configurations are used. In the continuous mode, membranes, agonist and quenching solution are displaced from separate syringes by a single pneumatically-driven piston unit (tower). A 100 msec stroke time is used to minimize turbulent flow. The time of reaction between AChR and agonist is determined by the length of the tube between the two mixers. Quenching occurs in the second mixer prior to manual separation of membranes from external ions. Reaction times of $10-100$ msec are conveniently obtained by varying the length of the tubing. For longer reaction times, the pulse mode is used. Membranes, tracer ion and agonist are mixed and permitted to stay in a holding tube. At any later time a second tower is fired and buffer is used to displace the reaction mixture from the holding tube and allow it to interact with the quencher. For more details on the design of the apparatus and the analysis of data see ref. 40. (Figure reprinted from ref. 40.)

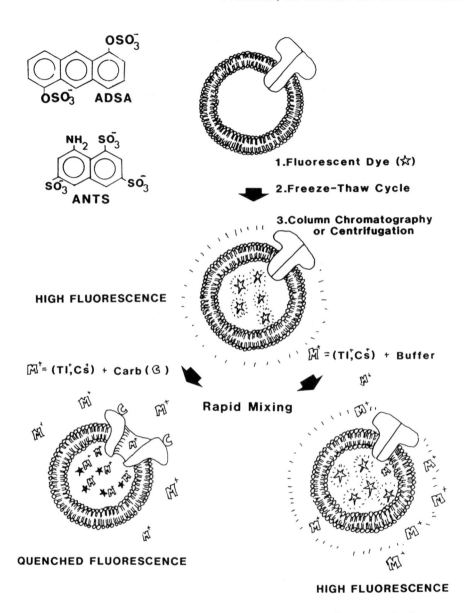

Figure 7. Schematic representation of the fluorescence quenching – stopped flow system for the measurement of rapid kinetics of ion flux in reconstituted vesicles containing the AChR. The mechanical design of the system is similar to that used for quench-flow. The major difference is that the reaction mixture is continuously monitored for fluorescence changes within a cuvette after mixing. Thus a single sample gives rise to an entire time dependence curve. (Figure reproduced from ref. 40.)

The first technique involves the agonist-stimulated uptake of a radioactive tracer ion. The binding of agonist to the receptor opens the ion channel allowing ions to move across the membrane and approach an equilibrium distribution. External ions are separated from internal ions using a cation-exchange column to trap the external ions. The membranes could also be separated from the external solution by filtration or centri-

Table 11. Test assays for AChR ion flux activity using ion-exchange columns (see *Figure 5*).

1. Prepare Dowex columns.

   a. Wash 500 g of Dowex 50W-X8 (50 − 100 mesh) cation-exchange resin (Sigma Stock No. 50X8-100) with distilled water until the supernatant is clear. To 1 vol. of resin plus 1 vol. of water add 220 g of Tris base and incubate with gentle stirring at room temperature for 1 h. Check the pH (it should be 10.6). Wash the resin extensively with distilled water (at least 25 batch washes) until pH 7. Wash several times with doubly distilled water and store at 4°C.

   b. Prepare 2 ml disposable columns (e.g. Isolab QS-Y) by adding resin to a level just below the reservoir area and washing with 170 mM sucrose containing 3.3 mg/ml BSA (to block non-specific binding sites). Drain the columns so that the buffer level is just above the resin and store capped at 4°C until needed.

2. Prepare solutions for influx assays. The solutions described below are suitable for approximately 50 − 60 assays, corresponding to about 10 different samples. Volumes can be scaled up or down.

   a. IFA-1: mix together 50 μl of stock Rb and 250 μl of DB-1 (*Table 5*).
   b. Carb[a]: dissolve 18.2 mg of carbamylcholine in 1 ml of water 0.1 M Carb). Add 100 μl of 0.1 M Carb to 700 μl of DB-1 (12.5 mM Carb).
   c. IFA-1/Carb: mix 125 μl of IFA-1 with 125 μl of 12.5 mM Carb.
   d. IFA-1/Buff: mix 125 μl of IFA-1 with 125 μl of DB-1.

3. Assay samples.

   a. To 50 μl of membrane sample on ice, add 10 μl of IFA-1/Carb or IFA-1/Buff.
   b. Incubate for 30 sec.
   c. Remove 50 μl and apply immediately to top of Dowex column.
   d. After the solution has entered the resin immediately add 3 ml of 175 mM sucrose. Allow the eluate to collect directly into a scintillation vial.
   e. Count vials in the liquid scintillation counter in a 100 − 900 window without fluor.

   Each sample should be assayed at least in duplicate with and without agonist. Each column may be re-used two or three times without additional treatment.

   f. Standards: add 10 μl of IFA-1/Buff to 1 ml of water and count 10 μl aliquots.
   g. Background: add 50 μl of DB-1 instead of membranes in one set of assays.
   h. Internal volume: Incubate 200 μl of membranes with 40 of IFA-1/Carb or IFA-1/Buff for 36 − 48 h. Then pass 50 μl aliquots through a Dowex column as described above.

[a]Other agonists can be substituted for carbamylcholine.

fugation techniques, but the ion-exchange method is rapid and avoids problems with loss of the small reconstituted membranes through the pores of the filters. In the scheme shown in *Figure 5*, a quencher (curare) is added to stop the influx reaction after a specified time. The quenching step becomes essential when the assay is carried out on a very fast time scale using rapid mixing techniques (109) as described in *Figure 6*. The addition of quencher is not necessary when the reaction is allowed to proceed nearly to equilibrium (10 − 30 sec) as illustrated in *Table 11*. By pre-treating the receptor with an agonist before adding tracer cations, the time-dependent loss of channel activity can be monitored.

A related assay (*Figure 7*) works on a very different principle. If a fluorescent dye is trapped inside sealed vesicles, its fluorescence can be quenched by the influx of heavy monovalent cations into the vesicles. Raftery and colleagues (110) have developed a stopped-flow method for monitoring the influx of thallium ions into vesicles loaded with fluorescent dye using freeze − thaw techniques. The time course of fluorescence quenching can be related to the rate of influx of the monovalent cations. Both this assay and one developed by Hess and colleagues (111) take advantage of the fact that the AChR allows nearly all monovalent cations to pass easily through the ion channel.

4.2.2 *Voltage-gated channels*

A more challenging application of ion flux assays arises in the case of voltage-gated channels, such as those for sodium and calcium. A fully reconstituted system requires that a transmembrane potential be established and that methods for effecting controlled changes in the potential be developed. No one has yet achieved a reconstitution that fully mimics the native protein. However, there are toxins, such as veratridine, that selectively bind to the sodium channel protein and stabilize it in an open state. Using a combination of toxin binding and ion flux assays, several groups have shown that purified sodium channels can be functionally reconstituted into vesicles (112).

4.3 **General comments**

The strategies outlined above could be applied to almost any membrane system in which a membrane translocation event takes place. The integrity of the membrane vesicle is probably the limiting factor in obtaining good results. Even if a reconstitution experiment leads to vesicles with random inside-outside distributions of the protein, a vectorial assay can be designed in which one or more of the necessary substrates is restricted to one side. A complication can arise if the compound being transported binds to the membrane. For example, divalent cations like calcium bind with relatively high affinity to both lipids and the proteins. In order to measure calcium fluxes, considerable care must be taken to correct for non-specific binding.

Using a combination of ligand binding, enzyme and permeability/transport assays, it should be possible to obtain useful quantitative information about the structure and function of many membrane proteins both in the solubilized and membrane-bound states.

5. NEW APPLICATIONS OF RECONSTITUTED MEMBRANES

In addition to their role in defining protein activity, it appears that solubilization and reconstitution strategies can play an important role in elucidating aspects of protein structure. The most striking example is the use of solubilization and reconstitution to prepare ordered samples for analysis of protein tertiary structure, some examples of which are given below.

5.1 **Reconstituted membrane proteins for study of protein structure**

Structural information on membrane proteins, at the resolution required to position peptide backbones, is available only through diffraction techniques. These methods require a repeating array of membrane proteins. Two-dimensional membrane lattices can be studied at low resolution by electron or optical diffraction (113), while three-dimensional crystals are required for high-resolution X-ray diffraction (2). The use of reconstituted membranes for protein structural analyses is in its infancy. Only bacterio-rhodopsin (for review, see 114) and recently the light-harvesting chlorophyll *a/b* protein (115) have formed two-dimensional crystals from reconstituted membranes. Preliminary studies in the Stroud laboratory demonstrate that three-dimensional microcrystals can be grown from reconstituted AChR (S.Choe, personal communication). Although the value of using a reconstituted membrane is clear, successful crystallization, whether in two or three dimensions, requires screening many conditions, including type and concentration of detergent, the presence of small amphiphiles to 'plug' holes in the crystalline lattice, and phase changes in membrane lipids. Reconstituting a purified protein into a defined lipid not only simplifies the starting system, but also makes it possible to manipulate what may well be the most important variable for membrane proteins: the lipid–protein interface. Another advantage of using a reconstituted membrane protein is that membranes with low lipid-to-protein ratios, which normally form vesicles in the native state, often form two-dimensional sheets when reconstituted (116). Image reconstruction is simplified when only one layer of ordered protein is present.

The method employed by Li (115) to obtain well-ordered lattices of a reconstituted membrane protein is outlined below. The light-harvesting chlorophyll *a/b* complexes were isolated from pea seedlings in 1.5% (v/v) Triton X-100, and reconstituted into asolectin vesicles by freeze–thawing (65). Excess Triton X-100 was removed by stirring with BioBeads SM-2, and lattice formation was induced by incubating the membranes in 2 mM $MgCl_2$. Membranes were fractionated on a 0–40% (w/v) sucrose gradient in buffer which contained 0.03% Triton X-100. The small amount of detergent facilitated removal of excess lipids from the lattice and allowed infiltration of stain into the lipid region. The reconstituted membranes were centrifuged at 95 000 *g* for 20 h, sucrose was removed by dialysis or dilution, and lattices or vesicles were collected by centrifugation. Lattices were stained for EM with 2% (w/v) uranyl acetate. Image reconstruction of the electron micrographs showed the three-dimensional structure of the proteins to within 30 Å resolution.

6. CONCLUDING REMARKS

In spite of major advances in all aspects of membrane biology we are still forced to concur (at least partly) with Racker (118) that 'we have no intellectual guidelines to the approach of reconstitution when faced with a new task. We cannot recommend an all-purpose detergent for the solubilization of membrane proteins nor can we predict a reconstitution procedure that will work with all proteins'. Consequently, this chapter has been written with the aim of highlighting the major principles and practices which should be applied in any practical solubilization and reconstitution strategy. Additionally, an attempt has been made to give some flavour of the possible uses of reconstitution systems in membrane biology.

7. REFERENCES

1. Singer,S.J. and Nicolson,G.L. (1972) *Science,* **175**, 720.
2. Michel,H. (1983) *Trends Biochem. Sci.,* **8**, 56.
3. Bennett,A.B., O'Neill,S.D. and Spanswick,R.M. (1984) *Plant Physiol.,* **74**, 538.
4. Hjelmeland,L.M. and Chrambach,A. (1984) In *Methods in Enzymology.* Jakoby,W.B. (ed.), Academic Press, New York and London, Vol. 104, p. 305.
5. Klausner,R.D., Van Renswoude,J., Blumenthal,R. and Rivnay,B. (1984) In *Receptor Biochemistry and Methodology Vol. 3.* Venter,J.C. and Harrison,L.C. (eds), Alan R.Liss, New York, p. 209.
6. Lichtenberg,D., Robson,R.J. and Dennis,E.A. (1983) *Biochim. Biophys. Acta,* **737**, 285.
7. Helenius,A. and Simons,K. (1975) *Biochim. Biophys. Acta,* **415**, 29.

8. Reynolds,J.A. (1982) In *Lipid−Protein Interactions*. Jost,P.C. and Griffith,O.H. (eds), Wiley, New York, Vol. 2, p. 193.
9. Andersen,J.P., LeMaire,M., Kragh-Hansen,U., Champeil,P. and Moller,J.V. (1983) *Eur. J. Biochem.*, **134**, 205.
10. Earnest,J.P., Stroud,R.M. and McNamee,M.G. (1986) *Biophys. J.*, **49**, 522a.
11. Jackson,M.L. and Litman,B.J. (1985) *Biochim. Biophys. Acta*, **812**, 369.
12. Roux,M. and Champeil,P. (1984) *FEBS Lett.*, **171**, 169.
13. Tietz,N.W. ed. (1976) *Fundamentals of Clinical Chemistry*. Saunders, Philadelphia.
14. Akiyama,T., Takagi,S., Sankawa,U., Inari,S. and Saito,H. (1980) *Biochemistry*, **19**, 1904.
15. Siakatos,A.N. and Rouser,G. (1965) *J. Am. Oil Soc.*, **42**, 913.
16. Morrison,W.R. and Smith,L.M. (1964) *J. Lipid Res.*, **5**, 600.
17. Small,D.M. (1971) In *The Bile Acids*. Nair,P.P. and Kritchevsky,D. (eds), Plenum Press, New York and London, p. 249.
18. Giotta,G.J. (1976) *Biochem. Biophys. Res. Commun.*, **71**, 776.
19. Garewal,H.S. (1973) *Anal. Biochem.*, **54**, 319.
20. Griffith,J.D. (1957) *Chem. Ind.*, **30**, 1041.
21. Terland,O., Slinde,E., Skotland,T. and Flatmark,T. (1977) *FEBS Lett.*, **76**, 86.
22. Bar,H.P. and Kulshrestha,S. (1975) *Can. J. Biochem.*, **53**, 472.
23. Tanford,C. (1980) *The Hydrophobic Effect*. 2nd Edition, Wiley, New York.
24. Mittal,K.L. and Lindman,B. (1984) *Surfactants in Solution Vol. 3*. Plenum Press, New York.
25. Rosen,M.J. (1984) *Structure−Performance Relationships in Surfactants*. ACS Symposium Series, Vol. 23, ACS, Washington.
26. Rivnay,B. and Metzger,H. (1982) *J. Biol. Chem.*, **257**, 12800.
27. Mukerjee,P. and Mysels,K.J. (1971) *Critical Micelle Concentration in Aqueous Surfactant Systems*. National Bureau of Standards, NSRDS-NBS 36. Natl. Bur. Stds., Washington DC.
28. Mukerjee,P. (1967) *Adv. Coll. Int. Sci.*, **1**, 241.
29. Bordier,C. (1981) *J. Biol. Chem.*, **256**, 1604.
30. Huganir,R.L. and Racker,E. (1982) *J. Biol. Chem.*, **257**, 9372.
31. Noel,H., Goswami,T. and Pande,S.V. (1985) *Biochemistry*, **24**, 4504.
32. Devaux,P.F. and Seigneuret,M. (1985) *Biochim. Biophys. Acta*, **822**, 63.
33. Warren,G.B., Toon,P.A., Birdsall,N.J.M., Lee,A.G. and Metcalfe,J.C. (1974) *Proc. Natl. Acad. Sci. USA*, **71**, 622.
34. Robinson,N.C., Strey,F. and Talbert,L. (1980) *Biochemistry*, **19**, 3656.
35. Jones,O.T., Eubanks,J.H. and McNamee,M.G. (1985) *Biophys. J.*, **47**, 493a.
36. Chang,H.W. and Bock,E. (1977) *Biochemistry*, **16**, 4513.
37. Talvenheimo,J.A., Tamkun,M.M. and Catterall,W.A. (1982) *J. Biol. Chem.*, **257**, 11868.
38. Epstein,N. and Racker,E. (1978) *J. Biol. Chem.*, **253**, 6660.
39. Anholt,R., Fredkin,D.R., Deerinck,T., Ellisman,M., Montal,M. and Lindstrom,J. (1982) *J. Biol. Chem.*, **257**, 7122.
40. McNamee,M.G., Jones,O.T. and Fong,T.-M. (1986) In *Ion Channel Reconstitution*. Miller,C. (ed.), Plenum Press, New York, p. 231.
41. Rooney,M.W., Lange,Y. and Kauffman,J.W. (1985) *J. Biol. Chem.*, **259**, 8281.
42. Boyd,N.D. and Cohen,J.B. (1980) *Biochemistry*, **19**, 5344.
43. Schreiber,G., Henis,Y.I. and Sokolovsky,M. (1985) J. Biol. Chem., **260**, 8789.
44. Skou,J. and Esmann,M. (1979) *Biochim. Biophys. Acta*, **567**, 436.
45. De Meis,L. (1981) *The Sarcoplasmic Reticulum*. Wiley, New York.
46. Aldrich,R.W., Corey,D.P. and Stevens,C.F. (1983) *Nature*, **306**, 436.
47. Criado,M., Eibl,H. and Barrantes,F.J. (1984) *J. Biol. Chem.*, **259**, 9188.
48. Simmonds,A.C., Rooney,E.K. and Lee,A.G. (1984) *Biochemistry*, **23**, 1432.
49. Baldwin,P.A. and Hubbell,W.L. (1985) *Biochemistry*, **24**, 2633.
50. Nedivi,E. and Schramm,M. (1984) *J. Biol. Chem.*, **259**, 5803.
51. Pick,U. and Bassilian,S. (1983) *Eur. J. Biochem.*, **133**, 289.
52. Ackrell,B.A.C., Kearney,E.B. and Coles,C.J. (1977) *J. Biol. Chem.*, **252**, 6963.
53. Holloway,P.W. (1973) *Anal. Biochem.*, **53**, 304.
54. Racker,E. (1972) *J. Membr. Biol.*, **10**, 221.
55. Jackson,M.L. and Litman,B.J. (1982) *Biochemistry*, **21**, 5601.
56. Kaback,H.R. (1986) *Ann. Rev. Biophys. Chem.*, **15**, 279.
57. Popot,J.-L. and Changeux,J.-P. (1984) *Physiol. Rev.*, **64**, 1162.
58. Furth,A. (1980) *Anal. Biochem.*, **109**, 207.
59. Helenius,A., Sarvas,M. and Simons,K. (1981) *Eur. J. Biochem.*, **116**, 27.

60. Klausner,R.D., Bridges,K., Tsunoo,H., Blumenthal,R., Weinstein,J.N. and Ashwell,G. (1980) *Proc. Natl. Acad. Sci. USA*, **77**, 5087.
61. Radany,E.W., Bellet,R.A. and Garbers,D.L. (1985) *Biochim. Biophys. Acta*, **812**, 695.
62. Deamer,D.W. and Uster,P.S. (1983) In *Liposomes*. Ostro,M.J. (ed.), Marcel Dekker, Inc., New York, p. 27.
63. Racker,E., Violand,B., O'Neal,S., Alfonzo,M. and Telford,J. (1979) *Arch. Biochem. Biophys.*, **198**, 470.
64. Kasahara,M. and Hinkle,P.C. (1976) *Proc. Natl. Acad. Sci. USA*, **73**, 396.
65. McDonnel,A. and Staehelin,L.A. (1980) *J. Cell Biol.*, **84**, 40.
66. Schullery,S.E., Schmidt,C.F., Felgner,P., Tillack,T.W. and Thompson,T.E. (1980) *Biochemistry*, **19**, 3919.
67. Muscatello,U. and Horne,R.W. (1968) *J. Ultrastruct. Res.*, **25**, 73.
68. Henderson,R. and Unwin,P.N.T. (1975) *Nature*, **257**, 28.
69. Dubochet,J., Adrian,M., Lepault,J. and McDowell,A.W. (1985) *Trends Biochem. Sci.*, **10**, 143.
70. Carlson,S.S., Wagner,J.A. and Kelly,R.B. (1978) *Biochemistry*, **17**, 1188.
71. Fong,T.-M. and McNamee,M.G. (1986) *Biochemistry*, **25**, 830.
72. Mimms,L.T., Zampighi,G., Nozaki,Y., Tanford,C. and Reynolds,J.A. (1981) *Biochemistry*, **20**, 833.
73. Popot,J.-L., Cartaud,J. and Changeux,J.-P. (1981) *Eur. J. Biochem.* **118**, 203.
74. Goldin,S.M. and Tong,S.W. (1974) *J. Biol. Chem.*, **249**, 5907.
75. Goldin,S.M. (1977) *J. Biol. Chem.*, **252**, 5630.
76. Racker,E., Chien,T.F. and Kandrach,A. (1975) *FEBS Lett.*, **57**, 14.
77. Ochoa,E.L.M., Dalziel,A.W. and McNamee,M.G. (1983) *Biochim. Biophys. Acta*, **727**, 151.
78. Strauss,G. and Hauser,H. (1986) *Proc. Natl. Acad. Sci. USA*, **83**, 2422.
79. Seto-Young,B., Chen,C.-C. and Wilson,T.H. (1985) *J. Membr. Biol.*, **84**, 259.
80. Van der Steen,A.T.M., Taraschi,T.F., Voorhout,W.F. and De Kruijff,B. (1983) *Biochim. Biophys. Acta*, **733**, 51.
81. Walker,J.W., Lukas,R.J. and McNamee,M.G. (1981) *Biochemistry*, **20**, 2191.
82. Tamkun,M.M. and Catterall,W.A. (1981) *J. Biol. Chem.*, **256**, 11457.
83. Carroll,R.C. and Racker,E. (1977) *J. Biol. Chem.*, **252**, 6981.
84. Parise,L.V. and Phillips,D.R. (1985) *J. Biol. Chem.*, **260**, 10698.
85. Anholt,R., Lindstrom,J. and Montal,M. (1981) *J. Biol. Chem.*, **256**, 4377.
86. Criado,M., Hochschwender,S., Sarin,V., Fox,J.L. and Lindstrom,J.L. (1985) *Proc. Natl. Acad. Sci. USA*, **82**, 2004.
87. Saito,A., Wang,C.T. and Fleischer,S. (1978) *J. Cell Biol.*, **79**, 601.
88. Herbette,L., Scarpa,A., Blasie,J.K., Bauer,D.R., Wang,C.T. and Fleischer,S. (1981) *Biophys. J.*, **36**, 27.
89. Chen,Y.S. and Hubbell,W.L. (1973) *Exp. Eye Res.*, **17**, 517.
90. Wickner,W. (1976) *Proc. Natl. Acad. Sci. USA*, **73**, 1159.
91. Agnew,W.S.A. (1984) *Annu. Rev. Physiol.*, **46**, 517.
92. Klymkowsky,M.W. and Stroud,R.M. (1979) *J. Mol. Biol.*, **128**, 319.
93. Levitzki,A. (1985) *Biochim. Biophys. Acta*, **822**, 127.
94. Schmidt,J. and Raftery,M.A. (1973) *Anal. Biochem.*, **52**, 349.
95. Weber,M. and Changeux,J.-P. (1974) *Mol. Pharmacol.*, **10**, 15.
96. Heidmann,T. and Changeux,J.-P. (1984) *Proc. Natl. Acad. Sci. USA*, **81**, 1897.
97. Dunn,S.M.J. and Raftery,M.A. (1982) *Proc. Natl. Acad. Sci. USA*, **79**, 6757.
98. Ross,E.M. and Gilman,A.G. (1980) *Annu. Rev. Biochem.*, **49**, 533.
99. Smigel,M., Katada,T., Northup,J.K., Bokoch,G.M., Ui,M. and Gilman,A. (1984) *Adv. Cyclic Nucleotide Protein Phos. Res.*, **17**, 1.
100. Brandt,D.R. and Ross,E.M. (1985) *J. Biol. Chem.*, **260**, 266.
101. Brandt,D.R., Asano,T., Pedersen,S.E. and Ross,E.M. (1983) *Biochemistry*, **22**, 4357.
102. Citri,Y. and Schramm,M. (1982) *J. Biol. Chem.*, **257**, 13257.
103. Keenan,A.K., Gal,A. and Levitzki,A. (1982) *Biochem. Biophys. Res. Commun.*, **105**, 615.
104. Shorr,R.G.L., Strohsacker,M.W., Lavin,T.N., Lefkowitz,R.J. and Caron,M.G. (1982) *J. Biol. Chem.*, **257**, 12341.
105. Pedersen,S.E. and Ross,E.M. (1982) *Proc. Natl. Acad. Sci. USA* **79**, 7228.
106. Fleming,J.W. and Ross,E.M. (1980) *J. Cyclic Nucleotide Res.*, **6**, 407.
107. Curtis,B.M. and Catterall,W.A. (1984) *Biochemistry*, **23**, 2113.
108. Racker,E. and Stoeckenius,W. (1974) *J. Biol. Chem.*, **249**, 662.
109. Cash,D.J. and Hess,G.P. (1981) *Anal. Biochem.*, **112**, 39.
110. Moore,H.-P.H. and Raftery,M.A. (1980) *Proc. Natl. Acad. Sci. USA*, **77**, 4509.

176

111. Karpen,J.W., Sachs,A.B., Cash,D.J., Pasquale,E.B. and Hess,G.P. (1983) *Anal. Biochem.*, **135**, 83.
112. Tanaka,J.C., Furman,R.E. and Barchi,R.L. (1986) In *Ion Channel Reconstitution*. Miller,C. (ed.), Plenum Press, New York, p. 277.
113. Amos,L.A., Henderson,R. and Unwin,P.N.T. (1982) *Prog. Biophys. Mol. Biol.*, **39**, 183.
114. Stoeckenius,W. and Bogomolni,R.A. (1982) *Annu. Rev. Biochem.*, **52**, 587.
115. Li,J. (1985) *Proc. Natl. Acad. Sci. USA*, **82**, 386.
116. De Grip,W.J., Olive,J. and Bovee-Geurts,P.H.M. (1983) *Biochim. Biophys. Acta*, **734**, 168.
117. Tanford,C., Nozaki,Y. and Rhode,M.F. (1977) *J. Chem. Phys.*, **81**, 1555.
118. Racker,E. (1983) In *Liposome Letters*. Bangham,A.D. (ed.), Academic Press, London, p. 179.
119. Levitzki,A. (1984) *Receptors — A Quantitative Approach*. Benjamin/Cummings Co., Menlo Park, CA.

CHAPTER 6

The isolation and labelling of membrane proteins and peptides

JOHN B.C.FINDLAY

1. INTRODUCTION

Depending on their method of association with the bilayer, membrane proteins have been categorized as either integral (intrinsic) or peripheral (extrinsic). The first group are presumed to integrate deeply into the hydrophobic phase of the bilayer and to present an essentially hydrophobic face to the surrounding fatty acid/cholesterol milieu. Regions of these polypeptides exposed to the aqueous environment resemble so-called 'water-soluble' proteins, hence they are usually described as amphipathic. Peripheral proteins are associated with the membrane surface through interactions either with phospholipid head groups or with other membrane proteins. Although regions of their surfaces may be hydrophobic, this class of polypeptide presents an essentially hydrophilic outer surface (see front cover).

In this chapter, central aspects of protein and peptide labelling, solubilization and purification will be described. Since there are almost as many solvents, detergents, resins and procedures as there are different proteins, only general guidelines to possible purification strategies can be given, along with important principles and observations which should be carefully considered. However, the information provided should allow useful approaches to be readily devised. The methods used in protein modification are more routine and many of the more successful are detailed.

2. SOLUBILIZATION OF MEMBRANE PROTEINS

The structural classification outlined above can readily be translated into operational behaviour. Peripheral proteins can be solubilized without substantially disrupting the bilayer but liberation of integral species is only usually achieved by destroying the integrity of the membrane. The principles and practices used in solubilizing integral membrane proteins with detergents or chaotropic agents have been discussed in the previous chapter and will not be substantially repeated here. Prior to such solubilization, however, it is often useful, particularly if subsequent fractionation is envisaged, to remove or separate the population of peripheral proteins.

2.1 Peripheral proteins

Depending on the particular problem, the membrane can be subjected to either mild or harsh conditions.

Mild conditions vary from low ionic strength solutions such as $0.1-1$ mM EDTA which will also remove complexed divalent cations (1,2) to high ionic strength buffers containing greater than 1 M NaCl or KCl (2,3), with or without EDTA. Anions such as iodide or diiodosalicylate should not be used since they are chaotropic agents and will tend to act like detergents. The pH employed can vary from about 6.0 to 8.0. Under these conditions it is unlikely that either integral or peripheral proteins will be irreversibly denatured.

More aggressive solubilization methods often liberate greater amounts of protein (>50% of the total membrane content) but usually result in denaturation which, in many cases, is irreversible. Examples of such conditions include 6 M guanidinium chloride (4), 8 M urea, 1 mM *p*-chloromercuribenzoate (5), dilute acids pH 2.0−3.0 (6) or alkali, pH 9.5−11.0 (7). To prevent cleavage of peptide bonds, 'stripping' of the membrane in this way is usually carried out at 0−4°C (7). Acidic conditions sometimes result in precipitation of the solubilized proteins and so the alkaline approach is employed more often.

Since considerable amounts of protein are often removed by these procedures, it should be borne in mind that the membrane may invert and/or vesiculate during such treatment. Care must be taken, therefore, to adjust the centrifugation conditions to ensure good recovery of membrane (see Chapter 1). It is advisable as quickly as possible thereafter, to restore the pH to neutrality.

2.2 Integral proteins

The principles and practices used for the dissolution of membranes with detergents or chaotropic agents are provided in Chapter 5. It is important to appreciate that this dissolution can release/activate proteases. Careful consideration should, therefore, be given to the addition of protease inhibitors at this point even if they were added during the initial stages of membrane isolation. Numerous complex cocktails of these inhibitors can be used depending on the sensitivity of the system. A useful, but not comprehensive, agent is the serine protease inhibitor, phenylmethylsulphonyl fluoride (PMSF) but bear in mind that this reagent, stored at 100 mM in 2-propanol or ethanol and added to 100 μM, has a relatively short half-life in aqueous systems [110 and 35 min at pH 7.0 and 8.0, respectively (25°C), longer at lower temperatures (8)]. For -SH proteases, 10 mM sodium tetrathionate can be useful. Other protease inhibitors, their specificities and their uses are given in Chapter 1, Table 3.

2.2.1 Phase partitioning

A quite different but sometimes very useful way of separating integral and peripheral species is the application of phase partitioning, particularly successful with Triton X-114 (9). To obtain as homogeneous a preparation of Triton X-114 as possible, some pretreatment can be carried out.

(i)   Add 10 g of Triton X-114 and 8 mg of butylated hydroxytoluene to a final volume of 200 ml in 20 mM potassium phosphate pH 7.5, 0 15 M KCl.

(ii)   Cool the mixture to 0°C to ensure complete solubilization, then incubate overnight at 35°C.

Above 20−25°C, the purified detergent separates out as a distinct phase of smaller volume which can be recovered following gentle centrifugation and re-subjected, if required, to the above procedure. For membrane solubilization carry out the following steps.

(i)   Add up to 2.0% Triton X-114 to samples of membrane (1−3 mg/ml protein) at 0−4°C in 10 mM Tris-HCl or phosphate buffer pH 7.4, containing up to 150 mM NaCl or KCl.

(ii)   Incubate at 0−4°C for up to 1 h before centrifuging the mixture at 100 000 *g* for 1 h to remove insoluble material.

(iii) Overlay the soluble fraction on a cushion of buffered 6% sucrose in a suitable centrifuge tube.

(iv)   Warm to 25−30°C for 5 min and then centrifuge for about 5 min at 100 *g* and above 20°C.

(v)   The detergent phase appears as an oily layer at the bottom of the tube. The aqueous fraction can be re-equilibrated with 0.5% (w/v) Triton X-114 and the detergent phase with aqueous buffer, and the process repeated.

The usefulness of this procedure lies in the unusual properties of the detergent at 0°C where it is soluble in aqueous buffers and above 20°C where it separates out as a distinct detergent phase. In general, hydrophilic (i.e. peripheral) proteins are then found in the aqueous phase and hydrophobic species in the detergent phase. As with all detergents, the protein of interest may not be soluble, native or active in Triton X-114 and this should be ascertained at the start. If insoluble, it can be recovered by centrifugation at 0°C. Integral proteins with pronounced hydrophilic character, for example species with a single transmembrane helix and/or substantial carbohydrate content may sometimes be found in the aqueous phase either due to self-association or to binding of detergent, some of which always remains in the aqueous phase. It has also been suggested that channel-forming proteins (e.g. acetylcholine receptor, ATPases) may also partition anomalously (10,11). Applications of this technique can be found in ref. 12.

2.2.2 Delipidation

Solubilization of the membrane in detergent usually gives rise to mixed micelles containing both phospholipid and detergent. Most integral polypeptides exist, with or without other proteins, as complexes with a monomolecular belt of detergent (see Figure 1, Chapter 5). A small amount of residual phospholipid may still persist. This lipid is often removed during susequent purification procedures but in certain avid associations, it may be necessary to extract with organic solvents or to denature the protein in SDS or organic mixtures.

(i) *Organic solvents.* Pure organic solvents such as chloroform; chloroform:methanol (or ethanol or butanol); 1-butanol; diethylether:ethanol 3.1 (v/v) have been used to solubilize proteins (13−16). The type of polypeptides soluble in these conditions are often classed as proteolipids. They are very hydrophobic and less amphipathic and do do not represent a major fraction of most membranes. Good examples are the DCCD-binding protein from the ATP synthase complex (17) and the proteolipid proteins from myelin, all of which are soluble in chloroform:methanol, 2:1, v/v (13). Under the extraction conditions used in the latter case (chloroform:methanol:aqueous buffer, 8:4:3 by vol.) most membrane proteins form a precipitated layer at the aqueous:organic interface, and any proteolipids present are in the organic phase. Some heavily glycosylated species, for example glycophorin, glycolipid, can be found in the aqueous phase. Varying the chloroform:methanol ratio can slightly alter the spectrum of proteins extracted (18). It is also of some advantage to extract the membrane with chloroform:methanol (2:1 v/v) acidified with up to 50 mM HCl (19) or 5 mM *p*-toluene sulphonic acid (20). In these cases, the range and amount of proteins partitioning into the organic phase can be increased somewhat.

Recovery of proteins from these organic solvents can readily be achieved by rotary evaporation but it is possible in some cases to precipitate them out using 10−20 volumes of cold acetone, 5−10 volumes of cold diethyl ether or various combinations of diethyl ether and ethanol (1:1−3:1 v/v). These conditions can also be used to separate protein from detergent (which remains in solution). Where neutral organic mixtures have been used as solvents, biological activity of the soluble component(s) may still be preserved but this is much less likely with acidified solvents or where the protein has been recovered by precipitation.

More recently, two very useful solvent systems have been introduced which maintain virtually all integral membrane proteins (indeed nearly all water-soluble species too) in a denatured but soluble form. The first of these is *formic acid:ethanol (3:7 v/v or 1:2 v/v)* used in the study of bacteriorhodopsin (21) and rhodopsin (22). It is most useful for peptides where no lipid or detergent is present. The second, *formic acid:acetic acid:chloroform:ethanol, 1:1:2:1 by vol.* (23,24), renders most membrane proteins soluble and has the additional advantage of stripping away all phospholipid and preventing aggregation of protein and self-association of lipid. This latter property substantially facilitates subsequent fractionation. The second solvent (abbreviated FACE) is particularly useful when solubilizing intact membranes. It is best added to pelleted membranes where the volume of water is low and can be readily accommodated by the solvent. In both solvents, methanol can be substituted for ethanol but the Sephadex LH resins which can be used with these mixtures, pack better in the latter. Acetic acid is necessary in FACE to prevent phase separation on standing. Note that once solubilized and fractionated using these systems, many integral membrane proteins are reluctant to return to aqueous solutions containing detergent, with the exception of SDS where fairly good solubilization can be achieved. Subsequent purification of the soluble extracts can be carried out as below.

3. CHROMATOGRAPHIC SEPARATION OF PROTEINS

Peripheral proteins removed from the membrane by the conditions identified above behave essentially as water-soluble polypeptides and can be fractionated by any of the procedures used for protein purification. These methods, which rely on variations of size, charge and specificity, are described in detail in other volumes in the Practical Approach Series dealing with h.p.l.c. and affinity chromatography. This account will concentrate, instead, on specific applications relevant to integral membrane proteins and peptides. The procedures and solvents described, however, can readily be used or adapted for the fractionation of peripheral species.

3.1 Size fractionation

Provided the protein of interest is stable, useful initial fractionation can be achieved by gel filtration in aqueous buffers containing detergent [usually at or above its critical micellar concentration (CMC)]. Integral species surrounded by their monomolecular belt of detergent can exhibit apparent molecular weights up to twice that anticipated from their amino acid compositions/sequences. In addition, many functional complexes are not dissociated by non-denaturing detergent. For these reasons, it is usual to employ the high porosity Sepharoses, agaroses or polyacrylamides for column chromatography (see also Chapter 5). In some difficult cases, proteins may aggregate, or be irreversibly adsorbed to the resin, but generally, a reasonably successful crude fractionation will result if the detergent is chosen carefully. Where irreversible adsorption is suspected, the resin type may be changed successfully; polyacrylamide based materials tend to be the most inert. This type of resin can be used with both mild and harsh (i.e. denaturing) detergents but *not* with purely organic solvents, which require modified supports (see later).

The use of h.p.l.c. and f.p.l.c. based systems for fractionation of integral membrane proteins can also be successful, particularly when yields and time are limiting factors. There are an ever-expanding range of gel filtration h.p.l.c. and f.p.l.c. supports utilizing silica or synthetic polymer-based materials (e.g. hydroxylated polyethers). The molecular weight range of the proteins under investigation will largely govern which pore size should be used. Manufacturers claim ranges from 500 to 2×10^7 daltons. The presence of non-denaturing detergents has a tendency to reduce the resolution which can be achieved. Most of these resins can be successfully used with neutral organic solvents, detergents and denaturing conditions such as guanidine hydrochloride and SDS. We have found the TSK resins (e.g. LKB Ultrapac SW) quite successful for SDS-solubilized proteins.

Particular caution must be exercised when extremes of pH are employed. There are two areas for concern. The degradation of silica-based supports is accelerated, some more than others, by the very acidic conditions often used in conjunction with organic solvents. (Alkaline conditions above pH 8.0 are even more destructive.) In these situations, irreversible adsorption of polypeptide is an increasing problem as the resin is stripped of its protective groups. Secondly, there is the possibility that high levels of formic acid may cause damage to the stainless steel components of the h.p.l.c. apparatus. Some manufacturers claim that the quality of their stainless steel makes it refractory to such damage but others have resorted to the introduction of ceramic or titanium-lined mixing chambers.

Fortunately there are alternatives. If sufficient protein is available, hydroxypropylated dextrans (Sephadexes LH-60 and LH-20) can be used with almost all conceivable mixtures of organic solvents and acid. In this respect, the FACE and formic acid:ethanol solvents are particularly useful, especially since the former will usually remove all detergent and lipid components which then chromatograph as their low molecular weight monomers. *Remember that these organic solvent systems require the use of glass and Teflon for all components.* Several proteins and many peptides are soluble (some as aggregates) in formic acid up to about 90% v/v. Under these extreme conditions the Sephadex and polyacrylamide range of supports can be used for fractionation on the basis of size. These are potentially destructive conditions for both resin and peptide, however, so low temperatures and 'snappy' run times are recommended. Obviously, the resins should not be stored in these solvents.

With aggressive organic mixtures, the Pharmacia f.p.l.c. system can also be useful. The apparatus relies more on Teflon and glass and hence is much less liable to damage by acid. The silica-based resins so far introduced appear more resistant to these solvents but the agarose-based supports (Superoses) tend to irreversibly adsorb membrane proteins. The latter are more useful for gel filtration but should be used in detergent conditions. The silica resins are best employed for reverse-phase chromatography (see later). The development times for h.p.l.c. and f.p.l.c. systems are much faster (30−60 min) than conventional glass column systems and can often give better resolution depending on the particular problem. The capacity of the columns is limited, however, and the chromatography may need to be repeated many times on identical aliquots of original sample. It is important not to overload the resins or to employ too fast a flow-rate. As a very rough guide, one should not load more than 1 mg of average total protein per ml of packed resin nor use a run time shorter than 30 min or longer than 90 min.

3.2 Charge fractionation

3.2.1 *Ion-exchange chromatography*

All the classical ion-exchange systems used in protein purification can be adapted for membrane proteins provided the native charge characteristics of the polypeptide have not been obliterated by the addition of charged detergents or by their dissolution in strong acid or alkali. The weakly ionic DEAE and CM-derivatized resins (e.g. cellulose and dextrans) still remain the most popular but when used in the presence of detergent, care must be exercised that the detergents themselves do not interfere with the interaction between protein and resin. The success of these supports in producing effective purifications depends largely on the nature of the protein. Where a large proportion of its structure is in the aqueous phase, conditions for firm interaction with an ion-exchange resin can be devised. In contrast, proteins which have most of their mass embedded in the bilayer and hence are smothered in the solubilizing detergent have much less scope for interaction with ionic supports.

Bearing in mind these considerations, DEAE columns are normally equilibrated in a relatively low ionic strength buffer (e.g. <0.1 M Tris-HCl or phosphate) containing detergent (e.g. 0.1%) and perhaps also phospholipid (0.1%) at pHs from about 6.5 to 8.0. They are developed with an increasing concentration gradient of salt (e.g. 0−1 M NaCl or KCl) still containing detergent and/or phospholipid (25). CM resins are usually equilibrated at slightly acidic pHs (5.5−7.0) and often are developed with a pH rather than a salt gradient. There have been applications of DEAE- (26) and CM- (27) cellulose in organic mixtures (e.g. chloroform:methanol containing water, where the proportion of water has been varied but this is an 'in extremis' rather than a first choice method for the great majority of applications.

Finally, ion-exchange silicas and synthetic resins in a range of pore sizes have been developed for h.p.l.c. and f.p.l.c. (28−30). These possess at least 20−30 times higher capacity for total protein than gel filtration systems. They can be used in much the same way as conventional columns and have the usual advantages and disadvantages of h.p.l.c. and f.p.l.c. systems. An additional resin, often difficult to use in large columns due to flow-rate and recovery problems, but more malleable in h.p.l.c., is hydroxylapatite which has been successful for membrane systems (31). There are also a variety of weak and strong ion-exchange resins which have the potential for success with membrane proteins but they have not yet been used widely (see manufacturers' catalogues, e.g. Biorad).

3.2.2 *Isoelectric focusing*

In their native form, membrane proteins exhibit a range of isoelectric points potentially as wide as water-soluble proteins. This characteristic can be harnessed to produce effective purifications using the technique of isoelectric focusing in which the proteins migrate under the influence of an electric current. This migration occurs through a pH gradient and when the isoelectric point of a particular protein is reached, further movement stops and the protein 'focuses'. Very high resolution can be achieved to the point where the phosphorylated and non-phosphorylated versions of the same protein can be distinguished (32). Some recent examples of its use for membrane proteins can be found in references 33–36.

Various techniques (i) for setting up the pH gradients with ampholytes using a number of stabilizing supports, polyacrylamide and agarose gels, glycerol, sucrose or sorbitol density gradients; (ii) for loading the protein, and (iii) for carrying out the electrophoresis can be found in references 37, 38 and in manufacturers' handbooks (e.g. Pharmacia or LKB). Since membrane proteins require detergents for solubility, care must be taken to use detergents (usually non-ionic but theoretically also zwitterionic) which do not alter the charge characteristics of the protein, interfere with the pH gradient or focus themselves. In general it is useful to begin with a fairly wide pH range (pH 3–10) eventually narrowing down, in the refined conditions to a range of 2 pH units roughly in the middle of which the protein of interest focuses. For the examination of the more basic membrane proteins (pI ~7), it is often advisable to employ, in addition, non-equilibrium pH gradient electrophoresis (NEPHGE) where samples are applied in the acidic region of the gel (39). In this system, the pH gradient does not reach complete equilibrium and thus the position of the protein does not indicate its precise pI. Separations are usually carried out at $0-4°C$ and continued for at least a further 25% of the period (usually 1–2 h) required to attain constant voltage conditions. The technique can be used analytically or on a preparative scale by pumping the support carefully to limit mixing, into a fraction collector after the proteins have focused. Both operations depend for maximum efficiency on a good assay for the solubilized protein, on good solubilization in detergent and, very importantly, on the prevention of aggregation and/or precipitation.

One problem which can be irksome, particularly if protein analysis and sequencing is to be carried out, is the difficulty of removing all the ampholytes. The usually recommended method is precipitation of the protein with 80–100% saturated $(NH_4)_2SO_4$. If no precipitation occurs the protein may be adsorbed on a hydrophobic column, for example 1 ml/mg protein of phenyl or octyl-Sepharose equilibrated with 80–100% $(NH_4)_2SO_4$. The resin is then washed with 3–5 column volumes of the equilibration buffer to remove the ampholytes, and the protein eluted at low ionic strength, for example 0.1 M ammonium bicarbonate. Unfortunately neither of these methods, nor straightforward gel filtration or prolonged dialysis is routinely successful for integral membrane proteins in detergent-containing solution. Where retention of biological activity is not required, one successful method utilizes gel filtration with Sephadex LH60, equilibrated and developed with the formic:ethanol or FACE solvents as described before. Effective removal of ampholytes will be achieved with a 100 ml column, the protein (mol. wt >15 000) occurring in the void volume and the ampholytes in the included volume.

3.2.3 *Chromatofocusing*

For preparative rather than analytical purposes, chromatofocusing promises to be one of the most effective techniques for the purification of integral membrane proteins (40,41). The method is related to isoelectric focusing except that the pH gradient is formed on the immobile column resin rather than in the buffer. These supports are stable between pH 3 and pH 12 but are generally most useful in the pH 5−9 range. They can be employed with neutral organic solvents, with denaturants such as 8 M urea and 6 M guanidine hydrochloride and with detergents which are either uncharged (e.g. digitonin, Triton X-100) or zwitterionic (CHAPS). The method is rapid, efficient and of moderate cost. Polybuffer may be removed as described above for ampholytes. Pharmacia can provide a useful booklet describing the theoretical and practical aspects of this potentially powerful procedure.

3.3 **Hydrophobic interaction chromatography**

Increasing use is being made of hydrophobic interaction chromatography, a process whereby proteins/peptides adhere to hydrophobic substituents on the resins. Interaction of protein with resin, for example octyl- or phenyl-Sepharose or -Superose, occurs through hydrophobic influences and is enhanced by high salt concentrations, for example 50−80% ammonium sulphate, by low pH and by elevated temperatures. The protein is then eluted by the application of a reverse salt gradient. Inclusion of ethylene glycol, methanol or ethanol in the eluting buffer may improve fractionation and recovery yield, as will higher pHs and lower temperatures. It must be pointed out, however, that this approach has limited applicability for most integral membrane proteins. Firstly, the efficacy of the derivatized supports is drastically reduced by the presence of detergent (more so for organic solvents). Secondly, resolution from other integral membrane proteins is likely to be poor. It can sometimes be useful in separating integral and peripheral species.

3.3.1 *Reverse-phase chromatography*

This represents a specific form of hydrophobic interaction chromatography, particularly relevant to h.p.l.c. and f.p.l.c. systems. Macroporous (250 Å or greater) *supports*, both silica and synthetic (e.g. polystyrene-divinylbenzene), can be obtained derivatized with cyanopropyl or (di)phenyl groups or, more usefully with n-alkyl chains containing between two and 18 CH_2 groups. The general rule is the larger and more hydrophobic the protein or peptide the shorter the chain length which should be employed (C4 is particularly effective for large peptides, C18 with small). To protect the resin and reduce artefacts it is better to choose supports which have been capped, that is have had those silanol substituents not containing the C4-C18 chains, blocked with small side-chain silanes.

The choice of *solvent* (mobile phase) is important although only a limited number of systems have been developed so far. Alkaline pHs are to be avoided because of the instability of silica-based resins, while neutral pHs can often give broad peaks and poor resolution. The latter observation has been attributed to interactions between amino groups on the peptide and silanols on the resin. Under acid conditions, the silanols are protonated and hydrophobic interactions predominate. For the average peptide mixture, starting conditions of $0.01-0.1\%$ trifluoroacetic acid (TFA) or pentafluorobutyric acid are most popular. Alternatively, up to 0.1% phosphoric and perchloric acids may be used in order to increase the hydrophilicity of the peptides but they have only limited applicability to hydrophobic peptides. Inclusion of organic amines in the aqueous solvents, for example 0.1% triethylamine phosphate, have the effect of reducing ionic interactions between support and peptide which could be helpful. Small amounts of organic solvent can improve peptide solubility.

Samples should be applied to the resin in a small volume (i.e. 100 μl) of the aqueous solvent. Proteins/peptides of strong hydrophobic character may require up to 90% formic acid for solubilization. Although the latter conditions are not conducive to long resin life, they can be used for loading purposes. As the formic acid is diluted the peptides adhere (or precipitate?) on the support. They can then be eluted by gradients of increasing organic content, that is decreasing polarity. It is useful to bear in mind the following decreasing order of polarity: water > methanol > ethanol > acetonitrile > 1-propanol > 2-propanol > butanol. Acetonitrile is popular because of its low viscosity, good solubilizing properties and because it can be obtained free of u.v.-absorbing material at reasonable cost. A typical gradient might be 0.1% TFA containing 10% acetonitrile to 0.1% TFA containing 70% acetonitrile. It has been suggested that small amounts of additives such as monomethoxyethylene glycol can increase peptide solubility. Using such mobile phases, peptides applied in strong acids and which are effectively insoluble under the starting conditions, can become soluble and elute at high concentrations of organic solvent.

It is our unfortunate experience, however, that many of the larger and more hydrophobic peptides from integral membrane proteins, are recovered not at all or only in low yield using C18 resins and the solvents described above. More drastic conditions are usually necessary. Khorana *et al.* (42), for example, used a gradient from 5% formic acid in water to 5% formic acid in ethanol with C-18 resins. We have had some success in operating a gradient from 50% formic acid in water to 50% formic acid in isopropanol or ethanol (Medina and Findlay, unpublished). With both these systems it is advisable to use the f.p.l.c. apparatus and columns (Pharmacia) since they may be more resilient to the harsh conditions. Other solvent systems which have had some success are 0.24 M potassium acetate pH 6.5 in chloroform:methanol (1:2 v/v) containing 8% water and 1% benzene (43), different chloroform:methanol ratios (44), a gradient from 10% to 60% of 12 mM HCl, ethanol:*n*-butanol (4:1 v/v) in 12 mM HCl (45) and the inclusion of up to 1% dimethyl sulphoxide (DMSO) at the end of the gradient.

As a general operating rule of thumb, it may be best, first to fractionate proteins/peptides on LH60 and LH20, then subject pools containing material of similar molecular weight to reverse-phase chromatography as above. Drying down of peptide/protein should be carried out with care using rotary evaporation followed by lyophilization of the still wet sample. To aid re-solubilization and/or disaggregation, it can be useful to add a very small volume (100 μl) of anhydrous TFA to the dry sample, prior to the main solvent.

3.4 **Affinity chromatography**

This approach allows purification on the basis of some biological specificity or activity not possessed by most other proteins or peptides in the mixture Membrane-bound enzymes, for example, may be retained by insoluble supports to which substrate analogues, prosthetic groups or specific effectors have been covalently attached. Receptors will bind to immobilized drugs or hormones. Needless to say solubilization of the membrane such that the protein of interest retains its biological activity is usually an essential prerequisite to this approach. The protein is usually recovered from the resin by including an agent, such as excess drug, which successfully competes for the interaction site. The technique usually concentrates the protein as well as purifying it. A complete description of the theory and practical procedures in affinity chromatography is given in an earlier volume of this series (46). The discussion here will concentrate on applications related to membrane components.

3.4.1 *Supports*

Athough a number of materials, dextrans, cellulose, silicas, glass or synthetic resins, can be used as immobile supports, the vast majority of successful applications have utilized Sepharose, agarose, polyacrylamide and now, polystyrene resins. The size of the resultant protein−ligand complex will largely determine which porosity of gel to employ, always bearing in mind that the receptor protein may be oligomeric and that the associated detergent will substantially increase the size of the soluble species. These supports are then activated in such a way that they can then covalently couple to the specific ligand. It is often more convenient to start with many of the activated resins available commercially (e.g. Beckman, Bio-Rad, Calbiochem, Pharmacia, Pierce). They can be obtained with a number of reactive species attached close to the resin backbone or on extended arms (see *Table 1*). The advantage of these extensions is that they can increase the availability of the ligand and reduce steric hindrance, both of which are particularly important in detergent-containing solutions. Sometimes, however, they can be disadvantageous by participating in non-specific interactions. Increasingly, these affinity resins will become available for use in h.p.l.c. and f.p.l.c. systems.

3.4.2 *Ligands*

In virtually all cases, the coupling of ligand to resin is via primary amino, carboxylic acid or sulphydryl groups. It can easily be seen from *Table 1* that small ligands and proteins with suitable moieties can all be covalently attached. In some instances suitable reactive groups can be added to the ligand or protein (see protein modification and *Tables 4* and *5*). The choice of support will usually be governed by the structure and properties of the ligand.

There are a number of important miscellaneous considerations to bear in mind when proteins are employed as ligands. First, during the coupling procedure, care should be taken to preserve the activity of the protein. This will often involve inclusion during

Table 1. Affinity supports and specificities.

Resin/functional group	Ligand specificity	Coupling	References
Sepharose-CNBr activated	−NH₂	Spontaneous interaction	7
[a]Agarose ~~~O—C(=O)—N(pyrrole)	−NH₂	Spontaneous interaction	186
[b]Agarose ~~~~N(succinimide)	−NH₂	Spontaneous interaction	187
[b]Agarose ~~~COOH	−NH₂	Catalysed by water-soluble carbodiimide	188
[b]Agarose ~~~NH₂	−COOH	Catalysed by water-soluble carbodiimide	188
[b]Polyacrylamide ~~~NH₂	−COOH	Catalysed by water-soluble carbodiimide	190
[a,b]Agarose ~~~SH (Note that thiol resins may contain a protective grouping)	−SH	Resin may require activation by sulphydryl reagent	191
[a,b]Agarose ~~~Hg or ⬡—HgCl	−alkylating agents −SH	Spontaneous interaction Spontaneous interaction	192
[b]Polyacrylamide ⬡—B(OH)₂	Cis-diols,e.g. sugars/coenzymes	Spontaneous interaction	193
[a]Polystyrene ~~~NH—⬡—B⁻(OH)₃	Cis-diols, e.g. sugars/coenzymes	Spontaneous interaction	194

[a]From Pierce.
[b]From Biorad.

the coupling steps, of the substrate, analogue, prosthetic group or any ions essential for activity. It may also be necessary during re-equilibrium of the affinity resin to add these components again. Second, excess reactive groups on the support should be blocked with relatively inert species (e.g. ethanolamine) before the resin is thrown into combat. Blocking agents such as amino acids are less satisfactory since they can introduce ion-exchange properties. Third, as well as possessing specific binding sites, proteins also provide larger surfaces which can act as foci for hydrophobic interactions or, much more commonly, as ion-exchange resins. The detergent present often masks or occupies hydrophobic pockets, so reducing non-specific hydrophobic interactions. To reduce non-specific ionic interactions it is often worthwhile including sodium chloride (e.g. 0.1 M) in the buffer.

Highly specific affinity resins containing peptide hormones or antibodies as ligands can often be generated in this way (see Chapter 4). They do, however, suffer from a few disadvantages, the most troublesome being low capacity, low affinity and the harsh conditions (e.g. low pH) which are often required to effect elution of the bound fractions. They tend, therefore, to be used later in a purification sequence. Instead, initial fractionation can often be achieved very successfully with carbohydrate-binding proteins, called *lectins*, which attach to support very easily and which retain activity in mild detergent (47,48, for general review see 49).

Because of their different sugar specificities, lectins can be employed to differentiate between glycoproteins (48,50) but more often they retain a range of carbohydrate-containing membrane components (for a review of lectins and their apparent specificities see 51). Although elution of bound glycoprotein is readily obtained with specific sugars, the nature of the carbohydrate with which some lectins interact suggests that they can have overlapping/ill-defined specificities. Recoveries are generally very high (>90%), especially with lectins such as those from *Lens culinaris* (lentil) but it has been a common observation that Concanavalin A is less good in this respect (47,48,52). Due to variability in the side-chains of many membrane glycoproteins, individual lectins may only retain a certain proportion of the total population of a particular protein (48). This can be circumvented to some extent by using lectins of similar but wider specificity. Another useful trick where the protein of interest contains terminal sialic acid is to use a relatively non-specific lectin or combination of lectins (e.g. *Lens culinaris* and wheat germ agglutinin), to bind the vast majority of the membrane glycoproteins. The unbound sialic acid-containing fraction is then treated with neuraminidase (see Section 6.1.2) in order to generate a new terminal carbohydrate moiety which can now be recognized by the affinity resin (53). Finally, for reasons that are not entirely clear, small carbohydrate-containing peptides are often not retained by lectin-affinity supports even though the parent protein binds well.

As an alternative to lectin systems, carbohydrate-containing proteins can be retained non-specifically by two boronate resins (see *Table 1*) and eluted by the inclusion of sorbitol, citrate or acetate in the buffer or, in desperation, by low pH (e.g. 25 mM HCl, 100 mM formic acid). This approach is relatively new and has not yet been widely used for membrane proteins.

3.4.3 *Operating guidelines*

Samples should be applied to columns containing affinity supports (e.g. 10−20 ml packed resin) at a slow flow-rate (2−3 ml/h), to allow maximum equilibration and interaction. Elevated temperatures (20−35°C) may enhance binding. In some cases, it may also be useful to temporarily stop buffer flow altogether once the sample has been applied. The resin should then be washed with 5−10 column volumes of buffer to remove unbound or weakly retained material. At this stage, the ionic strength of the buffer could be increased to reduce non-specific retention. The eluting buffer is applied once some criterion (e.g. A_{280} nm) has been achieved suggesting that the resin has been adequately washed. The specifically-bound components are usually eluted rapidly in an asymmetric peak with a sharp leading edge followed by trailing tail. The use of sugar gradients with lectin resins to capitalize on variations in binding affinities has had only very limited success. If the column has been loaded at an elevated temperature 20−30°C, elution is often facilitated by cooling to 4°C.

After washing with high salt, the resin may be stored for up to 1−2 years in a solution containing protease inhibitors, anti-bacterial agents and substrate or substrate analogue. The resins are made ready for use by extensive washing (up to 100 bed volumes) with the loading buffer. With some supports, washing with 0.01% SDS is sometimes carried out prior to re-equilibration but this step is only recommended where the covalently-attached protein is stable to SDS or readily renatured. Note that precipitation results if SDS is mixed with potassium ions or a detergent such as dodecyltrimethylammonium bromide (DTAB).

4. GEL ELECTROPHORESIS

Modern techniques in recombinant DNA technology and protein sequencing have obviated the need in some cases for classical purification procedures. Instead, if the protein of interest can be identified on a one or two-dimensional gel system, it can be eluted from the gel in conditions whereby its amino acid composition and partial protein sequence can be obtained. In an earlier volume in this series (54) the principles and practices of gel electrophoresis are described in detail. For membrane proteins, systems containing non-denaturing detergent have been devised, for example Triton X-100 (55) and deoxycholate (56), but the resolution obtained is often poor. Charge-shift electrophoresis can, however, be a very productive approach (see Appendix V). Denaturants have been of great value, especially when used in conjunction with polyacrylamide gel electrophoresis. The most successful detergents, because of their strong solubilizing properties, have been sodium and lithium dodecyl sulphates (57). Chloral hydrate, although potentially very useful (58), is to be avoided due to its toxic properties, particularly at the high concentrations required.

As a general rule, the quantity of SDS added to a membrane must be at least twice the total dry weight of the preparation while maintaining the SDS concentration at 1% or above. The pH of the buffer can vary widely, for example 5−9. Reducing agents (2-mercaptoethanol, dithiothreitol) are generally added but this is not obligatory. Our experience with 2-mercaptoethanol indicates that the high concentrations often used (1−5%) can induce proteolysis and that lower levels of dithiothreitol (2 mM) are preferable. Sucrose or glycerol are not necessary at this stage and can be included in the gel loading buffer. The sample can now be subjected to electrophoretic analysis immediately or it may first be incubated at various temperatures (e.g. 60−90 min at 20−30°C) to aid solubilization as monomeric species. The sample may also be boiled for 3 min, principally to destroy proteases, but great care must be exercised because some membrane proteins, such as rhodopsin, a lens plasma membrane protein, and the β-adrenergic, muscarinic and nicotinic receptors, are aggregated by boiling. However, proteolysis can be a problem since denaturated membrane proteins provide excellent substrates for SDS-resistant proteases, some of which may even be activated during membrane dissolution. For these reasons it is usually best to prepare membranes in buffers containing one or a cocktail of protease inhibitors (see Appendix I). If further precautions are required, 1 mM PMSF and 10 mM EDTA added along with SDS should prove successful. Finally the sample may be concentrated prior to electrophoresis by lyophilization, ultrafiltration (Amicon Centricon microconcentrators) or by precipitation with 50−90% acetone (latter not generally recommended since re-solubilization in monomeric form may be difficult).

4.1 Recovery of protein

Following electrophoresis, proteins can be visualized by a number of techniques. From the point of view of subsequent procedures, for example protein sequencing, the use of Coomassie blue R, sodium acetate or KCl (59−61) or reversible fluorescent probes is recommended (62) — reagents such as dansyl chloride, flucrescamine, *o*-phthalaldehyde or those used in silver and glycoprotein staining procedures, all of which can react irreversibly with the amino groups in the proteins, should be avoided. Best of all are procedures whereby the protein can be identified by previous covalent attachment of specific ligands (radioactive, fluorescent, etc), by modification of a limited number of sites, such as Cys, Tyr, or even by specific recognition with ^{125}I-labelled antibodies or lectins. The most successful method for recovery of the protein from the gel, electroelution with dialysis, is described in *Table 2*. Simple dialysis can also be

Table 2. Electroelution of proteins from SDS−PAGE.

Equipment/Reagents

1. Flat-bed gel tank of type in common use for DNA separations or adapted tube gel apparatus (see *Figure 1*)
2. Dialysis tubing — boiled for 15 min in 0.1 M sodium bicarbonate with 20 mM EDTA, then thoroughly rinsed with distilled water — use Spectropor for small proteins/peptides.
3. Coomassie stain stolution: 0.1% w/v Coomassie brilliant blue in 50% v/v aqueous methanol, 7% v/v acetic acid. The de-stain solution is identical, minus dye.

Procedure

1. Stain the gel very briefly with Coomassie blue (5−10 min) then rinse in de-stain solution until the protein bands become visible (10−15 min). Keep stain and de-stain times as short as possible for maximum yields.
2. Excise the band of interest. Do *not* break up or homogenize the gel slice.
3. Fill the flat-bed tank with buffer (25 mM Tris-Gly pH 8.5 0.1% w/v SDS, 50 ml Tris−acetate pH 7.8, 0.1% SDS, or 0.1 M sodium phosphate pH 7.8, 0.10% SDS) to a level approximately 1 cm above the platform. 0.1% v/v 2-mercaptoethanol or 2−5 mM dithiothreitol can be added to the buffer if required. For membrane proteins, the SDS concentration of the buffer can be raised to 1% w/v to ensure solubility.
4. Cut a length of dialysis tubing long enough to take the gel slice plus approximately 2 cm at either end. Clip one end with a Mediclip and fill the tubing with buffer from the tank. Place the excised gel slice in the tubing and gently squeeze out most of the liquid before sealing. Position the gel slice on one side of the dialysis bag. Care should be taken to avoid trapping any air bubbles.
5. Place the dialysis bag onto the platform of the electrophoresis tank. The level of buffer should be adjusted so that it just covers the bag.
6. Electrophorese for 3−20 h at 25−100 V constant current (20−150 mA) then reverse the current for approximately 30 s (in order to electrophorese the protein off the dialysis membrane surface). Longer staining/de-staining periods require the more extended electroelution times.
7. Remove the gel slice from the bag. Re-stain to check that the protein has been eluted. Any small fragments of gel that remain *must* be removed. If necessary, decant the solution and centrifuge briefly or filter to remove any small pieces. Re-seal the bag and dialyse the protein solution against at least five changes of distilled water (5 litres each) over 2−3 days at 0−4°C. The dialysis solutions can contain up to 0.25% w/v SDS to keep the protein in solution, but make the last change against distilled water only. The Coomassie dye stays with the protein throughout and can thus be used as an indicator of the progress of the electroelution.

Modifications

The procedure can be used with all forms of staining and with any apparatus based on tube gels where the dialysis bag is fixed onto the bottom of the tube (see *Figure 1*).

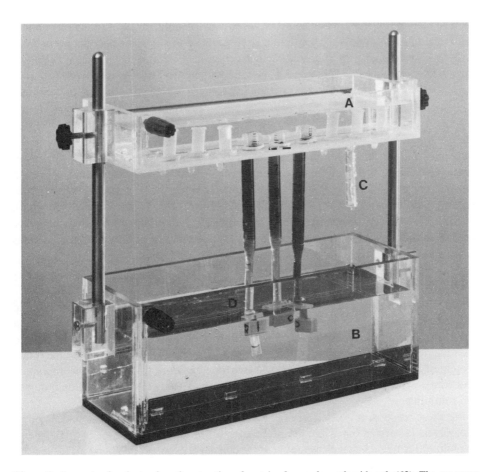

Figure 1. Apparatus for electrophoretic extraction of proteins from polyacrylamide gels (63). The apparatus consists of a height adjustable upper reservoir A and a lower reservoir B, both containing a lengthwise strip of platinum wire with appropriate plugs for connection to a power supply. Buffer is circulated between the reservoirs using a peristatic pump; an overflow pipe in the upper reservoir is shown (c). Up to 18 extractions of gel strips (two rows of nine holes) can be carried out simultaneously. Shown are three gel strips, contained in 10 ml pipettes cut off at the 5 ml mark. Remaining holes are closed off with plastic centrifuge tubes. The top of the pipette (pre-cleaned in acid) is sealed by a square sheet of dialysis membrane secured by an 'O'-ring. Proteins are electro-eluted (50 V for 12−18 h) into a length of dialysis tubing (D), fastened using a rubber 'O'-ring over the tip of the pipette and closed with a Mediclip (Spectrum Medical Industries Inc., 60916 Terminal Annex, Los Angeles, CA 90054). Almost identical systems can be constructed using circular electrophoresis apparatus (Medina and Findlay, unpublished). It is very useful to fabricate the glass tubes such that they contain small grooves at the top and bottom to accommodate the 'O'-rings and so ensure tight seals.

effective. Both methods are suitable for rehydrated gels but the yields are lower. In our experience, procedures which involve substantial crushing or dissolution of the gel produce unsatisfactory preparations either because they are contaminated or because the proteins are irreversibly modified or degraded.

Following recovery of the protein, it may be necessary to remove SDS. A number of procedures have been reported employing either resins to bind the detergent (64−67), or precipitation (60,68) or simply dialysis for an extended period (5 days) with numerous buffer changes. In most cases some SDS remains (if only for the proteins to remain soluble), unless a great excess of other detergents or in some cases denaturants such as 2−8 M urea, and 2−6 M guanidine hydrochloride are added to compete off residual SDS. The use of organic solvent (e.g. FACE) with gel filtration using LH60 or LH20 has been the most effective in removing all detectable traces of detergent (see Section 3.2.2).

5. ANALYSIS OF MEMBRANE PROTEINS

Once purified, all the usual analytical methods used in the study of protein structure, amino acid composition, N- and C-terminal residues, carbohydrate content and protein sequencing, etc., can be employed but, in most instances, care has to be taken to present the protein in a solubilized non-aggregated form. This is particularly true where enzymic digestion is to be carried out. These general methods are all detailed in another volume in this series (69). There are, however, a number of important considerations to bear in mind.

5.1 Protein cleavage

The purification of peptides from integral membrane proteins presents considerable problems due to their increased tendency to aggregate and their insolubility in most aqueous solvents. When soluble in detergent they generally exist as mixtures associated with the detergent micelles rather than in the monomeric forms exhibited by the intact protein or large fragments. In organic solvents the number of different chromatographic procedures which can be employed are limited since extremes of pH are often required for solubility.

For these reasons, it is often advisable to dissect the polypeptide in two stages. First, proteolysis is carried out while the protein is still in the membrane or in a detergent-solubilized form, such that cleavage occurs at only a very few peptide bonds. This restriction is best achieved by maintaining the protein in its native state, the lipid bilayer being the best environment of all. The choice of protease is largely governed by the protein under investigation. Good results have been achieved with the enzyme from *Staphylococcus aureus* V8 which, under the following conditions, cleaves at the C-terminal side of glutamic acid residues (22).

(i)   50−100 mM phosphate pH 7.0;
(ii)   1 mg/ml membrane protein (in membrane or detergent);
(iii)   up to 2% w/w V8 protease;
(iv)   25−37°C for up to 6 h.

The experimental conditions can be varied somewhat to get efficient cleavage at only a few sites. The enzyme can even be used in 0.1−1.0% SDS. Other proteases worth considering include trypsin, chymotrypsin, thermolysin and endoproteases Lys-C and Arg-C, incubated in phosphate or bicarbonate buffers pH 7.0−8.0. The large fragments which then result can be separated by chromatography in organic solvents or SDS−PAGE. When the latter is used it is often helpful to carry out pre-electrophoresis for a few hours to remove agents which may block N-termini but this often results in broader protein bands. Including 1% thioglycollic acid in the running buffer helps protect methionine residues. More extensive cleavage is obtained with enzymes such as pronase, subtilisin and papain. This type of proteolytic/PAGE approach can be used very successfully for peptide mapping (70,71).

The second stage of the strategy involves further degradation of these purified large fragments to give peptide mixtures simpler than would be obtained by degrading the whole protein. Proteolysis at this stage is often incomplete because of the insoluble or aggregated nature of the fragments or of the denaturing effects of the solvent. However, chemical cleavage, particularly at methionine residues by cyanogen bromide (CNBr) can be very effective. A simple protocol is given below.

(i)   Dissolve protein/peptide in 70% formic acid or TFA (the latter is better for Met-Thr and Met-Ser bonds but will oxidize tryptophan residues - ref. 4).
(ii)   Add 50- to 100-fold molar excess of CNBr over methionine (take care, CNBr breaks down to very toxic cyanides, weigh out and manipulate in a fume hood).
(iii)   Incubate in a N_2 atmosphere, in the dark at $20-25°C$, with gentle stirring, for 24 h.
(iv)   Rotary evaporate to near dryness.
(v)   Add water and lyophilize.

Other useful chemical cleavages can be obtained at tryptophan residues by *o*-iodosobenzoic acid in the presence of tyrosine (72) or by BNPS-skatole (73), at asparagine−glycine bonds by hydroxylamine (74) and, less successfully, at cysteine residues using 2-nitro-5-thiocyanobenzoic acid (75). The resulting peptides can then be purified in organic and/or acid solvents using gel filtration, reverse-phase chromatography (see above) or, in a few rare cases, by using affinity supports capable of interacting with -SH groups.

5.2 Determination of protein concentration

Because of the presence of detergents or phospholipid, often compounded by the small quantities of protein available, the determination of protein concentration can often be a particular problem. *Table 3* lists a number of possible assay methods which can be used. Unfortunately none of these are so free of problems that they can be recommended generally but we have found that the fluorescamine (Fluram, Roche; 76) and manual ninhydrin (77) methods and particularly amino acid analysis, are the most reliable. The last-named relies on having an amenable operator and machine (which is increasingly problematic) but the manual ninhydrin method, while a little tedious and time-consuming for large numbers of samples, is relatively sensitive, free of interference and can deal with difficult proteins/peptides, such as those soluble only in organic mixtures. For ease of operation and sensitivity the fluorescamine, Bradford (78), BCA (Pierce) and microLowry (79) procedures are very useful. In most cases serum albumin can be used as the standard but some variability should be expected.

5.2.1 *Fluorescamine*

(i)   Add 0.2 M sodium borate pH 9.0−9.5 (proteins), 8.0−9.0 (peptides) to the sample to give a volume of about 2 ml and a buffer concentration of over 0.1 M.
(ii)   Add 0.5 ml of 0.2 mg/ml fluorescamine in acetone; mix thoroughly.
(iii)   Leave for 10 min at room temperature before reading with excitation and emission wavelengths of 390 and 480 nm, respectively.

Table 3. Protein assays.

Assay	Range	Comments
Absorbance 280 nm	0.1−3 mg	Affected by aromatic content of protein and by some absorbing detergents, e.g. Tritons. Generally gives only rough estimate.
Absorbance 212 nm	1 μg−1mg	Depends on absorbance by peptide bond, therefore sensitive and relatively invariable, *but* subject to substantial interference (e.g. acids).
(Micro) Biuret	0.1−1 mg	Employs peptide bond therefore little variation with protein composition. Turbidity in detergents can be reduced with 1,2-propanediol (82).
Bradford	1−100 μg	Some variation with protein composition since relies on interaction with amino groups. Detergents can interfere. Variants on this scheme could be more sensitive or useful in some cases (83).
[³H]Dansyl chloride	0.1−1 μg	Reactive with amines, phenols, thiols and imidazoles. Sensitive but slow and complex procedure. Some detergents can interfere (84).
Fluorescamine	0.1−50 μg	Fluorescent on reaction with primary amines under alkaline conditions. Relatively free of interference, simple procedures, rapid results (Roche).
(Micro) Lowry	1−50 μg	Some variability with proteins, interfering detergents can be solubilized by SDS (79).
Manual ninhydrin	1−50 μg	No protein variability since proteins are degraded to amino acids. High levels of detergents can interfere. Time consuming (77).
BCA (Pierce)	0.5−50 μg	Very sensitive, easy to use and relatively free from interference.

(iv)   Sensitivity can be increased by hydrolysing protein/peptide with NaOH (see manual ninhydrin below), neutralizing with acetic acid and adding sodium borate buffer to pH 9.0.

5.2.2 *MicroLowry*

(Adapted from reference 80.)

(i)   To 200 μl of the protein solution in detergent, add 1 ml of fresh Lowry reagent (1 ml each of 2.0% sodium potassium tartrate and 1% $CuSO_4$ into 100 ml of 2% Na_2CO_3 in 0.1 N NaOH), mix thoroughly and allow to stand for 10 min.
(ii)   Add 10% SDS to a final level of 1%; mix thoroughly.
(iii)   Add 100 μl of 1:2 diluted Folin−Ciocalteau reagent, mix immediately.
(iv)   Read at 660 nm after 30 min and within 2 h.

5.2.3 *MicroLowry B*

(Adapted from reference 81.) Since not all detergents are soluble in SDS (e.g. DTAB), an alternative procedure involving precipitation of protein can be employed.

(i)   Add 30 μl of 1% sodium deoxycholate to protein solution (25 μg in 5−200 μl).
(ii) Add 1 ml of cold 12% trichloroacetic acid.
(iii) Incubate for 10 min. Spin at 1000 g for 20 min.
(iv) Add 1 ml of Lowry reagent to the pellet and solubilize (10−15 min).
(v)   Add 100 μl of 1:2 diluted Folin−Ciocalteau reagent; mix immediately.
(vi) Read at 660 nm after 30 min.

5.2.4 *Manual ninhydrin*

(i)   Dry down the sample in *polypropylene tubes*.
(ii) Add 0.15 ml of 9 M NaOH and incubate for 90 min at 110°C.
(iii) Carefully add 0.25 ml of glacial acetic acid while samples are still warm.
(iv) Add 0.5 ml of fresh ninhydrin reagent (2 g of ninhydrin, 40 mg of $SnCl_2$ in 75 ml of 2-methoxyethanol and 25 ml of 4 M acetate pH 5.5 bubbled with N_2) and incubate at 100°C for 15−20 min.
(v)   Cool, add up to 2.5 ml of ethanol and read at 570 nm.

6. COVALENT LABELLING OF PROTEINS

Covalent modification of individual amino acid side-chains can be used to facilitate the purification of membrane proteins which cannot be tagged with a specific ligand or to investigate the topographical arrangement of the polypeptide chain in the bilayer. Many of the reagents which have been developed for water-soluble proteins can be equally effective for integral membrane proteins. *Tables 4* and *5* contain a long but not complete, list of suitable compounds, concentrating particularly on those reagents which are readily available, can be easily synthesized and have been used with membranes. An instructive example of the way such reagents (and proteases) can be used to characterize membrane protein topography is seen in the studies on rhodopsin (reviewed in ref. 85). Appendix IV lists a further selection of reagents which could be used with membranes but which have not so far been employed in such studies.

6.1 Surface reagents

A particularly important consideration in many membrane studies is the need to guard against the possibility of the reagent penetrating the bilayer or in some cases being transported across the membrane. For these reasons the reagents have been classified as permeable where there is positive proof or even the possibility of their being able to enter or traverse the bilayer. However, in many instances the nature of the biological system may prevent this whilst in others (e.g. anhydrides) the reactivity of the reagent is so great that it is either inactivated relatively quickly in water or reacts rapidly with moieties present in the phospholipid headgroups. Carrying out the labelling reaction at low temperatures (0−4°C) can also help by reducing or inhibiting altogether the activity of transport systems. The net result is that many modifying agents are effectively impermeable. To ensure that there are no ambiguities in the data, however, it is advisable, at the end of the labelling period, to inactivate and remove by washing

Table 4. Membrane labelling reagents.

Amino acid	Potentially permeable	Impermeable	Reactions/conditions/comments	References
Cysteine	Acetic anhydride		See other anhydrides under lysine	
		Bromo-, chloro- and iodo-acetic acid/acetamide	pH 8.5 for maximum specificity, any suitable buffer, up to 10 mM, 30–60 min, 4–30°C. Iodo-derivatives most active.	86
		p-Chloromercuri-benzoate or benzene sulphonic acid	Non-covalent and reversible, slowly permeant. pH 7.0, up to 1 mM, 20–30°C for up to 3 h.	87
	Dansyl and dabsyl chlorides		Fluorescent, used dispersed in lipid vesicles; 0.1 M phosphate, 0.15 M sucrose pH 9.0, 4°C for up to 3 h.	88
	N-Ethyl and phenyl maleimides		pH 7.0–8.0, up to 1 mM, 20–37°C for up to 1 h. Can react with His and Lys under extreme conditions.	89
		3,5-Diiodo-4-diazobenzene sulphonic acid	See text Section 6.1.1	90
		NAP-Taurine (see also Table 5) Impermeable at 0°C	pH 7.0–8.0, up to 1 mM, 0–20°C for up to 30 min.	91,92
Glutamic acid	Carbodiimides	Some water-soluble forms	pH 4.5–7.0, mM, 20–30°C for up to 1 h.	93, 94
Glutamine	Transglutaminase		Catalyses incorporation of labelled amino-containing compounds, e.g. [14C]putrescine. pH 7.4, 25–37°C for up to 18 h.	95
Histidine	Dansyl and dabsyl chlorides		As for cysteine	
	Iodine		See text Section 6.1.1	
		3,5-Diiodo-4-diazo benzene sulphonic acid	As for cysteine	

Table 4. continued

Residue	Reagent	Conditions	Ref.
Lysine	Methyl and ethyl acetimidates	pH 8.0, 1–10 mM, 20–30°C for up to 2 h.	96,97
	Acetic and succinic anhydrides	Essentially impermeable due to high reactivity. pH 7.0–7.5, in aliquots up to mM, 4–25°C for 10–60 min.	90,98
	3-Azido-2,7-naphthalene disulphonate (see Table 5)	pH 6.0–8.0, up to 1 mM, 4–37°C for up to 1 h.	99
	3,5-Diiodo-4-diazo-benzene sulphonic acid	See Section 6.1.1.1	90
	Dansyl and dabsyl chlorides	As for cysteine, amino groups more readily labelled.	
	1-Fluoro-2,4-dinitrobenzene (Sanger's reagent)	pH 8.0–9.0, up to 10 mM, 20–30°C for 1–3 h. Also reacts with cysteines and tyrosines.	100
	N-formyl methionyl sulphone methyl phosphate ([35S]-FMMP)	pH 9.5–10.0, 50 μM, 20–30°C for 10–30 min.	101
	N-Hydroxy succinimyl biotin and iminobiotin, sulphosuccinimyl biotin	Donate free biotinyl moieties, thereby facilitating purification using avidin-affinity resins. Can introduce spacer arms and cleavable bonds; pH 8.0, 1 mM, 4°C for 2 h.	102 Pierce
	Dimethyl aminoazobenzene- and phenyl isothiocyanates	pH 7.0–8.0, up to 1 mM, 4–30°C for up to 1 h. Show particular specificity for the anion transport protein (of erythrocytes)	103 104

continued overleaf

Table 4. continued

Amino acid	Potentially permeable	Impermeable	Reactions/conditions/comments	References
Lysine (cont.)	N-Succinimidyl propionate and its 3-(4-hydroxy phenyl), 3-(4-hydroxy-5 iodo phenyl) and 3-(4-hydroxy-3,5 diido-phenyl) derivatives (Bolton and Hunter Reagent).	Pyridoxal phosphate plus tritiated borohydride	pH 7.5, up to 1 mM, 4–25°C for up to 30 min.	105
		t-Butoxy-L-[35S]methionine, N-hydroxy-succinimidyl ester ([35S]LR)	pH 6–9 (hydrolyses at alkaline pH)	Amersham
		Sulpho succinimidyl-3-(4-hydroxy phenyl) propionate (water-soluble Bolton and Hunter Reagent).	μM (10- to 100-fold molar excess over -NH$_2$), 0–30°C for 10–30 min (see *Table 7*). Widely used to ^{125}I-iodinate proteins — see Section 6.1.1	106 107
		Sulphosuccinimidyl acetate	Use as for succinimyl esters (*Table 7*)	Pierce
		2,4,6-Trinitro-benzene sulphonate	pH 8.0–9.0, up to 1 mM 20–40°C, for up to 24 h	108
Tyrosine (-OH)	Acetic anhydride Dansyl and dabsyl chlorides	3,5-Diido-4-diazobenzene sulphonic acid	See under Lysine	
			See under Lysine	
	Free iodine	Iodination with iodogen, iodobeads, lacto-peroxidase	See Section 6.1.1	
		p-Nitrobenzene sulphonyl chloride	pH 7–8.0, 1 mM, 25°C; 30 min.	

Table 5. General photosensitive labelling reagents.

Permeable	Impermeable	λ_{max}/comments/references
Adamantane diazirine		337−372 nm (118)
1-Azido-4-iodobenzene		260 nm (119−122)
1-Azido-5-iodonaphthalene		310 nm (122)
5-Azido-1-naphthalene	3-Azido-(2,7)-naphthalene disulphonate	319,382 nm 0.1−1 mM, pH 7−8 4−30°C (99)
12-(4-azido-2 nitrophenyoxy) stearoyl glucosamine		1−2M slight detergent properties (123)
N-(4-Azido-2-nitrophenyl) 2-amino ethyl sulphonate (NAP-Taurine)		permeability can be substantially reduced at 0−4°C. 0.1−1 mM, may have detergent-like properties (91,92)
5-(4-Azido-2-nitrophenyl) [^{35}S]thiophenol		395 nm (129)
Diazofluorene	Diazotrifluorides (also permeable versions)	(124, 125)
	3,5-Diiodo-4-azido benzene sulphonic acid	(126)
Hexanoyldiiodo-N-(4-azido-2-nitrophenyl)-tyramine		460 nm (129)
Pyrene-1-sulphuryl azide		355 nm (130)
3-Trifluoromethyl-3-(m-iodophenyl) diazirine (TID)		diazo isomer inert 353 nm (127,128)

as much as possible of the reagent before proceeding. Since non-covalent associations with the membrane surface can persist, it is often advantageous to include a carrier protein (e.g. up to 1 mg/ml BSA) in the washing buffer.

6.1.1 *Iodination*

The use of ^{125}I as a radioactive label has been particularly successful because of its ease of detection (γ-radiation), its high specific activity and its relatively long half-life (60 days). ^{125}I can be introduced into a protein by direct iodination of tyrosine residues or by the attachment of reagents containing an iodo-substituent to a number of amino acid side-chains. Procedures using the enzyme lactoperoxidase are the gentlest.

(i) *Lactoperoxidase.* (109,110). This enzyme catalyses the incorporation of iodide ions into the meta position of tyrosine side chains in a reaction dependent on hydrogen peroxide. In the presence of high levels of H_2O_2, molecular iodine is rapidly generated and the enzyme can be inhibited. Since I_2 itself carries out iodination of both lipids and proteins and readily penetrates the bilayer, surface labelling can be compromised. The problem is substantially reduced by including an enzyme (glucose oxidase) which will generate H_2O_2 from glucose only at about the rate it is utilized by lactoperoxidase (i.e. no large excess is present at any time). Using lactoperoxidase bound to insoluble resins is thought to ensure that only surface labelling occurs. The simplest method employs beads containing both bound lactoperoxidase and glucose oxidase (Enzymobeads - Biorad). I_2 can also be generated from I^- by peroxidases present within cells. Therefore, it is often important to include a control lacking lactoperoxidase.

Procedure.

(1)   Mix substituents such that the final concentrations are: 200 mM phosphate pH 7.2 [or phosphate-buffered saline (PBS)]; membrane (cell) suspension at up to 1 mg protein/ml (max); 20 mM β-D-glucose; and 1 mCi carrier-free Na[^{125}I]. *Azides and Thiols should be rigorously excluded since they inhibit iodination.*

(2)   For every 100 μl of incubation mixture, add 50 μl of the Enzymobead preparation *or* 2 μg of lactoperoxidase plus 1 U of glucose oxidase.

(3)   Incubate for $10-30$ min at $4-25°C$.

(4)   Wash membranes (cells) three times in the phosphate buffer.

(5)   Pellet beads by centrifugation (100 g for 10 min). Membranes may be dissolved in detergent prior to centrifugation.

(6)   A useful test of whether any non-specific labelling has occurred is to measure the amount of ^{125}I in extracted lipids (see Chapter 3).

(ii) *1,3,4,6-Tetrachloro-3,6-diphenylglycouril.* (See ref. 111 for earlier references.) This reagent (IODO-GEN; Pierce) efficiently catalyses the iodination of proteins via the oxidation of I^-. It is water insoluble and normally used as a coating on the walls of the reaction vessel. It is hydrophobic, however, and potentially could penetrate the bilayer. As a vectorial label, therefore, it is less useful than other procedures.

Procedure.

(1)   Coat glass vials with up to 1 mg/ml Iodo-Gen dissolved in chloroform and evaporate the solvent in a vacuum or a N_2 stream.

(2)   Add membranes (1 mg protein) suspended in 1 ml of PBS pH 7.4 and 1 mCi Na[^{125}I]. (1 μg reagent/10 μg protein).

(3)   Incubate at $4-25°C$ for $5-30$ min with occasional gentle swirling.

(4)   Terminate reaction by removing membranes and washing in buffer containing 1 mg/ml BSA.

(iii) *Immobilized Chloramine T (112).* The Chloramine-T method for iodinating tyrosines is very effective but also quite harsh. Its worst effects can be mitigated by the use of Iodo-Beads (large polystyrene beads derivatized with N-chlorobenzenesulphonamide; Pierce). Good vectional labelling can be achieved. The reagent may be used in the presence of azide but its activity is effectively quenched by 2-mercaptoethanol. Recommended reaction conditions are:

$5-500$ μg protein/bead;

$0.1-0.2$ M phosphate pH $7.0-7.4$ (Tris slightly inhibitory);

1 mCi Na[^{125}I];

total volume of up to 1 ml/bead;

incubate at $4-25°C$ for $10-30$ min.

(iv) *3,5-Diiodo-4-diazobenzene sulphonate (DDISA).* Sections (i)−(iii) predominantly result in the iodination of tyrosine residues. It may be that the number of such residues is limited or that modification of tyrosines results in the destruction of biological activity. In such cases, use can be made of agents which have wider or different specificity (110). DDISA (90,113) is very useful since it can react with lysine, histidine, cysteine as well as tyrosine residues and the conditions can be simply modified to ensure preservation of biological activity.

Reaction conditions (90 and references therein):

membranes at up to 2 mg protein/ml;

0.1−0.2 M sodium phosphate pH 7.4;

up to 1 mM [^{125}I]DDISA (5 Ci/mol) or about 100-fold molar excess over protein;

incubate at 4−25°C for 5−30 min.

The reaction can be quenched with a large molar excess (100-fold e.g. 10 mM) of histidine and the membranes washed free of reagent by centrifugation and resuspension in phosphate buffer.

(v) *N-Succinimidyl-3(4-hydroxyphenyl)propionate (106,107).* This compound (Bolton and Hunter reagent) or its water-soluble sulphonated derivative (Pierce) can be obtained in iodinated form or be iodinated using the Chloramine-T method (Amersham). It covalently attaches to amino groups via the succinimidyl ester moiety under mild conditions (e.g. phosphate buffers, pH 7.0−7.5, 10- to 100-fold molar excess, 0−30°C for 10−30 min), thereby irreversibly iodinating proteins (see *Tables 4* and *7*). Substantial protein modification can occur and the reaction conditions should be carefully monitored, especially if biological activity is to be preserved.

6.1.2 *Cell surface carbohydrates*

If modification of cell surface proteins is undesirable, methods exist whereby carbohydrate side-chains can be labelled. The best of these uses galactose oxidase to convert terminal galactose residues to the corresponding 6-aldehyde, followed by reduction with NaB[$_3$H]$_4$ to regenerate the original galactose now bearing a tritiated grouping (114). Some prefer KB[$_3$H]$_4$ due to its increased storage stability.

(i)   Incubate membranes (cells) at 1 mg protein/ml PBS pH 7.4 with 10−20 U of galactose oxidase at 25−37°C for 30 min with gentle agitation.

(ii)   Wash membranes (cells) three times in the same buffer using centrifugation conditions that will not disrupt the cells.

(iii)   Resuspend membranes (cells) in the phosphate buffer, add 1 mCi of NaB[^{3}H]$_4$ dissolved in 10 mM NaOH and incubate at 25°C for 30 min with gentle agitation.

(iv)   Wash by centrifugation and resuspension as above.

It is important to include a control incubation which has not been subjected to the action of galactose oxidase, since NaB[^{3}H]$_4$ will react with some side-chains (e.g. sulphydryls).

Terminal galactose residues can be created/increased by removal of sialic acid using neuraminidase (1 U as above, prior to galactose oxidase). Alternatively, sialic acid itself may be labelled with borohydride following treatment of membranes with 1−2 mM sodium metaperiodate in phosphate buffer pH 7.4 for 5−10 min at 0−20°C (115,116). The reaction is quenched with 100 mM glycerol in buffer, the membranes washed and then exposed to [^{3}H]borohydride, as above.

6.2 Hydrophobic reagents

This class of compounds (*Table 5*; see 117 for review) has been specifically developed with membranes in mind, the object being to inculcate the inactive reagents into the bilayer and once there, to generate a reactive species which will react only with intra-membranous substituents. This activation is usually achieved by irradiation but care must be taken not to destroy biological activity. For this reason wavelengths above 300 nm are preferred. The two most important criteria for these reagents are their hydrophobicity and reactivity. Since the aim in many instances is to specifically label and perhaps then identify amino acid side-chains exposed to the fatty acid milieu of the bilayer, scavenging agents are often added to the aqueous phase to prevent any surface labelling. Thiol-containing compounds such as cysteine and reduced glutathione are the most effective in this respect. However, their use may sometimes reduce the totality of information that can be gained since they may competitively prevent labelling in the region of the membrane occupied by phospholipid head-groups.

Recent evidence suggests that the azides and diazirines (nitrene and carbene precursors, respectively), because of their different reactivities can give rise to information of different kinds. Activated products of the less reactive azides (*Table 5*) attack primarily nucleophilic groups, particularly cysteine and tyrosine (perhaps also methionine) residues followed by tryptophan, lysine and histidine (120,121). They have only a very limited ability to insert into hydrocarbon side chains. The activated products of the diazirines, on the other hand, whilst preferring nucleophilic residues, will also react with hydrocarbons (128). For general labelling, therefore, the latter, particularly 3-trifluoromethyl-3-(*m*-iodophenyl)diazirine (TID) may be best (127). Where there is a danger that so much labelling can occur that the positive identification of labelled sites by protein sequencing becomes a problem due to background contamination, the azides may be preferred.

On activation by u.v. irradiation, azides and diazirines break down to give the reactive nitrene and carbene species, respectively, together with a series of azirines and diazo derivatives which may also be reactive (that of TID is not). Since many of the species have increased hydrophilicity and longer life-times, the possibility of significant diffusion to surface regions is a potential problem. This can be guarded against by photoactivation of samples at below the lipid transition temperature but freezing should be carried out very rapidly to prevent protein aggregation (120,121). A typical protocol is given in *Table 6*. It is also worthwhile consulting the various references to evaluate individual variations. The controls indicated can be very useful.

6.3 Amphipathic reagents

This group includes reagents such as 12-(4-azido-2-nitrophenoxy)stearoyl glucosamine (123) which will integrate into bilayers such that the sugar residue is positioned at the aqueous surface. With sealed membranes this type of probe is thought to reside in only the exposed monolayer but with time some redistribution to the inner layer must occur, if only slowly. Care must be taken that the reagent does not exert its detergent properties.

Where it is possible to reconstitute the protein(s), the use of phospholipids containing photoactivatable moieties in the head group (132), glycerol (133) or fatty acyl chains (134–136) can have advantages. These molecules remain membrane bound (i.e. do

Table 6. Labelling with photosensitive hydrophobic reagents.

1.	Pre-incubate membranes (1−5 mg protein/ml) in a suitable buffer (e.g. phosphate, Tris-HCl) containing, if required, up to 10 mM scavenger, at a suitable temperature (4−37°C) and in a N₂-atmosphere. A glass vessel (capped) is preferred.
2.	*In the dark*, add the reagent dissolved in a miscible organic solvent (usually ethanol) such that the final volume of solvent is less than 1% (or at a level that does not affect the labelling or biochemical activity of protein/membrane) and the reagent is present at 10 μM to 1 mM. Depending on the partition coefficient, the intramembranous concentration will be at least 1000-fold higher so lower levels are preferable. Continue incubation for 10−30 min.
3.	Rapidly freeze the sample as a thin film by dropwise addition into a glass vessel suspended in liquid N₂. This step is optional and can be used when diffusion may be a problem.
4.	Irradiate the sample (frozen) for 1−90 min under a N₂ atmosphere, the time being dependent on the level of probe incorporation required and the need to avoid damage to the biological components. (Remember that glass has u.v.-absorbing properties.) A u.v. lamp emiting a range of wavelengths can be used but it is better to use one of defined wavelengths corresponding to the absorbance maximum of the reagent. *Wear u.v.-absorbing goggles, reduce exposure to a minimum and irradiate in a fume cupboard or well-vented room.* Continue the incubation in the dark for a further period of up to 90 min to ensure complete reaction of any long-lived species. This is particularly important if low temperature conditions are being employed.
5.	Wash the membranes extensively (after rapid thawing if necessary) in buffer containing up to 1 mg/ml BSA to remove non-covalently bound photolysis products. Alternatively, they may be immediately dissolved in SDS or FACE (120,121) or the proteins precipitated with 7−10 volumes of cold acetone.
6.	It is important to include two controls.
	(1) A membrane sample incubated with reagent *but not irradiated and* carried through the whole procedure in the dark.
	(2) A membrane sample incubated in the dark with *pre-irradiated* reagent (say up to 10 mM) then carried through the protocol in the dark.
	The first control guards against interactions not associated with the reactive species. The second gives some estimate of the degree of labelling that could result from long-lived reactive species and, depending on how soon after irradiation the probe is added to the membrane, the life-time of these reactive moieties.

not repartition to any significant extent) but it should not be assumed that the position of the photosensitive group is uniform or static. Membrane fluidity characteristics can mean that the group is found at different depths in the bilayer. Their use as 'depth probes' can, therefore, be problematic. These phospholipid analogues are generally reconsituted along with native phospholipids at molar ratios ranging from 1:10 to 1:1000. The experiment is usually critically dependent on reconstitution faithfully representing the native organization of the protein *in situ* (see Chapter 5). They can be particularly useful when introduced as modified fatty acids, and subsequently incorporated into the membrane in the course of *in vivo* biosynthesis (137).

6.4 Cross-linking agents

One of the more common approaches to help define the molecular organization of proteins is the use of bifunctional agents which can cross-link components which are in permanent association over the time course of the experiment (for general reviews see 117, 138). The number of reagents available has been growing exponentially over the last few years, and they now vary widely in size, reactivity and in the nature of their active moieties. Three chemical species are most common, the imidates, the N-hydroxy

Table 7. Cross-linking agents.

Specificity	Reagents/reaction	Comments/conditions	References
Arginine $\left(-NH-\overset{NH}{\underset{\parallel}{C}}-NH_2\right)$	*Phenyl glyoxals*	pH 7–8 but product slowly cleaved at neutral or alkaline pH mM 30–60 min 20–30°C	145 146
Aspartic and Glutamic acids $(-COOH)$	*Carbodiimides* $-N=C=N-$ $R-NH_2 + HOOC-R_1 \longrightarrow R-\overset{H}{\underset{}{N}}-\overset{O}{\underset{}{C}}-R_1$	Catalyse formation of peptide bonds with intrinsic or add amino containing compound pH 4.5–7.0 mM 30–60 min 20–30°C	147
Glutamine $\left(-\overset{O}{\underset{\parallel}{C}}-NH_2\right)$	catalysed by *transglutaminase* $R-\overset{O}{\underset{\parallel}{C}}-NH_2 + NH_2-R_1 \longrightarrow R-\overset{O}{\underset{\parallel}{C}}-\overset{H}{\underset{\parallel}{N}}-R_1 + NH_3$	pH 7.4 25–37°C 1–18 h	95 148
Cysteine $(-SH)$ (free or generated by reduction)	*Acyl halides or amino halides* $Y=\overset{O}{\underset{\parallel}{C}}-CH_2-Br\ (I)$ $R-SH \longrightarrow R-S-CH_2-\overset{O}{\underset{\parallel}{C}}-X$ *Cupric di-(1,10)-phenanthroline* $R-SH + SH-R_1 \longrightarrow R-S-S-R_1$	pH 7–9 µM–mM 4–25°C for up to 2 h. pH 7–9 up to 100 µM phenanthroline plus up to 10 µM CuSO$_4$ 0–25°C for 5–20 min Induces disulphide formation of suitably situated sulphydryls Quench with 1 mM EDTA	156 119 149 150

206

	Reagent	Structure	Conditions	Notes	Refs
Cysteine (cont.)	*Maleimides*		pH 7–9 μM 0–40°C 10–60 min	Under harsh conditions will react with lysine and histidine gives disulphide bonds with thiophthalimides	151 152
	Pyridyl disulphides		pH 8 μM 4–40°C 1–18 h.		153 156
	Sulphenyl halides		pH 8.0 μM 0–40°C, 10 min–1 h	for synthesis see 154. also reacts with tryptophans (155)	151 153 154
Lysine (-NH₂)	*Aryl azides* R₁=H,OH,NO₂		pH 7–9 mM −100°C to +25°C 10–60 min	Add in dark, irradiate above 300 nm.	138 152 154 155 156 157 158 159 160
	Aryl halide		pH 7–9 μM 0–40°C	has also been used to cross-link phospholipids	139 162 163 162

Reagent	Reaction scheme	Conditions	Ref.
Diazotrifluorides	$X-\overset{O}{\overset{\|}{C}}-\overset{N_2}{\overset{\|}{C}}-CF_3$; $R-NH_2 \longrightarrow$?	pH 7.0–8.0 μM irradiated at 254 nm for 2 min at 4°C Has wider specificity e.g. -SH.	125
N-Hydroxysuccinimide esters $R = -SO_3Na$ for water soluble *sulphonates*	$X-\overset{O}{\overset{\|}{C}}-O-N\text{(succinimide, R)}$; $R-NH_2 \longrightarrow R-\overset{H}{N}-\overset{O}{\overset{\|}{C}}-X$	pH 6–9, some hydrolysis at alkaline pH. μM concentrations, 10- to 100-fold molar excess over amino groups. 0–30°C 10–30 min Add in solutions of ethanol, acetone or DMSO. Keeping final solvent concentration below 1%. For synthesis see ref. 170	139. 164 165 166 167 168 169 170
Imidates	$X-\overset{NH_2^+Cl^-}{\overset{\|}{C}}-OCH_3$; $R-NH_2 \longrightarrow R-\overset{H}{N}-\overset{NH}{\overset{\|}{C}}-X$	pH 8–10, hydrolyse more rapidly at neutral pH μM concentrations or at least 100-fold molar excess over amino groups. 0–40°C, significantly more active at the higher temperatures. 10–60 min.	139 168 171
Isothiocyanates	$X-N=C=S$; $R-NH_2 \longrightarrow \text{Ⓧ}-\overset{H}{N}-\overset{S}{\overset{\|}{C}}-\overset{H}{N}-\text{Ⓨ}$	pH 7–9, but unstable at alkaline pH μM–mM 0–50°C 5, 30 min	104
Diazobenzenes (Tyrosine/Histidine)	$X-\text{benzene}-N_2^+Cl^-$; phenol(OH) $\longrightarrow N=N-\text{Ⓧ}$ (with OH)	pH 7.0–8.0 mM 4–40°C for up to 30 min also reacts with amines and thiols.	103 172
Trifluorodiazirines Non-selective	$CF_3-\overset{N=N}{\overset{\diagup\diagdown}{C}}$ (diazirine)	pH 6–8, μM, 0–37°C for up to 20 min activated by irradiation at 350 nm nucleophiles most reactive.	117

succinimide esters and their water-soluble sulphonate variants, and the photoactivatable aryl azides. The first two react almost exclusively with amino groups whilst the third is directed principally against nucleophiles. Some use has also been made of activated halogen reagents (e.g. iodo and bromo-acetyl groups) and of maleimides, especially where thiol groups are involved or suspected. *Table 7* summarizes the various reactive species and their specificities. Bifunctional reagents arise through combinations of similar (homobifunctional) or different (heterobifunctional) species and vary further in the length of chain between the active moieties, whether or not they can be cleaved and their hydrophobicity (for available agents see Amersham, Calbiochem and Pierce).

These reagents are usually stored at −20°C under anhydrous conditions since many are labile in aqueous conditions. They can then be added to the reaction mixture in buffer, ethanol, or dimethylsulphoxide. At the end of the experiment, excess reagent should be quenched and removed by washing the membrane. Since the aim in many experiments is to define the most intimate interprotein associations, it may be instructive to begin the investigation with reagents catalysing disulphide (Cu-phenanthroline) or peptide (carbodiimide, transglutaminase) bond formation or with reagents possessing very short distances between the reactive moieties (e.g. difluorodinitrobenzene, 139). Whether or not the geometry allows successful cross-linking with such so-called 'zero-length reagents', longer and more flexible cross-linkers and agents possessing different specificities should then be recruited to the study. *Table 7* summarizes the conditions under which these reagents can be employed.

6.4.1 *Nearest-neighbour interactions*

Where a membrane protein can be detected on a SDS−polyacrylamide gel either through specific labelling (usually covalent) or because it is essentially the sole or only one of very few species present [e.g. rhodopsin (139), myelin (140)], cross-linking agents can be employed to investigate its molecular organization. The experimental conditions are those which apply for water-soluble proteins and depend mainly on the nature of the cross-linker (see *Table 7*). However, there are a number of controls which ought to be considered if the results are to be unambiguous. Firstly, it is worthwhile using agents (e.g. those containing a disulphide bond) which are cleavable, because the ability to regenerate the monomeric polypeptides helps guard against any non-specific but irreversible aggregation which may result from modification of amino acid side-chains (see *Table 8*). Controls with no or inactivated cross-linking agent, or best of all, with a related monofunctional reagent, should also be included.

Secondly, it must be appreciated that the effective protein concentration in the membrane is very high (i.e. with a total protein concentration in the reaction vessel of 1 mg/ml, the *effective* concentration is nearer 100 mg/ml). The frequency of random collision is correspondingly that much greater and may give rise to considerable amounts of non-specific association thereby adding confusion to the interpretations. The situation can be improved by reducing three variables — the reagent concentration (from mM to μM), the incubation time (e.g. from 1 h to 10 min) and the incubation temperature. The degree of temperature reduction will depend, of course, on the activity of the functional group but there are some advantages to be gained from *rapidly* reducing it below 4°C or at least below the transition temperatures of the phospholipid and thereby effectively 'freezing the steady-state'.

Table 8. Cleavable substituents in cross-linking agents.

Groups		Conditions	References
Amidine	HN H $\|$ $\|$ -C-N-R	2 M methylamine pH 11.5 75% acetonitrile 3 h 37°C	173
Azo	-N=N-	0.1 M sodium dithionite 0.15 M NaCl pH 8.0 30 min 20−30°C	174
Disulphide	-S-S-	10 mM dithiothreitol or 2-mercaptoethanol pH 7.0−8.0 30 min 4−30°C	
Ester	O $\|\|$ -C-O-	1 M hydroxylamine pH 7−9 25 mM CaCl$_2$ 1 mM benzamidine 3−6 h 20−40°C	175
Glycol	OH OH $\|$ $\|$ -CH-CH-	15 mM sodium periodate pH 7.5 (not Tris) 4−5 h 20−30°C	176
Sulfone	O$_2$ $\|\|$ -S-	100 mM NaH$_2$PO$_4$ pH 11.5 6 M urea 2 mM dithiothreitol 2 h 37°C	177

The interpretation of cross-linking results is a hazardous occupation (141). By far the most popular procedure involves identification of the oligomeric species on SDS−PAGE. Where cross-linking *by more than one agent* gives rise to dimers and tetramers as the predominant species, it is usually reasonable to deduce a native dimeric (or perhaps tetrameric) organization. A geometrically decreasing progression of dimers, trimers, tetramers, pentamers, etc. more often implies a monomeric arrangement or a massive aggregate. Where one protein is present in high concentration, artefactual oligomeric arrangements can be seen with one or two bifunctional reagents, so it is important to experiment with several cross-linkers.

Sometimes clear confirmation of the interpretations can be obtained by first dissolving the membrane samples in detergents which preserve the native activity (and hence presumably also the organization) of the protein and then adding the cross-linking agent. Under these conditions, the effective protein concentration and the frequency of random collisions are reduced and with it any non-specific 'background' cross-linking (139).

6.4.2 *Receptor—ligand interactions*

The addition of cross-linking agents to a membrane suspension pre-incubated under equilibrium conditions with radiolabelled hormone or other protein ligand has successfully led to the identification of specific protein receptors (for review see 142). The experimental conditions are generally very similar to those used for nearest-neighbour studies (see Section 6.4.1) except that the ratio of receptor to ligand should be kept as high as possible (see later). These experiments, however, are prone to the same problems of background cross-linking identified above and for these reasons, an alternative strategy has been devized. In this case, the hormone/protein ligand is pre-labelled with a photosensitive heterobifunctional agent. Provided no significant disruption has occurred in its ability to react with the specific receptor, the derivatized ligand is now introduced to the membrane in the dark and in concentrations sufficient to promote interaction with specific high-affinity receptors. Once equilibrium has been reached, the sample is exposed to u.v. light and the dormant azide activated. Although the overall amount of cross-linking between ligand and receptor thereby obtained is significantly below the 20—25% optimal value often seen (at best 1—2%), the increased specificity of the interaction clarifies ambiguity. It can also help define more accurately which subunit in an oligomeric receptor possesses the binding site for the hormone/protein ligand.

For such experiments, bifunctional agents based on an aryl azide and the N-hydroxy succinimide ester with or without cleavable groups have proved successful (143,144). Two examples are:

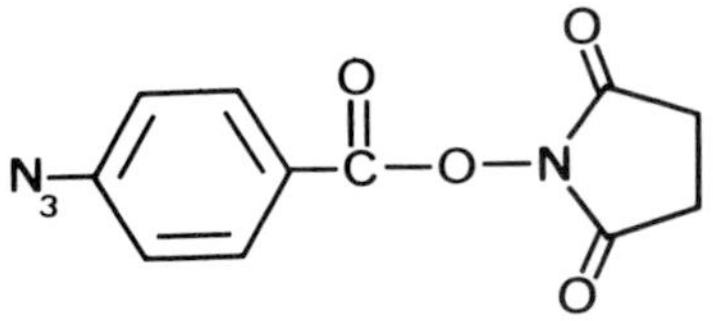

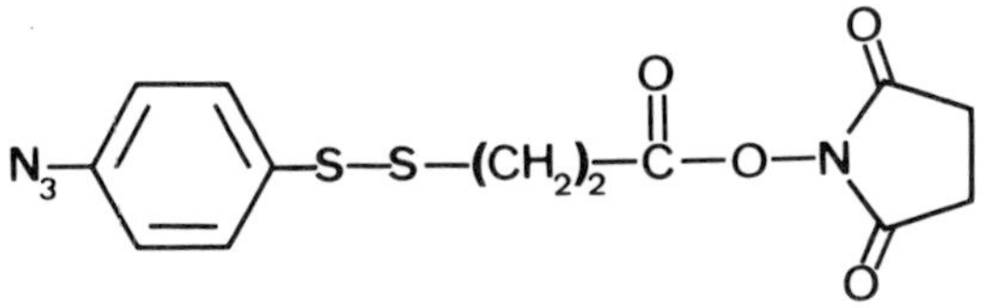

A typical procedure could be as follows:

(i)   Iodinate the protein ligand using the methods described in Section 6.1.1. The soluble protein should be separated from the components of the reaction by centrifugation or gel filtration.

Steps (ii)—(iv) should be carried out in the dark.

(ii) Incubate [^{125}I]ligand (10—20 µg) under N_2 in a solution containing 0.1 mM of the bifunctional cross-linker in 0.1 M phosphate pH 7.4 for 10−30 min at 4−25°C. The reaction may be terminated with 1 mM lysine.

(iii)   Purify the modified [^{125}I]ligand by gel filtration on a small column (e.g. 2 ml of Sephadex G-50) in buffer appropriate for incubation with membranes (cells). *Check that binding constants are not substantially different.*

(iv)   (a)   Add conjugated [^{125}I]ligand to membranes under conditions used in the receptor binding assays (0−37°C) and allow the establishment of equilibrium. In an ideal situation, sufficient ligand should be added such that receptor site occupancy is in the region of 50% while unbound ligand represents 60−70% of the total added. This ideal is achieved when K_d values are low (10^{-8} to 10^{-9} M) and receptor numbers are high (for discussion see ref. 117). If binding is very tight, membranes can be washed to reduce non-specific binding but it may be preferable to include a control (b).

        (b)   Set up a parallel incubation in which non-iodinated ligand is also added but in great excess (at least 200-fold).

(v)   Irradiate both samples under N_2 with u.v. light (e.g. mercury arc lamp) for 1−10 min at 4−25°C. The reaction may be terminated with 1 mM dithiothreitol. Wash the membranes (cells) in appropriate buffer.

(vi)   Specific receptor binding is identified on SDS−polyacrylamide gels by the presence and absence of labelled species resulting from incubations (a) and (b), respectively.

    Where it is inadvisable or impossible to render the ligand radioactive, steps (ii)−(vi) can be carried out using bifunctional reagents such as [cysteamine-^{35}S]N-succinimidyl -3-[2-nitro-4-azidophenyl)-2-aminoethyldithio]propionate ([^{35}S]SNAP; Amersham). This reagent has the advantage not only of identifying the cross-linking complex but also of identifying the receptor itself which retains the radioactive moiety after treatment of the complex with dithiothreitol to uncouple ligand and receptor.

    N-[4-(4′-azido-3′-[^{125}I]iodophenylazo)benzoyl]-3-aminopropyl-*N*-oxy-succinimide ester, available from New England Nuclear, is another reagent of this type. It is cleaved by sodium dithionite (see *Table 8*) and therefore does not leave a free -SH group as in the case of SNAP.

7. RADIATION INACTIVATION

Estimates of molecular weight based on gel filtration and subunit composition can be usefully supplemented by methods which do not rely on solubilization of the protein. Such an approach is represented by radiation inactivation analysis which may in certain cases be the only or the best means of determining native molecular weight (for review see 178). The method is based on the destruction of the biological activity of the protein *in situ* by X-irradiation. Protein inactivation is a function (usually logarithmic) of the total dose of radiation and this dose-dependency is in turn closely related to target size (i.e. the molecular volume of the fully functional protein). The relationship at its simplest, is expressed as:

$$\text{Molecular weight} = 6.4 \times \frac{10^{11}}{D}$$

where D is the dose in rads which reduces activity to 37% of the control value. This equation was drived empirically from experiments carried out at room temperature. It is recommended, however, that most studies on membrane-bound systems be carried out in the frozen state since sensitivity to irradiation increases as the temperature rises. Under these conditions, it is usual to construct a calibration curve of molecular weight versus sensitivity to irradiation for a range of known proteins. Data for the unknown system can then be applied directly to this reference standard. The approach works well with monomeric membrane proteins (179,180) and moderately well with oligomeric systems (181,182). Usually the derived molecular weight gives the native active molecular composition although this is not necessarily the same as the smallest entity capable of full activity. In some instances, target sizes representing only part of the molecular complex are obtained (183), presumably reflecting the absence of 'subunit-cross-talk'. It is important, therefore, to utilize other lines of investigation.

Radiation inactivation can prove intriguingly informative in the study of induced association — for example, the coupling of receptor and adenylate cyclase in the presence of hormone. In such situations, molecular weight estimates are capable of distinguishing between an uncoupled cyclase system and the coupled complex in the presence of hormone (184). Even more than before, however, independent verification of the interpretation is essential since, being equilibrium systems, the scope for variability is even greater with such complex assemblies (185). Wherever possible, SDS−polyacrylamide gel electrophoresis should be carried out at the end of the experiment to assess the effects of irradiation on the membrane protein(s).

8. CONCLUSION

Although the manipulation of integral membrane proteins is still a difficult and non-routine task, each year sees some progress in the development of new or improved procedures. There seems to be particular scope for advances in instrument and resin technology for the isolation of hydrophobic peptides/proteins. When coupled with modern efficient protein labelling and sequencing techniques, these improved isolation methods promise rapid progress in our understanding of protein structure, disposition and organization.

9. REFERENCES

1. Marchesi,V.T. and Steers,E., Jr. (1968) *Science,* **159**, 203.
2. Fairbanks,G., Steck,T.L. and Wallach,D.F.H. (1971) *Biochemistry,* **10**, 2606.
3. Tanner,M.J.A. and Gray,W.R. (1971) *Biochem J.,* **125**, 1109.
4. Steck,T.L. (1972) *Biochim. Biophys. Acta,* **225**, 553.
5. Carter,J.R., Jr. (1973) *Biochemistry,* **12**, 171.
6. Schiechl,H. (1973) *Biochim. Biophys. Acta,* **307**, 65.
7. Steck,T.L. and Yu,J. (1973) *J. Supramol. Struct.,* **1**, 221.
8. James,G.T. (1978) *Anal. Biochem.,* **86**, 574.

9. Bordier,C. (1981) *J. Biol. Chem.*, **256**, 1604.
10. Clemetson,K.J., Bienz,D., Zahno,M.-L. and Luscher,E.F. (1984) *Biochim. Biophys. Acta*, **778**, 463.
11. Mahar,PA. and Singer,S.J. (1985) *Proc. Natl. Acad. Sci. USA*, **82**, 958.
12. Pryde,J.G. (1986) *Trends Biochem. Sci.*, **11**, 160.
13. Folch,J., Lees,M. and Slone-Stanley,G. (1957) *J. Biol. Chem.*, **226**, 497.
14. DeRobertis,E. (1975) *Rev. Physiol. Biochem. Pharmacol.*, **73**, 11.
15. Altendorf,K., Lukas,M., Lohl,B., Muller,C. and Sandermann,H. (1977) *J Supramol. Struct.*, **6**, 229.
16. Phizackerly,P.J.R., Town,M. and Newman,G.E. (1979) *Biochem. J.*, **183**, 731.
17. Altendorf,K. (1977) *FEBS Lett.*, **73**, 271.
18. DeRobertis,E., Fiszerde Plazas,S., Llorente de Carlin, C., Aguilar,J. and Schlieper,P. (1978) *Adv. Pharmacol. Ther.*, **1**, 235.
19. Boyan-Salyers,B., Vogel,J., Riggan,L., Summers,F. and Howell, R. (1978) *Metab. Bone Dis. Rel. Res.*, **1**, 143.
20. Criado,H., Aquilar,J. and DeRobertis,E. (1980) *Anal. Biochem.*, **103**, 289.
21. Gerber,G.E., Anderegg,R.J., Herlihy,W.C., Gray,C.P., Biemann,K. and Khorana,H.G. (1979) *Proc. Natl. Acad. Sci. USA*, **76**, 227.
22. Findlay,J.B.C., Brett,M. and Pappin,D.J.C. (1981) *Nature*, **293**, 314.
23. Brett,M. and Findlay,J.B.C. (1983) *Biochem. J.*, **211**, 661.
24. Pappin,D.J.C. and Findlay,J.B.C. (1984) *Biochem. J.*, **217**, 605.
25. Sigel,E., Stephenson,F.A., Mamalaki,C. and Barnard,E.A. (1983) *J. Biol. Chem.*, **258**, 6965.
26. Fillingame,R. (1976) *J. Biol. Chem.*, **251**, 6630.
27. Siebald,W. and Wachter,E. (1980) *FEBS Lett.*, **122**, 307.
28. Welling,G.W., Groen,G. and Welling-Webster,S. (1983) *J. Chromatogr.*, **266**, 629.
29. Lundahl,P., Greijer,E., Lindblom,H. and Fagerstam,L.G. (1984) *J. Chromatogr.*, **297**, 129.
30. McGregor,J.L., Clezardin,P.L., Manach,M., Gronlund,S. and Dechavanne,M. (1985) *J. Chromatogr.*, **326**, 179.
31. Engel,W.D.S., Schagger,H. and Von Jagow,G. (1980) *Biochim. Biophys. Acta*, **592**, 211.
32. Aton,B.R., Litman,B.J. and Jackson,M.L. (1984) *Biochemistry*, **23**, 1737.
33. Gotti,C., Conti-Tonconi,B.M. and Raftery,M.A., (1982) *Biochemistry*, **21**, 3148.
34. Momoi,N.Y. and Lennon,V.A. (1982) *J. Biol. Chem.*, **257**, 12757.
35. Hammonds,R.G., Nicolas,P. and Li,C.H. (1982) *Proc. Natl. Acad. Sci. USA*, **79**, 6494.
36. Schneider,W.J., Beisiegel,U., Goldstein,J.L. and Brown,M.S. (1982) *J. Biol. Chem.*, **257**, 2664.
37. Schmitt-Ullrich,R. and Wallach,D.F.H. (1977) In *Biological and Biomedical Applications of Isoelectric Focussing*. Catsimpoolas,N. and Drysdale,J., (eds), Plenum Press New York, London, p. 191.
38. Lilley,L., Eddy,B., Schaber,J., Fraser,C.M. and Venter,J.C. (1984) In *Receptor Biochemistry and Methodology. Vol. 2 Receptor Purification Procedures*. Venter,J.C. and Harrison,L.C. (eds), Alan R. Liss, New York, p. 77.
39. O'Farrell,P.Z., Goodman,H.M. and O'Farrell,P.H. (1977) *Cell*, **12**, 1133.
40. Wakefield,L.M., Cass,A.E.G. and Radda,G.K. (1984) *J. Biochem. Biophys. Methods*, **9**, 331.
41. Aton,B.R. (1986) *Biochemistry*, **25**, 677.
42. Khorana,H.G., Gerber,G.E., Herlihy,W.C., Anderegg,R.J., Gray,C.P., Nihei,K. and Biemann,K. (1979) *Proc. Natl. Acad. Sci. USA*, **76**, 5046.
43. Blondin,G.A. (1979) *Biochem. Biophys. Res. Commun.*, **87**, 1087.
44. Tandy,N.E., Dilley,R.A., Hesmondson,H.A. and Bhatnagar,D. (1982) *J. Biol. Chem.*, **257**, 4301.
45. Van der Zee,R., Welling-Webster,S. and Welling,G. (1983) *J. Chromatogr.*, **266**, 577.
46. Dean,P.D.G., Johnson,W.S. and Middle,F.A., eds (1985) *Affinity Chromatography − A Practical Approach*. IRL Press, Oxford and Washington, D.C.
47. Allan,D., Auger,J. and Crumpton,M.J. (1972) *Nature New Biol.*, **236**, 23.
48. Findlay,J.B.C. (1974) *J. Biol. Chem.*, **249**, 4398.
49. Lotan,R. and Nicolson,G.L. (1979) *Biochim. Biophys. Acta*, **559**, 329−376.
50. Kahane,I., Furthmayer,H. and Marchesi,V.T. (1976) *Biochim. Biophys. Acta*, **426**, 464.
51. Goldstein,I.J. and Hayer,C.E. (1978) *Adv. Carbohydr. Chem. Biochem.*, **35**, 127.
52. Nachbar,M.S., Oppenheim,J.D. and Aull,F. (1976) *Biochim. Biophys. Acta*, **419**, 512.
53. Carter,W.G. and Sharon,N. (1979) *Arch. Biochem. Biophys.*, **180**, 570.
54. Hames,B.D. and Rickwood,D. eds, (1981) *Gel Electrophoresis of Proteins − A Practical Approach*. IRL Press, Oxford and Washington DC.
55. Dewald,B., Dulaney,J.T. and Touster,O. (1974) *Methods Enzymol*, **32**, 82.
56. Dulaney,J.T. and Touster,O. (1970) *Biochim. Biophys. Acta*, **196**, 490.
57. Gershoni,J.M., Palade,G.E., Hawrot,E., Klimowitz,D.W. and Leutz,T.L. (1982) *J. Cell Biol.*, **95**, 422.
58. Griffin,D.L. and Landon,M. (1981) *Biochem. J.*, **197**, 333.

59. Nelles,L.P. and Bamberg,J.R. (1976) *Anal. Biochem.*, **73**, 522.
60. Hager,D.A. and Burgess,R.R. (1980) *Anal. Biochem.*, **109**, 76.
61. Higgins,R.C. and Dahmus,M.E. (1979) *Anal. Biochem.*, **93**, 257.
62. Aebersold,R.H., Teplow,D.B., Hood,L.E. and Kent,S.B.H. (1986) *J. Biol. Chem.*, **261**, 4229.
63. Kelly,C., Totty,N.F., Waterfield,M.D. and Crumpton,M.J. (1983) *Biochem. Int.*, **6**, 535.
64. Weber,K. and Kuter,D.J. (1971) *J. Biol. Chem.*, **246**, 4504.
65. Lenard,J. (1971) *Biochem. Biophys. Res. Commun.*, **45**, 662.
66. Fox,J.L., Stevens,S.E., Taylor,C.P. and Poulser,L.L. (1978) *Anal. Biochem.*, **87**, 253.
67. Kapp,O.H. and Vinogradov,S.N. (1978) *Anal. Biochem.*, **91**, 230.
68. Henderson,L.E., Oroszlan,S. and Konigsberg,W. (1979) *Anal. Biochem.*, **93**, 153.
69. Findlay,J.B.C. and Geisow,J.B.C., eds (1987) *Protein Sequencing − A Practical Approach.* IRL Press, Oxford and Washington, DC in press.
70. Cleveland,D.W., Fischer,S.G., Kerschmer,M.W. and Laemmli,U.K. (1977) *J. Biol. Chem.*, **252**, 1102.
71. Hudson,T.H. and Johnson,G.L. (1981) *J. Biol. Chem.*, **256**, 1459.
72. Mahoney,W.C. and Hermodson,M.A. (1979) *Biochemistry*, **18**, 3810.
73. Hunziker,P.E., Hughes,G.J. and Wilson,K.J. (1980) *Biochem. J.*, **187**, 515.
74. Bornstein,P. and Balian,G. (1977) *Methods Enzymol.*, **47**, 132.
75. Degani,Y. and Patchornik,A. (1974) *Biochemistry*, **13**, 1.
76. Udenfriend,S., Stein,S., Bohlen,P., Dairman,W., Leingruber,W. and Weigele,M. (1972) *Science*, **178**, 871−872.
77. Hirs,C.H.W. (1967) *Methods Enzymol.*, **11**, 328.
78. Bradford,M.M. (1976) *Anal. Biochem.*, **72**, 248.
79. Wang,C.-S. and Smith,R.L. (1975) *Anal. Biochem.*, **63**, 414.
80. Markwell,M.A.K., Haas,S.M., Bieber,L.L. and Tolbert,N.E. (1978) *Anal. Biochem.*, **87**, 206.
81. Bensadoun,A. and Weinstein,D. (1976) *Anal. Biochem.*, **70**, 241.
82. Futterman,S. and Rollins,M.H. (1973) *Anal. Biochem.*, **51**, 443.
83. Rubin,R.W. and Warren,R.W. (1977) *Anal. Biochem.*, **83**, 773.
84. Schulz,R.M. and Wassarman,P.M. (1977) *Anal. Biochem.*, **77**, 25.
85. Findlay,J.B.C. and Pappin,D.J.C. (1986) *Biochem. J.*, **238**, 625.
86. Gurd,F. (1972) *Methods Enzymol.*, **25**, 424.
87. Sutherland,R.M., Rothstein,A. and Weed,R.I. (1967) *J. Cell Physiol.*, **69**, 185.
88. Schmidt-Ullrich,R., Knüfermann,H. and Wallach,D.F.H. (1973) *Biochim. Biophys. Acta*, **307**, 353.
89. Knauf,P.A. and Rothstein,A. (1971) *J. Gen. Physiol.*, **58**, 211.
90. Barclay,P.L. and Findlay,J.B.C. (1984) *Biochem. J.*, **220**, 75.
91. Staros,J.V. and Richards,F.M. (1974) *Biochemistry*, **13**, 2720.
92. Mas,M.T., Wang,J.K. and Hargrave,P.A. (1980) *Biochemistry*, **19**, 684.
93. Solioz,M. (1984) *Trends Biochem. Sci.*, **9**, 309.
94. Carraway,K.L. and Koshland,D.E. (1972) *Methods Enzymol.*, **25**, 616.
95. Pober,J.S., Iwanij,V., Reich,E. and Stryer,L. (1978) *Biochemistry*, **17**, 2163.
96. Whiteley,N.M. and Berg,H.C. (1974) *J. Mol. Biol.*, **87**, 541.
97. Nemes,P.P., Miljamich,G.P., White,D.L. and Dratz,E.A. (1980) *Biochemistry*, **19**, 2067.
98. Carraway,K.L., Kobylka,D., Summer,S.J. and Carraway,C.A. (1972) *Chem. Phys. Lipids*, **8**, 65.
99. Dockter,M.E. and Koseki,T. (1983) *Biochemistry*, **22**, 3954.
100. Poensgen,J. and Passow,H. (1971) *J. Membr. Biol.*, **6**, 210.
101. Bretscher,M. (1971) *J. Mol. Biol.*, **58**, 775.
102. Zeheb,R., Chang,V. and Orr,G.A. (1983) *Anal. Biochem.*, **129**, 156.
103. Maddy,A.H. (1964) *Biochim. Biophys. Acta*, **88**, 390.
104. Cabantchik,Z.I. and Rothstein,A. (1974) *J. Membr. Biol.*, **15**, 227.
105. Rifkin,D.B., Combans,R.W. and Reich,R. (1972) *J. Biol. Chem.*, **247**, 6432.
106. Bolton,A.E. and Hunter,W.M. (1973) *Biochem. J.*, **133**, 529.
107. Ciccimarra,F., Rosen,F.S. and Merler,E. (1975) *Proc. Natl. Acad. Sci. USA*, **72**, 2081.
108. Gordesky,S.E., Marinetti,G.V. and Love,R. (1975) *J. Membr. Biol.*, **20**, 111.
109. Morrison,M. (1974) *Methods Enzymol.*, **32(B)**, 103.
110. Hubbard,A.L. and Cohn,Z.A. (1976) In *Biochemical Analysis of Membranes.* Maddy,A.H. (ed.), Chapman and Hall (London), p. 427.
111. Tuszynski,G.P., Knight,L.C., Kornecki,E. and Srivastava,S. (1983) *Anal. Biochem.*, **130**, 166.
112. Markwell,M.A.K. (1983) *Anal. Biochem.*, **125**, 427.
113. Luthra,M.G., Friedman,J.M., and Sears,D.S. (1978) *J. Biol. Chem.*, **253**, 5647.
114. Morell,A.G. and Ashwell,G. (1972) *Methods Enzymol.*, **28**, 205.
115. Gahmberg,C.G. and Andersson,L.C. (1977) *J. Biol. Chem.*, **252**, 5888.

215

116. Hedo,J.A., Kasuga,M., Van Obberghen,E., Roth,J. and Khan,C.R. (1981) *Proc. Natl. Acad. Sci. USA*, **78**, 4791.
117. Bayley,H. (1983) *Laboratory Techniques in Biochemistry and Molecular Biology*. Work,T.S. and Burdon,R.H. (eds), Elsevier, Amsterdam, Vol. 12.
118. Bayley,H. and Knowles,J.R. (1978) *Biochemistry*, **17**, 2414, 2420.
119. Wells,E. and Findlay,J.B.C. (1980) *Biochem. J.*, **187**, 719.
120. Davison,M.D. and Findlay,J.B.C. (1986) *Biochem. J.*, **234**, 413.
121. Davison,M.D. and Findlay,J.B.C. (1986) *Biochem. J.*, **236**, 389.
122. Tarrab-Hazdai,R., Bercovici,T., Goldfarb,V. and Gitler,C. (1980) *J. Biol. Chem.*, **255**, 1204.
123. Wisnieski,B.J. and Bramhall,J.S. (1981) *Nature*, **289**, 319.
124. Anajaneyulu,P.S.R. and Lala,A.K. (1982) *FEBS Lett.*, **146**, 165.
125. Casanova,J., Horowitz,Z.D., Copp,R.P., McIntyre,W.R., Pascual,A. and Somnels,H.H. (1984) *J. Biol. Chem.*, **259**, 2084.
126. Booth,A.G. and Kenny,A.J. (1980) *Biochem. J.*, **187**, 31.
127. Brunner,J. and Semenza,G. (1981) *Biochemistry*, **20**, 7174.
128. Hoppe,J., Brunner,J. and Jorgensen,B.B. (1984) *Biochemistry*, **23**, 5610.
129. Cerletti,N. and Schatz,G. (1979) *J. Biol. Chem.*, **254**, 7746.
130. Sator,J., Gonzalez-Ros,J.M., Calvo-Fernandez,P. and Martinez-Carrion,M. (1979) *Biochemistry*, **18**, 1200.
131. Hebdon,G.M., Knott,J.C.A. and Green,N.M. (1980) *Biochemistry*, **19** 6216.
132. Ross,A.H., Radhakrishnan,R., Robson,R.J. and Khorana,H.G. (1982) *J. Biol. Chem.*, **257**, 4152.
133. Montecucco,C., Bisson,R., Dabbeni-Sala,F., Pitotti,A. and Gutweniger,H. (1980) *J. Biol. Chem.*, **255**, 10040.
134. Brunner,J. and Richards,F.M. (1980) *J. Biol. Chem.*, **255**, 3319.
135. Robson,R.J., Radhakrishnan,R., Ross,A.H., Takagaki,Y. and Khorana,H.G. (1982) *Lipid-Protein Interactions*, **2**, 149.
136. Brunner,J., Spiess,M., Aggeler,R., Huber,P. and Semenza,G. (1983) *Biochemistry*, **22**, 3812.
137. Leblanc,P., Capone,J. and Gerber,E.G. (1982) *J. Biol. Chem.*, **257**, 14586.
138. Ji,T.H. (1983) *Methods Enzymol.*, **91**, 580.
139. Brett,M. and Findlay,J.B.C. (1979) *Biochem. J.*, **179**, 215.
140. Harris,R. and Findlay,J.B.C. (1983) *Biochim. Biophys. Acta*, **732**, 75.
141. Peters,K. and Richards,F.M. (1977) *Annu. Rev. Biochem.*, **46**, 523.
142. Massague,J. and Czech,M.P. (1985) *Methods Enzymol.*, **109**, 179.
143. Eberle,A.N. and DeGraan,P.N.E. (1985) *Methods Enzymol.*, **109**, 129.
144. Yip,C.C. and Yeung,C.W.T. (1985) In *Methods Enzymol.*, **109**, 170.
145. Vanin,E.F., Burkhard,S.J. and Kaiser,I.I. (1981) *FEBS Lett.*, **124**, 89.
146. Ngo,T.T., Yan,C.F., Lenhoff,H.M. and Ioy,J. (1981) *J. Biol. Chem*, **256**, 11313.
147. Pennington,R.M. and Fisher,R.R. (1981) *J. Biol. Chem.*, **256**, 8963.
148. Dutton,A. and Singer,S.J. (1975) *Proc. Natl. Acad. Sci. USA*, **72**, 2586.
149. Steck,T.L., Ramos,B. and Strapazon,E. (1976) *Biochemistry*, **15**, 1154.
150. Wells,E. and Findlay,J.B.C. (1980) *Biochem. J.*, **187**, 719.
151. Vanin,E.F. and Ji,T.H. (1981) *Biochemistry*, **20**, 6754.
152. Moreland,R.B., Smith,P.K., Fujimoto,E.K. and Dockter,M.E. (1982) *Anal. Biochem.*, **121**, 321.
153. Kincaid,R.L. and Vaughan,M. (1983) *Biochemistry*, **22**, 826.
154. Kiehm,D.J. and Ji,T.H. (1977) *J. Biol. Chem.*, **252**, 8524.
155. Demoliou,C.D. and Epand,R.H. (1980) *Biochemistry*, **19**, 4539.
156. Moreland,R.B. and Dockter,M.E. (1981) *Biochem. Biophys. Res. Commun.*, **99**, 339.
157. Steiner,M. (1980) *Biochem. Biophys. Res. Commun.*, **94**, 861.
158. Goewert,R.R., Landt,H. and McDonald,J.M. (1982) *Biochemistry*, **21**, 5310.
159. Schmitt,M., Painter,R.G., Jesaitis,A.J., Preissner,K., Sklar,L.A. and Cochrane,C.G. (1983) *J. Biol. Chem.*, **258**, 649.
160. Zarling,D.A., Miskimen,S.A., Fan,D.P., Fujimoto,E.K. and Smith,P.K. (1982) *J. Immunol.*, **128**, 251.
161. Jung,S.M. and Moroi,M. (1983). *Biochim. Biophys. Acta*, **761**, 152.
162. Marfey,S.P. and Tsai,K.H. (1975) *Biochem. Biophys. Res. Commun.*, **65**, 31.
163. Kornblatt,J.A. and Lake,D.F. (1980) *Can. J. Biochem.*, **58**, 219.
164. Zarling,D.A., Watson,A. and Bach,F.H. (1980) *J. Immunol.*, **124**, 913.
165. Staros,J.V. and Kakkad,B.P. (1983) *J. Membr. Biol.*, **74**, 247.
166. Pilch,P.F. and Czech,M.P. (1979) *J. Biol. Chem.*, **254**, 3375.
167. Sen,I., Bull,H.G. and Sutter,R.L. (1984) *Proc. Natl. Acad. Sci. USA*, **81**, 1679.
168. Bragg,P.D. and Hon,C. (1980) *Eur. J. Biochem.*, **106**, 495.
169. Laburthe,M., Breant,B. and Ronyer-Fessard,C. (1984) *Eur. J. Biochem.*, **139**, 181.

170. Giedroe,D.P., Puett,D., Ling,W. and Staros,J.V. (1983) *J. Biol. Chem.*, **258**, 16.
171. Baskin,L.S. and Yang,C.S. (1982) *Biochim. Biophys. Acta*, **684**, 263.
172. Bender,W.W., Garan,H. and Berg,H.C. (1971) *J. Mol. Biol.*, **58**, 783.
173. Packman,L.C. and Perham,R. (1982) *Biochemistry*, **21**, 5171.
174. Jaffe,C.L., Lis,H. and Sharon,N. (1980) *Biochemistry*, **19**, 4423.
175. Abella,P.M., Smith,P.K. and Royer,G.P. (1979) *Biochem. Biophys. Res. Commun.*, **87**, 734.
176. Rinke,J., Meinke,M., Brimacombe,R., Fink,G., Rommel,W. and Fasold,H., (1980) *J. Mol. Biol.*, **137**, 301.
177. Huang,C.K. and Richards,F.M. (1977) *J. Biol. Chem.*, **252**, 5514.
178. Jung,C.Y. (1984) In *Receptor Biochemistry and Methodology.* Ventor,J.C. and Harrison,L.C. (eds), Alan R. Liss, Vol. 3, p. 193.
179. Hughes,S.M., Harper,G. and Brand,H.D. (1983) *Biochem. Biophys. Res. Commun.*, **122**, 56.
180. Sterr,C.J., Kempner,E.S. and Ashwell,G. (1981) *J. Biol. Chem.*, **256**, 5851.
181. Saccomani,G., Sachs,G., Cupoletti,J. and Jung,C.Y. (1981). *J. Biol. Chem.*, **256**, 7727.
182. Venter,J.C. (1983) *J. Biol. Chem.*, **258**, 4842.
183. Fewtrell,C., Kempner,E., Poy,G. and Metzger,H. (1981) *Biochemistry*, **20**, 6589.
184. Nielsen,T.B., Lad,P.M., Preston,M.S., Kempner,E., Schlegel,W. and Rodbell,M. (1981) *Proc. Natl. Acad. Sci. USA*, **78**, 772.
185. Simon,P., Swillens,S. and Dumont,J.E. (1982) *Biochem. J.*, **205**, 477.
186. Bethell,G.S., Ayers,J.S., Hancock,W.S. and Hearn,M.T.W. (1979) *J. Biol. Chem.*, **254**, 2572.
187. Dufau,M.L., Ryan,D.W., Baukal,A.J. and Catt,K.J. (1975) *J. Biol. Chem.*, **250**, 4822.
188. Brody,R.S. and Westheimer,F.H. (1979) *J. Biol. Chem.*, **254**, 4238.
189. Merrill,A.H., Jr. and McCormick,D.B. (1980) *Methods Enzymol.*, **66**, 28.
190. Lockshin,A., Moran,R.G. and Danenberg,P.V. (1979) *Proc. Natl. Acad. Sci. USA*, **76**, 750.
191. Carlsson,J., Axen,R. and Unge,T. (1975) *Eur. J. Biochem.*, **59**.
192. de Pinto,V., Tommasino,M., Valmieri,F. and Kadenbach,B. (1982) *FEBS Lett.*, **148**, 103.
193. Davis,C.W. and Daly,J.W. (1979) *J. Cyclic Nucleotide Res.*, **5**, 65.
194. Garlick,R.L. and Mazer,J.S. (1983) *J. Biol. Chem.*, **258**, 6142.

CHAPTER 7

Optical spectroscopy of biological membranes

C.LINDSAY BASHFORD

1. INTRODUCTION

Optical spectroscopy, spectrophotometry and fluorimetry, can be used to monitor processes occurring in biological membranes provided that suitable chromophores are present which 'report' on the events in which they participate. The main problem with this type of experiment is not the acquisition but rather the interpretation of the data obtained. Straightforward analysis of the results depends on the clarity of the experimental design and the appropriate choice of chromophore. This chapter describes some of the problems that can be addressed by spectroscopic techniques and attempts to give guidance on good experimental design. Examples of typical protocols of which we have experience, will be described. *Table 1* lists some membrane systems that can be assessed by fluorimetry or spectrophotometry and indicates which are discussed in more detail below.

Two categories of chromophore are commonly used: endogenous, or 'intrinsic', chromophores already present in the membrane and exogenous, or 'extrinsic', pigments provided by the observer. In the former category, pigments which absorb light in the visible region of the spectrum (400 — 700 nm) are the most useful because they are restricted both in abundance and in the range of membrane processes in which they participate. Porphyrins, flavins and carotenoids exhibit characteristic spectra with rather narrow, intense absorbance bands. Thus the role of cytochromes in mitochondrial electron transport was appreciated long ago by Keilin (1) as a result of his studies of insect flight muscle using a simple spectroscope. In addition the co-factors most commonly found in biological oxidation — reduction reactions, namely the pyridine nucleotides (NAD, NADP) and the flavins are fluorescent in either their oxidized or their reduced forms. In photosynthetic systems, in addition to chlorophyll, the carotenoids, secondary 'antennae pigments', serve as 'molecular voltmeters' since the position of their absorption peaks alters as a function of the electrical potential across the photosynthetic membrane (2). This property is useful for studying the mechanisms of energy transduction in photosynthesis and has stimulated an extensive search for exogenous pigments with similar properties which could be used for studies of membrane potential in a wide variety of membrane systems (3).

In addition to exogenous pigments which register membrane potential, dyes have been employed to monitor a wide variety of cellular properties: for example, cytoplasmic and organelle pH, cytoplasmic Ca^{2+} and Mg^{2+} concentration, membrane fluidity or microviscosity, fate of ingested particles. In these cases the molecule in question is usually specifically designed: (i) sensitively to register the parameter of interest; and

Optical spectroscopy

Table 1. Chromophores useful for studies of the structure and function of biological membranes.

Chromophore	Method of measurement	Information	Reference[a]
Intrinsic			
Chlorophyll	fluorescence/absorbance	mechanism of	49, 50
Carotenoid[b]	absorbance	photosynthesis	2, 32
NADH	fluorescence/absorbance		
Flavoprotein	fluorescence	electron	14−16
Cytochromes[b]	absorbance	transport	
Haemoglobin[b]	absorbance	tissue	10, 25,
Myoglobin[b]	absorbance	oxygenation	26, 30
Extrinsic			
Acridines[b]	fluorescence	pH of	7, 51
Quene-1[b]	fluorescence	cellular	9
Neutral red[b]	absorbance	compartments	36,52,53
Quin-2[b]	fluorescence	cytoplasmic	8
Aequorin, obelin	luminescence	Ca^{2+}	54
Murexide, arsenazo	absorbance	content	55
Cyanines	fluorescence		
Oxonols[b]	fluorescence/absorbance	membrane	3,13,31
Merocyanines	fluorescence	potential	
Naphthalene	fluorescence	energy-coupling	56−59
sulphonates		surface charge	43, 60
Diphenylhexatriene	fluorescence	microviscosity	61, 62
Anthroyl esters	polarization	membrane fluidity	63−65

[a]References are not necessarily the first description of the use of the chromophores but serve as useful entry points to the original literature.
[b]These chromophores are discussed in more detail in the text.

(ii) simply to reach its ultimate destination. Probes of cytoplasmic Ca^{2+} or pH exhibit characteristic fluorescence or absorbance depending on the cations in their environment. Membrane-permeant esters of the indicators diffuse to the compartment where endogenous enzymes catalyse their hydrolysis to generate the active species *in situ*. The range of possible probe experiments is so great that it is impractical to describe each in detail. However, it is not always appreciated how simple many of the experiments are to perform using equipment available in most laboratories. A practical guide to spectrophotometry and fluorimetry in biochemistry laboratories, which describes many of the procedures referred to in this chapter, is available (4).

2. EXPERIMENTAL DESIGN

2.1 **Apparatus**

Optical spectroscopy requires either spectrophotometers, to measure absorbance, fluorimeters, to measure fluorescence or microscopes, which, in principle, can measure fluorescence or absorbance by single cells or small groups of cells. Fluorimeters and spectrophotometers usually require suspensions of membranes in conventional cuvettes; microscopes provide two-dimensional images from smears, slices or surfaces.

220

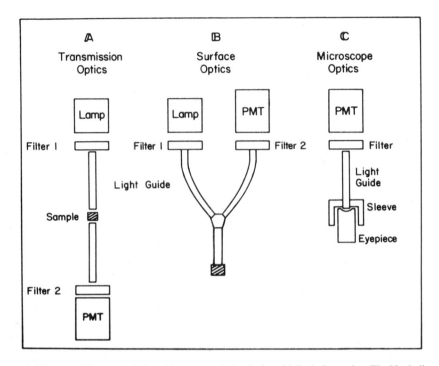

Figure 1. The use of fibre optic light guides to record signals from biological samples. The block diagrams indicate three configurations which we have found useful for recording absorbance or fluorescence signals. (A) Light is conducted to the sample by a flexible light guide which can be up to 2 m long. Light transmitted through the sample is collected by a second light guide and conducted back to a photomultiplier tube. Wavelength is selected by filters which have rather higher transmission than monochromators. For recording spectra the filters will be replaced by monochromators; for absorbance measurements only one filter is required either in position 1 or position 2; for fluorescence, filter 1 should pass only excitation light and filter 2 only emitted light. It is very important to check (despite any manufacturers' claims) that in the absence of fluorophore no signal is recorded! Because of the low intensity of fluorescence emission any excitation light that 'leaks' through the emission filter will normally mask the fluorescence entirely. To avoid stray light other than from the lamp, the apparatus may need to be set up in a dark room, and the sample shrouded to prevent room light reaching the collecting light guide. (B) In this configuration signals are recorded from the surface of the sample. Similar criteria should be applied as for case A above, except that in this case fluorescence and/or reflectance will be recorded. Bifurcated light guides are commercially available with fibres either randomly mixed at the common terminal (the most useful arrangement) or arranged in segments at the common terminal. More complete descriptions of apparatus using light guides for optical coupling can be found in refs. 25, 37 and 66. (C) In this case the collecting light guide is held in position over the eyepiece of a microscope. To prevent stray ambient light entering the light guide the fit of the brass sleeve both to the light guide and to the eyepiece should be snug.

Fluorimeters and spectrophotometers can be adapted to record from surfaces by making use of fibre optic 'light guides'. In this case, the guide (a bundle of optical fibres) conducts the appropriate illumination to the surface and both reflected and fluorescent light are collected by other fibres and conducted to a detector, usually a photomultiplier (*Figure 1*). The reflected light has two components: specularly reflected light, this is the mirror-like reflection, and diffusely reflected light. Only the latter contains information (analogous to absorbance) concerning the chromophores present at the surface. As in all types of photometry the proper use of filters and/or monochromators will

Table 2. Protocol to record exposure times with an automatic camera attached to a fluorescence microscope.

1.	Set the sensitivity by selecting the appropriate film speed (ASA/DIN). The greater the film speed the greater the sensitivity; an exposure of 20 sec at ASA 100 will take 10 sec at ASA 200.
2.	Locate a representative field of view using bright field (transmission) or phase contrast optics.
3.	Switch the microscope to fluorescence optics.
4.	Operate the camera and time the exposure with a stop-watch. Start the stop-watch when you open the shutter and stop it when you hear the shutter close. If the exposure time is less than 10 sec select a lower ASA setting, and if it is greater than 100 sec select a higher ASA setting.
5.	Repeat steps 1−4 at least five times with randomly selected fields of view.
6.	The mean exposure time (corrected to constant ASA setting) is inversely proportional to the fluorescence intensity of the image.

discriminate between the contributions from stray light and the signal of interest.

Light guides can also be used to convert microscopes into photometers or fluorimeters: light from the eyepiece is collected with a light guide and conducted to a suitable detector. The light guide, which is held in place by a snug-fitting brass ferrule, machined to exclude room light, monitors light arising from the image. Alternatively light output from the image can be assessed with the light metering system of a camera attached to the microscope. The time taken correctly to expose a film of given speed (ASA/DIN) is inversely related to the brightness of the image; cells that have twice the fluorescence of control cells will require half the exposure time. In some cases (e.g. Leitz Orthomat) the metering system can be operated in the absence of film. The exposure time, that is the interval between the opening and the closing of the shutter, is measured with a stop-watch. For accurate measurements the exposure time should lie in the 10−100 sec range; shorter exposures are difficult to time and longer exposures are compromised by stray light entering the objective (see Section 2.4.1 for discussion of stray light) and by photobleaching. The general procedure to record exposure times is set out in *Table 2*. The correlation between intensity and reciprocal exposure time is linear over a very wide range (4), that is, the exposure time is inversely proportional to the emission intensity. At high magnifications photobleaching, that is a time-dependent loss of intensity, usually occurs and it is necessary to arrange for short exposure times and minimal exposure of the field to the excitation beam.

2.2 Chromophore selection

The primary decision is the choice of a suitable chromophore with which to study the problem under consideration. In the case of intrinsic chromophores this must be a choice of system, namely the one in which the chromophore occurs and in which it exhibits the desired properties. For example, the carotenoid pigments record the light-dependent membrane potential associated with photosynthesis. Such pigments occur in many photosynthetic membranes but their use as 'molecular voltmeters' is most easily accomplished in preparations, known as chromatophores, from certain photosynthetic bacteria such as *Rhodopseudomonas sphaeroides* or *Rhodopseudomonas capsulata*. Again, studies of oxygen consumption and delivery *in vivo* require highly vascularized tissues which contain many mitochondria, for example from brain, liver or cardiac muscle (although in the latter case the myoglobin contribution cannot be resolved from that of haemoglobin). In each case, however, the oxygen carriers tend to mask the much less abundant mitochondrial pigments.

For extrinsic chromophores three factors strongly influence the final choice.

2.2.1 *Delivery of chromophore*

Most optical probes used in studies of membranes are lipophilic and can be stored as stock solutions in solvents such as ethanol or dimethyl sulphoxide. The stock solution is diluted at least 200-fold with the membrane preparation and the dye reaches its destination by diffusion. Such labelling is non-specific and in multi-membrane systems such as intact cells, the probes partition unevenly among the various compartments. This is helpful if the dye concentrates in the region of interest as cyanine dyes do in mitochondria (5) and weak bases do in acidic endosomes (6, 7). However, if cytoplasmic pH (Section 3.3.2) or plasma membrane potential (Section 3.2.2) are the object of interest, these dyes cannot be used in a straightforward way unless measures are taken to suppress responses from other compartments.

Slightly more specific labelling can be achieved by delivering an inactive (usually an ester) form of the dye which is locally converted to the active form by endogenous esterases. This is the technique used to probe cytoplasmic Ca^{2+} and H^+ levels with fluorescent analogues of EDTA (8, 9). Specificity is achieved by the location of active esterases in the compartment of interest (see Section 3.3.2).

Specific labelling can be achieved by coupling the chromophore covalently to a vehicle, usually an immunoglobulin or a hormone, that binds specifically to cellular receptors.

2.2.2 *Competition from endogenous pigments*

Membranes contain components which absorb near u.v. and scatter blue light strongly. Probes with fluorescence/absorbance in the red region of the spectrum are usually the most useful. Remember, however, that many photometers are less sensitive to red light than they are to blue light; so that the advantages gained by choosing a probe with little optical 'overlap' with native pigments are offset by the decreased sensitivity of the apparatus. The fluorescence of extrinsic probes will be severely quenched if the emission of the fluorophore overlaps with intense absorbance bands of intrinsic chromophores. Cellular autofluorescence (probably due to flavins) is green/yellow and can interfere with experiments using fluorescein-based chromophores. There is little autofluorescence in the orange/red region of the spectrum where rhodamine derivatives emit. In favourable circumstances, changes in fluorescence may be observed even when absorbance changes are masked by endogenous pigments (10).

2.2.3 *Toxic actions of chromophores*

Chromophores which are non-toxic at low concentrations may become potent inhibitors if they are concentrated in specific compartments. Cyanine dyes severely inhibit respiration at site I of the mitochondrial respiratory chain providing that the inner mitochondrial membrane potential is substantial (inside negative) (11). In living cells the inner mitochondrial membrane potential is about 180 mV and the plasma membrane potential is about 60 mV (both inside negative) and extracellular cyanine at 10^{-7} M is at electrochemical equilibrium with mitochondrial cyanine when the latter reaches 10^{-3} M. Very low concentrations of cyanines may thus adversely affect cellular respiration and energy-dependent processes (11).

2.3 Characterization and calibration of optical signals

Once the choice of chromophore and system has been made it is essential to characterize their optical properties. The spectrum of the chromophore (absorbance or fluorescence) *in situ* may contain unexpected contributions which will limit the scope of the experiment. Such investigations establish the signal-to-noise ratio for the intended observations (noise being used in its most general sense).

It is unfortunately the case that the calibration of optical signals, be it with pH, transmembrane potential or cation concentration, usually destroys the experimental system. For example, ionophores irreversibly modify the membrane potential during the calibration of dye indicators. Likewise, entrapped indicators of cell pH or Ca^{2+} are calibrated only after the cells have been completely permeabilized. Once a suitable calibration procedure is established, it is possible to record the desired information from the experimental system.

2.4 Artefacts

Optical investigations of biological membranes are beset with hazards and it cannot be too strongly emphasized that very rigorous criteria must be applied before interpreting the experimental data. Artefacts are more common in fluorescence experiments and a few of the most easily avoided ones are considered here.

2.4.1 *Stray light and turbid solutions*

Stray light is a general term for any light that reaches the detector which is either not of fluorescence origin or, in an absorbance experiment, of a wavelength different from that intended. It may include ambient light and leaks through the monochromators and filters from the source. It is avoided by correct apparatus construction, the correct choice of blocking filters (to exclude higher order diffraction and specular reflection from the grating in the monochromator) and the appropriate use of dark rooms.

Suspensions of biological membranes are turbid. It is especially important in such circumstances to resolve the contributions to the signal that arise from light scattering. Since scattered light usually exceeds fluorescence by a substantial margin, careful choice of filters and slit widths is essential to record just fluorescence (12). It is important to be aware that substantial changes in turbidity, and hence in the absolute value of fluorescence or absorbance, occur when membrane vesicles are exposed to sudden changes in osmolarity or are exposed to permeant salts (13). In order to maximize the chances of 'capturing' the emitted/transmitted photons the cuvette is usually placed as close as possible to the detector.

Turbid solutions can be useful when the chromophore has low extinction or can only be used at very low concentration. This arises because the multiple internal reflections amongst the particles greatly increases the effective path length of the absorbance experiment and hence increases the amount of light absorbed (a consequence of Lambert's Law). In such circumstances the *dual wavelength* optical arrangement is strongly recommended: absorbance at one wavelength is measured *in the same cuvette* with respect to light of a slightly different wavelength; changes in turbidity affect both beams equally and are thus cancelled out (4, 14).

2.4.2 *Inadequate excitation*

The excitation light may be absorbed by non-fluorescent material such as occurs when monitoring the surface fluorescence of tissues or organs. Blood completely absorbs the light used to excite fluorescence due to pyridine nucleotides and flavoproteins, hence an increase in the amount of blood in the field of view, due to vasodilation, will decrease fluorescence and lead to an erroneous conclusion that the oxidation−reduction state of the tissue mitochondria had changed. This problem is obviated by recording the ratio of the pyridine nucleotide (PN) and flavoprotein (FP) fluorescence: as mitochondria become reduced PN fluorescence increases and FP fluorescence decreases, as the mitochondria become oxidized the changes of fluorescence are reversed. Hence the FP/PN fluorescence ratio is a sensitive indicator of the mitochondrial oxidation−reduction state (15). It is not severely affected by masking problems as both signals are affected equally so that their ratio remains unaltered. An alternative procedure for correcting for the interference of blood is to obtain the 'corrected' fluorescence, namely the difference between the measured fluorescence and the measured reflectance (of the excitation light). In this mode, loss of excitation, due to absorbance by the blood pigments, leads to a loss of reflectance which compensates for the loss of fluorescence. The adequacy of this procedure has been demonstrated in perfused systems where the blood can be replaced by a transparent perfusate (16) but its applicability *in vivo* is less certain.

The excitation light may be absorbed so strongly by the fluorophore that fluorescence is reduced along the excitation path, one aspect of the inner filter effect (4, 12). Such an effect should be suspected if dilution of a sample leads, initially, to a fluorescence increase. This could arise, for example, if the detector 'observes' a part of the sample reached by the excitation beam only in dilute solutions. *Figure 2* illustrates this effect in a simple experiment that can be performed with all types of fluorimeter. The expected linear relationship between fluorescence and concentration occurs only with dilute ($< 10^{-3}$ M) preparations (4). Remember that it is not only the fluorophore that may absorb the excitation and that it is the *overall* absorbance that must be kept low to avoid any inner filter effects.

Inner filter effects can be reduced by altering the geometry of the fluorescence experiment: fluorescence from the surface of highly absorbing material is much less sensitive to the effect because neither the excitation nor the emission has to traverse an opaque medium in order for a fluorescence signal to be recorded. Indeed if a microscope with an epifluorescence attachment is used to excite the material the effective pathlength is of the order of a few microns and little filtering occurs. This can lead to confusion if results obtained with a microscope are compared with those obtained in a conventional fluorimeter. Cyanine dyes are avidly accumulated by mitochondria *in vivo* (5) with a resultant quenching of their fluorescence when assessed in cell suspension (17); however, in the fluorescence microscope the mitochondria appear as brightly fluorescent, filamentous structures (5). Similarly acridine orange and 9-aminoacridine are ac-

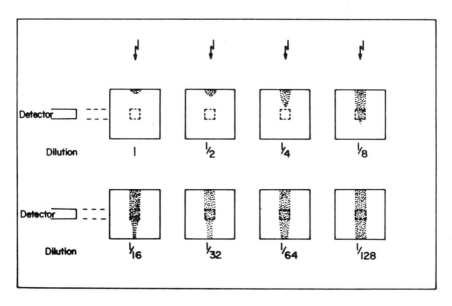

Figure 2. Fluorescence emission as a function of fluorophore concentration. 1 mM 9-aminoacridine in distilled water was placed in a quartz cuvette (all the faces of cuvettes for fluorescence measurements are usually polished), in a Perkin-Elmer MPF 44A fluorescence spectrophotometer. The excitation wavelength was set to 410 nm and the slitwidth to 6 nm. The cuvette was observed with the sample compartment and excitation shutter open and the emission shutter closed; for best effect it is preferable to switch off the room lights. The blue fluorescence of 9-aminoacridine arose from the stippled area as indicated. Note that the photodetector only monitors fluorescence emission arising within the central area of the cuvette indicated by the inner dashed box. As the concentration of 9-aminoacridine is reduced by serial dilution, the blue fluorescence is detected from progressively more of the excitation light path (the direction of incidence is indicated by the arrow).

cumulated by acidic endosomes with a quenching of their blue fluorescence; in the fluorescence microscope the vesicles appear as yellow or orange vesicles, this being the characteristic emission of very concentrated solutions of these dyes (4).

2.4.3 *Complex formation*

The dye chosen may interact with elements of the system with a consequential change in its properties. Thus the pK_a of a pH indicator in solution may not necessarily have the same value when the indicator is bound to cellular or other constituents (see, for example, refs. 18, 19); indeed the spectra of the bound dye may differ from those of the free dye, making any simple evaluation of pH problematical. It may be that the dye is useful as a qualitative indicator of the parameter under investigation but cannot be calibrated because it forms complexes with agents required in the calibration protocol. It is always good practice to use several, independent calibration procedures to check for internal consistency of the dye response. Thus when calibrating optical probes of membrane potential (20, 21) the same values should be obtained when valinomycin/K$^+$ or uncoupler carbonyl cyanide *p*-trifluoromethoxyphenylhydrazone (FCCP)/H$^+$ are introduced.

2.5 **Low temperature spectroscopy**

As the ambient temperature falls both absorption and fluorescence change, very often in a manner that simplifies spectroscopic investigations. At liquid nitrogen temperatures (77K), the absorption bands of most haemoproteins are blue shifted (by $1-4$ nm) relative to their positions at room temperature, and have significantly decreased spectral bandwidth (4, 22), that is they are 'sharper', and are increased in intensity. The latter effect arises both from intrinsic changes in extinction and an increase in the effective pathlength of the experiment (the same effect as that described for turbid suspensions in Section 2.4.1).

Improved spectral resolution and the virtual absence of chemical change at very low temperatures, allow the individual components of complex mixtures, such as intact electron transfer chains, to be resolved. Surprisingly some oxidation−reduction and ligand exchange reactions still occur at 4K and this allows the study of a number of otherwise very rapid reactions. Suitable reagents are frozen together and the reaction of interest is initiated photochemically once the system has reached its target temperature (4, 23, 24).

Quenching processes are much more temperature-sensitive than is fluorescence emission so that fluorescence yields almost always increase as the temperature is lowered. At 77K, the fluorescence of the mitochondrial flavoproteins is enhanced by at least an order of magnitude allowing a sensitive fluorometric determination of mitochondrial oxidation−reduction state in freeze-trapped tissue (15).

Increasingly spectrophotometers offer low-temperature accessories as an optional (usually expensive!) extra. Simpler 'home-made' devices have also been described (4, 23) and in our laboratory we use light guides to monitor signals at 77K simply by immersing the tip of the light guide into the liquid nitrogen that surrounds the sample (25). Thus, for example, the reflectance spectra of freeze-trapped rat kidney show clearly the disoxygenation of haemoglobin and the reduction of cytochrome aa_3 that occur during ischaemia. Note also that in ischaemic kidney, reduced cytochrome aa_3 and oxyhaemoglobin appear to be present in the same field of view implying a substantial metabolic heterogeneity in this organ (*Figure 3*).

Some reactions of haemoproteins can only be resolved kinetically at 4K (liquid helium temperatures); unfortunately the apparatus for such specialized measurements is not widely available.

3. EXAMPLES

3.1 **Oxidation−reduction state of tissue mitochondria**

The properties of mitochondrial suspensions have long been the centre of intense study and the characteristics of the pigments of the respiratory chain have been resolved by dual-wavelength spectroscopy (14; Section 2.4.1). However, there is always some uncertainty as to whether properties observed in an isolated preparation genuinely reflect the properties of the unperturbed system *in vivo*. Direct recording of the respiratory pigments *in situ* provides one way of resolving this dilemma. In anaesthetized animals spectroscopy of exposed tissues (25, 26) reports the oxidation−reduction state of the

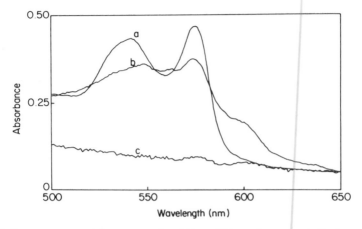

Figure 3. Reflectance spectra of freeze-trapped rat kidney. Kidneys from anaesthetized rats were freeze-trapped (27) before, 1 min and 10 min after tying off the renal blood supply. Reflectance spectra were recorded with a bifurcated light guide attached to a scanning, dual-wavelength spectrophotometer (*Figure 1B*, see ref. 25) using 700 nm as the reference wavelength. The spectrum of white polystyrene was stored and used as the baseline. Spectra labelled **a** and **b** were recorded from the normally perfused and 1 min ischaemic kidneys, respectively. The spectrum labelled **c** is the *difference* between the 1 and 10 min ischaemic kidneys. Reproduced with permission from ref. 29.

electron transport chain and the oxygenation of haemoglobin/myoglobin. Unfortunately the rapid fluctuations of O_2 delivery and consumption make it difficult to establish convenient steady states. An alternative procedure is to freeze the tissue rapidly and then to obtain spectra more leisurely at low temperatures where optical resolution is enhanced (Section 2.5).

Figure 3 shows reflectance spectra of freeze-trapped rat kidneys.

(i)   Remove the kidneys from the anaesthetized animal and freeze them rapidly by compression between aluminium tongs chilled to 77K (27).

(ii)   Keep the frozen tissue at 77K and record spectra from the surface using a bifurcated light guide (*Figure 1*) whose common terminal is chilled to 77K and applied directly to the frozen tissue.

Three different kidneys were assessed: a normally perfused kidney, and kidneys rendered ischaemic by tying a ligature round their arterial supply, and freeze-trapped either 1 min or 10 min later. The main features of the visible reflectance spectra arise from oxyhaemoglobin (peaks at 542 and 577 nm), haemoglobin (peak at 555 nm) and ferrocytochromes aa_3 (maximum at 605 nm) and c (maximum at 550 nm). The excess of haemoglobin means that in absolute spectra the ferrocytochrome contributions appear as shoulders rather than discrete maxima. In normal kidney (*Figure 3*, trace a) only oxyhaemoglobin is observed indicating adequate oxygen supply and adequate mitochondrial respiration. After 1 min of ischaemia (*Figure 3*, trace 2), the oxyhaemoglobin content is much reduced and the shoulders at 605 and 550 nm indicate reduction of the mitochondrial respiratory chain as the supply of oxygen to the point of consumption, fails. The general features of the spectrum after 10 min of ischaemia are similar to those found after 1 min and the *difference* between these two states (*Figure 3*, trace c) is featureless and gives an indication of the baseline for this experiment. Interesting-

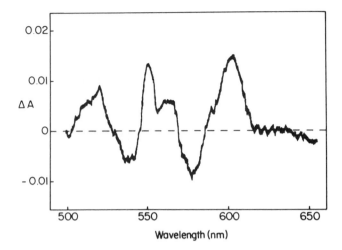

Figure 4. Reflectance difference spectrum of freeze-trapped cerebral cortex. Gerbil brains were freeze-trapped (67) 1 min after the anaesthetized animal began breathing either 5% O_2, 90% N_2, 5% CO_2 or 95% N_2, 5% CO_2. Reflectance spectra were recorded using a scanning spectrophotometer fitted with a bifurcated light guide (see *Figure 1B*) from the right cortex at 77K and the difference reflectance spectrum was computed from the memorized reflectance spectra recorded from each sample. Redrawn from data originally published in ref. 30.

ly, oxyhaemoglobin is not absent from the ischaemic tissue suggesting that there are metabolic compartments where oxygen is delivered but not consumed (28, 29).

Similar experiments with freeze-trapped gerbil cerebral cortex (25, 30; and see *Figure 4*) indicate that only in extreme anoxia does reduction of the mitochondrial pigments occur. *Figure 4* shows the difference reflectance spectrum obtained when the reflectance spectrum from a sample obtained from an animal after breathing 5% oxygen for 1 min is subtracted from that taken from an animal which had been breathing 0% oxygen for 1 min. In both situations the haemoglobin was largely deoxygenated and the difference spectrum of the whole tissue closely resembles that of reduced *minus* oxidized mitochondria. This shows that the respiratory chain still has enough oxygen under conditions where tissue haemoglobin is almost completely deoxygenated; or put another way the large difference in affinity for oxygen exhibited by mitochondria and haemoglobin *in vitro* is still apparent *in vivo*.

The experiments illustrated in *Figures 3* and *4* employed a microprocessor-controlled, scanning, dual-wavelength spectrophotometer which was capable of storing and manipulating spectra. An instrument with similar capacity for reflectance measurements is available from Perkin Elmer, the Lambda 5 UV/vis Spectrophotometer equipped with an External Integrating Sphere, and from Applied Photophysics Ltd., London.

3.2 Membrane potential of organelles and cells

Most biological membranes have the ability to sustain electrostatic potential differences between the aqueous phases which they separate. Indeed this potential, usually called the membrane potential, is often a significant component of membrane function; it is, for example, critical for signalling in excitable cells and makes a contribution to the accumulation of nutrients driven by the electrochemical Na^+ gradient across the plasma membrane of many cells. In subcellular organelles such as chloroplasts and mitochondria the membrane potential is an important 'high energy' intermediate in the synthesis of ATP. It is often impractical to record the potential directly with microelectrodes and of the other methods available for measuring transmembrane potential the use of dyes with specific sensitivity to potential is one of the most popular (31).

While it is clear that very many dyes change either their absorbance or their fluorescence as a response to a change in membrane potential (3) it is often difficult accurately to calibrate the signal with known potentials. In the simplest procedure the membrane permeability to a particular cation, usually K^+ or H^+, is increased by adding a reagent called an ionophore. In the presence of sufficient ionophore, such that all other permeabilities are relatively small, the membrane potential (V) will approach that predicted by the Nernst equation:

$$V = (-RT/F)\ln([M^+]_{in}/[M^+]_{out}) \qquad \text{Equation 1}$$

where M^+ is the relevant ion, R the gas constant, T the absolute temperature and F Faraday's constant. At 37°C this relation simplifies to:

$$V(mV) = -61.5 \log ([M^+]_{in}/[M^+]_{out}) \qquad \text{Equation 2}$$

which shows that, for a 10-fold change in cation gradient, you obtain a 61.5 mV change in potential. The cation gradient can be manipulated by altering the medium concentration of M^+ and hence a calibration curve of optical signal and potential can be obtained in the presence of ionophore.

3.2.1 *Light-induced membrane potential in chromatophores*

Chromatophores prepared from photosynthetic bacteria provide an ideal model system for studying changes in membrane potential. Electron transport in the photosynthetic reaction centre is initiated by infra-red light and short, intense flashes can be used to activate the photosynthetic apparatus just once: 'single turnover' flashes. As a consequence of electron transport, H^+ ions are pumped into the chromatophore with the generation of a substantial membrane potential (up to 300 mV inside positive), which induces a shift in the absorption spectrum of the carotenoid pigments. The potential-dependent shift of carotenoid absorption can be calibrated with K^+ in the presence of the ionophore valinomycin (ref. 2; see Section 3.2.2).

Figure 5 illustrates the light *minus* dark difference spectrum of a suspension of R. *sphaeroides* chromatophores containing a membrane potential indicating dye, oxonol-VI. The trough at 590 nm and the peak at 630 nm arise from the red shift of the oxonol absorbance (this region is featureless in the absence of dye at this sensitivity) and the peaks and troughs between 400 and 520 nm arise from the red shift of carotenoid absorbance. Thus both the intrinsic and the extrinsic chromophores provide signals and the former will reflect the properties of the latter. A detailed study of this system (32) revealed that the mechanism by which oxonols respond to potential is by migration into the chromatophore milieu (the distribution of the dye across the membrane obeys Equation 1) and subsequent binding of the dye to hydrophobic sites. Many of the most useful optical indicators of membrane potential work by this accumulation *plus* bin-

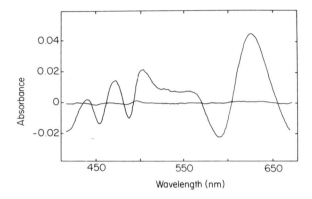

Figure 5. The light−dark difference spectrum of *R. sphaeroides* chromatophores in the presence of oxonol-VI. Chromatophores were prepared from *R. sphaeroides* GA (68) and suspended (2.1×10^{-5} M bacteriochlorophyll, final concentration) in 100 mM KCl, 20 mM morpholinopropanesulphonic acid (Mops), 1 mM $MgCl_2$, 0.5 mM ascorbate, 1.5×10^{-6} M oxonol-VI, pH adjusted to 6.9 with KOH at 23°C (32). The absorbance spectrum of the suspension was recorded and stored in the digital memory of a Johnson Foundation Scanning double beam spectrometer (25, 69) using 670 nm as the reference wavelength; the 'measuring' light is of insufficient intensity to initiate photosynthesis. A second spectrum of the suspension was recorded during illumination, at right angles to the 'measuring' beam, provided by a 15 W tungsten filament filtered through a Kodak Wratten Gelatin filter, number 88A, which transmits light of wavelengths >750 nm. The infra-red light activates photosynthesis and does not hamper the recording of spectra between 400 and 670 nm because of the low sensitivity of the photomultiplier to infra-red light. A third spectrum of the suspension was recorded and stored after extinguishing the infra-red illumination. Two *difference* spectra are presented: spectrum 1 *minus* spectrum 3, which gives the featureless baseline and suppresses any residual stimulation by the 'measuring' beam; and spectrum 2 *minus* spectrum 1, the 'light' *minus* 'dark' difference spectrum of the preparation. The maximum and minimum between 550 and 670 nm arise from the energy-dependent red-shift of oxonol-VI absorbance; the maxima and minima between 400 and 520 nm arise from the membrane potential dependent shift of the endogenous carotenoid pigments (32). Reproduced with permission from ref. 32.

ding mechanism. Calibration of the oxonol signal indicates that the dye has a logarithmic response to potential in this system.

3.2.2 *Plasma membrane potential of animal cells*

The membrane potential of whole cells is more difficult to assess than that of isolated organelles with optical indicators because cells contain many subcellular compartments and it is necessary to isolate the response of the plasma membrane from that of any other cell membrane. For example, the intense staining of mitochondria by cyanine dyes (5), which are membrane-permeant cations, means that if you choose a cyanine dye, or other lipophilic cation, to measure plasma membrane potential it is important to establish conditions where the mitochondrial potential is disabled. If an anionic dye is chosen, this problem is overcome but there is, then, very little entry of dye into the cells unless the dye is itself reasonably lipophilic. Thus, for studies of whole cells, we employ an oxonol dye similar to that used in the chromatophore studies but with phenyl rather than propyl substituents. The procedures for measuring plasma membrane potential with this dye, oxonol-V (33), are illustrated in *Figure 6*.

(i)   Suspend cells in a physiological saline (do not use serum or other albumin-containing media because albumin/protein will sequester all the oxonol and none

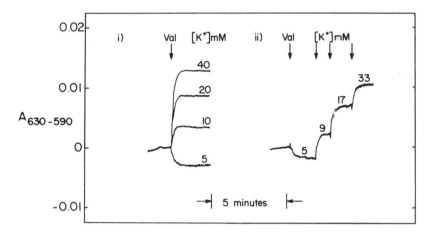

Figure 6. Measurement of Lettre cell membrane potential using oxonol-V. (i) 5×10^6 Lettre cells/ml were suspended in 5 mM Hepes, 1 mM MgCl$_2$, 2.2 × 10^{-6} M oxonol-V, pH adjusted to 7.4 with NaOH at 33°C and 155 mM NaCl *plus* KCl with the final K$^+$ concentration indicated. Valinomycin was added as indicated to give a final concentration of 1 × 10^{-6} g/ml. Traces from four separate experiments are superimposed such that the signal before the addition of valinomycin is identical. The extracellular concentration of K$^+$ at which valinomycin would give no change in absorbance was estimated to be 7.1 mM by interpolation, and the intracellular K$^+$ concentration was 73.3 mM giving a membrane potential (see Equation 2) under these conditions of −62 mV. (ii) 5 × 10^{-6} Lettre cells/ml were suspended in 150 mM NaCl, 5 mM KCl, 5 mM Hepes, 1 mM MgCl$_2$, 2.2 × 10^{-6} M oxonol-V, pH adjusted to 7.4 with NaOH at 33°C. Valinomycin (1 × 10^{-6} g/ml) and KCl to give the concentration indicated were added as shown by the arrows. The extracellular K$^+$ concentration at which valinomycin would give no change in absorbance was 6.9 mM, the cellular K$^+$ concentration was 63.3 mM and thus the membrane potential was −60 mV. Reproduced with permission from ref. 21.

of the dye will interact with the cells) in the presence of dye at 37°C. After an equilibration period of about 10 min the absorbance settles to a steady value.

(ii)   At this point, add valinomycin which moves the potential towards the K$^+$ equilibrium potential (defined in Equation 1) and the absorbance of the dye changes accordingly. In whole cells which have a large reservoir of internal K$^+$ the potentials in the presence of oxonol and valinomycin seem to be quite stable. If the K$^+$ equilibrium potential happens to equal the membrane potential then the addition of valinomycin will have no effect on the potential or the dye signal, this is the so-called K$^+$ 'null point' (34) and solution of Equation 1 using the medium concentration of K$^+$ for [K$^+$]$_{out}$ and the cytoplasmic K$^+$ concentration for [K$^+$]$_{in}$ will yield the value of the membrane potential. Should valinomycin affect the potential under the conditions employed, the K$^+$ null point can be obtained by either of two procedures both of which are illustrated in *Figure 6*.

(iii)   After valinomycin has been added to the cells and a new steady reading obtained, increase the concentration of K$^+$ in the medium by adding small aliquots of concentrated KCl, 4 M seems to be the highest manageable concentration (*Figure 6*).

(iv)   After taking readings at a few different extracellular K$^+$ concentrations, obtain the value of the null point by interpolation. It is usually sufficient to interpolate by joining data points rather than by using sophisticated least squares protocols,

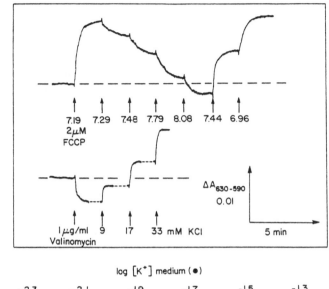

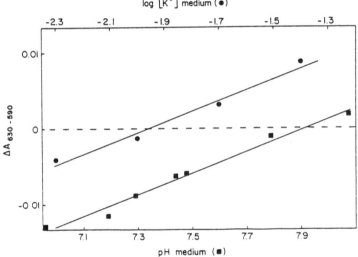

Figure 7. Measurement of lymphocyte membrane potential with oxonol-V. 10^7 lymphocytes/ml suspended in 150 mM NaCl, 5 mM KCl, 5 mM Hepes, 1 mM $MgCl_2$, 2×10^{-6} M oxonol-V, pH adjusted to 7.4 with NaOH at 37°C. FCCP and NaOH or HCl (**upper panel, upper trace**) or valinomycin and KCl (**upper panel, lower trace**) were added to give the final concentration or pH indicated. **Lower panel:** the difference between the absorbance recorded before and after the addition of FCCP at the relevant pH values (■) or of valinomycin at the relevant KCl concentrations (●) are plotted on equivalent axes. The concentrations of H^+ or K^+ at which the addition of FCCP or valinomycin, respectively, would have given no change in absorbance is obtained from the point of intersection of the solid lines with the base line (dashed). In these experiments the K^+ 'null-point' was 11 mM and the pH null point was 7.91; intracellular K^+ was 102 mM and intracellular pH was 7.0, giving membrane potentials of -59 and -56 mV, respectively. Reproduced with permission from ref. 21.

especially as the theoretical form of the dye absorbance versus potential plot is unknown.

This experimental protocol, using a single sample of cells, has the advantage of speed and uses little of a precious biological preparation. The alternative protocol is to incubate the cells in media of differing K^+ content and to assess the effect of valinomycin addition (*Figure 6*). Again the value of the K^+ null point is obtained by interpolation. This second procedure only works with cells that have rather low K^+ permeabilities; cells more permeable to K^+ depolarize in high K^+ media and the subsequent addition of valinomycin has little effect. In either procedure a knowledge of cytoplasmic K^+ concentration is required to calculate the potential accurately. This can be achieved by pelleting the cells through a mixture of di-*n*-butylphthalate (2 parts) and dinonylphthalate (1 part) with a density of 1.02; cells pellet through such an oil within 10 sec in a Beckman microcentrifuge B. Cell cations can be determined (after the water content; wet−dry weight) by atomic absorption spectrometry. The spillover of medium into the cell pellet is usually less than 0.05%, but this can be verified by including an extracellular space marker, such as [^{14}C]inulin, in the incubation medium.

There is a danger that the lipophilic valinomycin/K^+ complex may precipitate lipophilic anions such as the oxonol dyes and it is advisable to check on the calibration described above with another ionophore. The uncoupling agent FCCP is useful in this respect as it selectively increases the H^+ permeability of membranes. The experimental protocol for measuring the potential using FCCP is identical to that using valinomycin except that medium pH rather than K^+ is varied by addition of acid or base (*Figure 7*). Calculation of the membrane potential using the pH null point requires a knowledge of cell pH (usually close to 7.1) which may have to be determined separately, for example by using indicators (see below).

3.3 pH of cellular compartments

3.3.1 *pH of endosomes*

Most eukaryotic cells internalize components of their plasma membrane and associated exogenous material by endocytosis. The subsequent fate of the endosomes may be complex but it usually involves a step in which the vesicle milieu is acidified (35). The pH of such an endosome can be monitored directly if a pH-sensitive dye can be attached either to the relevant piece of membrane or to a suitable extracellular vehicle which

Table 3. Labelling of endosomes with fluorescein conjugates.

1.	Incubate the cell cultures in culture medium containing the fluorescein conjugate (present at ~1 mg/ml) at 37°C. The conjugate enters cells rapidly (within 10 min) and is subsequently transferred to secondary lysosomes[a]. Endocytosis and intracellular membrane traffic are markedly reduced at lower than physiological temperatures. At 4°C only binding of ligands to the cell surface occurs, so receptors can be 'loaded' at low temperature and endocytosis subsequently 'triggered' by raising the temperature.
2.	Remove the extracellular conjugate by washing the cells with cold 4°C), unlabelled medium. The residual fluorescence arises from within cells.
3.	Place the labelled cells in a fluorimeter/fluorescence microscope and record the fluorescence intensity. Cells attached to a solid substratum can be monitored with light guides (see *Figure 1*), or by fitting the substrate diagonally in a fluorescence cuvette and using a conventional fluorimeter (the final optical arrangement resembles the 'front face' system used to record fluorescence from opaque or strongly absorbing samples). It is important not to move the substrate during signal calibration.
4.	Calibrate the signal with extracellular pH using monensin and the protocol illustrated in *Figure 8a*.

[a]See ref. 35.

is endocytosed. The dye most commonly used is fluorescein because it is easily visualized with a fluorescence microscope and because its isothiocyanate derivatives can be covalently conjugated to the amino groups in proteins (for receptor-mediated endocytosis) or to dextran. Furthermore the fluorescence of fluorescein and its conjugates is sensitive to pH in the range 4.5−7.5 (18). The endosomes can be loaded with the fluorescein derivative by incubation in a suitable physiological medium and the extracellular dye is subsequently removed by washing (*Table 3*). The calibration of the fluorescence signal is illustrated in *Figure 8a*: after recording the signal, the pH of all the cellular compartments is set equal to the bulk pH by adding the ionophore monensin; the bulk pH is then systematically reduced until the original signal is restored and the final value

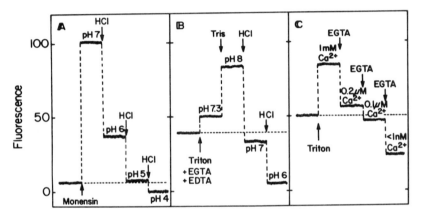

Figure 8. Calibration of intracellular probes of pH and Ca^{2+}. (**a**) Cells labelled with fluorescein−dextran (*Table 3*; ref. 18) in Hanks balanced salts solution buffered with 20 mM Hepes, pH 7.0 (NaOH) were excited at 493 nm (6 nm slitwidth) and fluorescence monitored at 520 nm (10 nm slitwidth) in a Perkin-Elmer MPF 4 fluorimeter operating in the 'ratio' mode. The initial signal (6 units) was calibrated by adding, as indicated, 1 μg/ml monensin (final concentration) to equalize the pH of all compartments in the system (18) followed by HCl to bring the bulk pH to the values indicated. The original signal from the cells corresponds to that found at pH 4.8 in the presence of monensin indicating that the fluorescein−dextran occupied a compartment whose pH was 4.8. Records should be corrected for autofluorescence (fluorescence of unlabelled cells) and, at the end of the calibration, the supernatant fluorescence should be measured and found negligible, <10% of that found in the presence of cells. (**b**) Thymocytes labelled with quene-1 (*Table 4*; ref. 9) were suspended at 5×10^6 cells/ml in a medium containing the inorganic salts of RPMI 1640 without phenol red and supplemented with 11 mM glucose and 10 mM Hepes, pH 7.3 (NaOH) at 37°C. Fluorescence at 530 nm (10 nm slitwidth) was excited at 390 nm (10 nm slitwidth) in a Perkin-Elmer model 44B spectrofluorimeter. The signal (49 units) was calibrated by permeabilizing the cells with 0.05% Triton X-100 (Triton) in the presence of 0.5 mM EGTA and 0.5 mM EDTA [the chelators reduce the free Ca^{2+} and Mg^{2+} to levels at which they do not affect the pH-sensitive fluorescence of quene-1 (9) and prevent quenching by other metal ions which may be present]. More than 90% of the dye is released by the Triton and it is titrated by the addition of Tris and HCl to give the bulk pH values indicated. The signal before permeabilization corresponds to that found at pH 7.1 in the presence of Triton indicating that cytoplasmic pH was 7.1 when the medium pH was 7.3. (**c**) Lymphocytes labelled with quin-2 (*Table 4*; ref. 8) were suspended at 5×10^6 cells/ml in 130 mM KCl, 20 mM NaCl, 1 mM $CaCl_2$, 1 mM $MgCl_2$, 5 mM glucose, 10 mM Mops, pH 7.4 (NaOH). Fluorescence at 500 nm (10 nm slitwidth) was excited at 339 nm (4 nm slitwidth) in a Perkin-Elmer MPF 44B spectrofluorimeter operating in the 'ratio' mode. The signal was calibrated by permeabilizing the cells with 0.05% Triton X-100 (Triton) and subsequent titration of the medium with EGTA, $MgCl_2$ and KOH to give 0.2 and 0.1×10^{-6} M Ca^{2+} as indicated, with 1 mM free Mg^{2+}, pH 7.05 (8). 'Zero' Ca^{2+} (<1 nM) was achieved by adding 2 mM EGTA and Tris to give a pH >8.3. Autofluorescence from unlabelled cells has been subtracted from each reading and the fluorescence of unpermeabilized lymphocytes corresponds to a Ca^{2+} level of 150 nM. All the traces in this figure are redrawn according to data and information originally published in refs 8, 9 and 18.

Table 4. Labelling of cytoplasm with Ca^{2+}- or pH-sensitive indicators starting with the acetoxymethyl ester.

1. Make a 10 mM stock solution of the acetoxymethyl ester of quin-2 (for Ca^{2+} measurement) or quene-1 (for pH measurement) in dimethyl sulphoxide (DMSO).
2. Incubate the cells in isotonic, buffered physiological medium at 37°C and add stock indicator solution to give a final concentration in the range $1-5 \times 10^{-6}$ M ($\sim 0.1-0.5$ nmol indicator/10^6 cells), the final DMSO content is $<0.5\%$ v/v.
3. Follow the progress of ester hydrolysis by recording fluorescence emission spectra from the cell suspension. The acetoxymethyl ester emission is blue-shifted (lower wavelength) with respect to the free acid form[a].
4. When ester hydrolysis is $>80\%$ complete, usually after $30-90$ min, pellet the cells, resuspend them in unlabelled medium and incubate for a further 30 min at 37°C.
5. Pellet the cells once more and resuspend them in medium for fluorescence measurement.
6. Calibrate the fluorescence signals with extracellular pH or Ca^{2+} as illustrated in *Figure 8b,c*.

If the labelling is slow, include bicarbonate in the incubation media as this helps to maintain intracellular pH near neutrality, the optimal value for the relevant cytoplasmic esterases.

[a]See refs 8 and 9.

of bulk pH equals that originally present in the endosomes. The major difficulty of this technique arises from cellular autofluorescence which may, itself, be pH-dependent.

A more qualitative estimate of endosome pH can be obtained by monitoring the distribution of a permeant, fluorescent amine such as 9-aminoacridine or acridine orange whose fluorescence varies with concentration. Such amines accumulate in acid compartments because the protonated form of the dye is much less membrane permeant than the free base. Indeed, if the latter equilibrates readily across membranes then the pH within the vesicle is given by the relationship (7):

$$pH_{in} = pH_{out} - \log([AH^+]_{in}/[AH^+]_{out}) \qquad \text{Equation 3}$$

normally $pH_{out} < pK-1$ so that $[AH^+] \cong [AH^+]+[A]$. 10 mM 9-aminoacridine exhibits a yellow (4) and acridine orange a red rather than the blue fluorescence seen at low dye concentrations. Cells incubated with 0.01 mM 9-aminoacridine may have many yellow vesicles when examined by fluorescence microscopy using the fluorescein settings; this indicates that such vesicles have accumulated the dye by at least two orders of magnitude and that the internal pH must be in the range $5-6$.

3.3.2 pH or [Ca²⁺] of cytoplasm or isolated membrane vesicles

Trapped dyes will also record the pH or calcium ion concentration in the cytoplasm or sealed compartments. For this purpose indicators of a type similar to quene-1 (9) or quin-2 (8) are convenient.

(i) Load the cells by adding the acetoxymethyl ester of the dye which permeates the plasma membrane and is subsequently hydrolysed by cellular esterases to the impermeant acid form (*Table 4*).
(ii) Follow the progress of esterification by recording the progressive change in the fluorescence emission spectrum from that of the ester to that of the acid. It is important to maintain cytoplasmic pH near neutrality, the optimum value for the relevant esterases, especially as the hydrolysis generates acid; this is best done by including bicarbonate in the incubation medium (9).

(iii) At the end of the incubation, wash the cells to reduce the medium fluorescence to a minimal value. The remaining signal arises from the cytoplasmically located dye. Overzealous washing may also remove intracellular dye, particularly if the cell is relatively leaky. It has been reported that a short, sharp spin in a microcentrifuge, followed by rapid resuspension and calibration (8) is the optimal protocol.

(iv) Once the steady-state has been established, apply stimuli and observe any fluorescence changes.

(v) At the end of the experiment, calibrate the fluorescence (*Figure 8b,c*) by permeabilizing the cells with digitonin or Triton-X 100 (in the presence of 0.5 mM EGTA and 0.5 mM EDTA to chelate metal ions, which are potent quenchers of fluorescence, for pH measurement) and titrating the suspension with H^+ or Ca^{2+} until the fluorescence observed in the absence of detergent is regained.

A word of caution about the cleaning of cuvettes is appropriate here; the use of chromic acid may introduce fluorescence quenchers. These can usually be removed by soaking the cuvettes in solutions of chelating agents such as EDTA.

Unfortunately some cells seem not to retain the acid form of these pH and Ca^{2+} indicating dyes very long even though the free acid is supposed to be membrane impermeant. It is very important, therefore, to check that the signal observed does indeed originate from within cells; this is best done by pelleting a sample of the suspension and verifying that the supernatant lacks dye (check the supernatant fluorescence). If leakage is particularly troublesome, reasonable data can be obtained by loading the cells for only a short time (much less than the time taken for complete hydrolysis of the ester), rapidly washing the cells and immediately making the pH or Ca^{2+} determination. Unfortunately this leads to an inevitable wastage of rather expensive starting material.

Permeant indicators record cytoplasmic pH provided that the contribution from extracellular dye is suitably compensated. Neutral red is ideally suited for this purpose. It readily enters cells and, provided that external buffering is strong, any perturbations of the signal can be ascribed to pH changes in the closed compartment (36). A typical experiment in which the absolute value of cellular pH was monitored with neutral red is shown in *Figure 9*.

For this experiment carry out the following procedure.

(i) Incubate cells (in this case an ascites tumour cell line known as Lettre cells) with Hepes-buffered physiological saline in the presence of 0.025 mM neutral red at room temperature (with 5×10^6 cells/ml approximately 50% of the dye is cell associated) and record the spectrum (spectrum a).

(ii) Pellet the cells and record the spectrum of the supernatant (spectrum b). Spectrum c is the difference between the two former spectra, that is the spectrum of cell-associated dye.

(iii) Calculate the pH of each compartment from a standard curve established using solutions of neutral red of known pH.

The parameter chosen was the ratio of the absorbance due to the acid form (A_{530}) to that of the isosbestic wavelength (A_{477}) since this obeys the Henderson−Hasselbalch relationship and is independent of dye concentration (4). The pH of Lettre cells under these conditions was 7.01. We have assumed that the pK_a and the spectral properties of cell-associated dye are identical to those of free dye, an assumption that remains

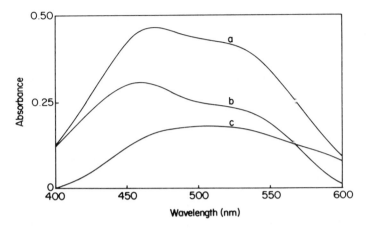

Figure 9. Measurement of Lettre cell pH with neutral red. The absorbance spectra of four samples were recorded at 23°C and stored in the digital memory of a scanning spectrophotometer (Double Wavelength Spectrometer, Applied Photophysics Ltd. London): (1) medium alone comprising 150 mM NaCl, 5 mM KCl, 5 mM Hepes, 1 mM $MgSO_4$, pH adjusted to 7.3 with NaOH, the baseline for **spectrum b**; (2) medium plus 3×10^{-6} Lettre cells/ml, the baseline for **spectrum a**; (3) medium plus 3×10^6 Lettre cells/ml plus 4×10^{-5} M neutral red; (4) supernatant remaining from medium 3 after pelleting the cells at 16 000 g for 1 min in a Beckman microcentrifuge B. The following difference spectra are presented: **spectrum a:** scan 3 − scan 2, the spectrum of extracellular plus intracellular neutral red; **spectrum b:** scan 4 − scan 1, the spectrum of extracellular neutral red; **spectrum c:** (scan 3 − scan 2) − (scan 4 − scan 1), the spectrum of intracellular neutral red, that is the difference between **spectrum a** and **spectrum b**. The pH of the intracellular and extracellular compartments was calculated from the ratio of the absorbance at 530 nm to that at 477 nm (4). For **spectrum b**, extracellular space, $A_{530}/A_{477} = 0.768$, pH $= 7.32$. For **spectrum c**, intracellular space, $A_{530}/A_{477} = 1.032$, pH $= 7.01$.

to be validated. However, the value obtained by the indicator technique described here agrees well with that obtained by two independent techniques, namely ^{31}P nuclear magnetic resonance spectroscopy and the direction of cation leakage from Sendai virus-permeabilized cells (37).

3.4 Surface potential of membranes

Biological membranes usually carry a net negative charge due to the presence of acidic groups (phosphate, sulphate, carboxylate) attached to lipids and proteins. The absolute magnitude of the negative charge (the surface potential) is not directly accessible but electrophoretic measurements (38−40) allow an estimate of the zeta-potential which is the magnitude of the surface potential at the 'plane of shear' between the membrane and the aqueous medium and subsequently of the surface charge density. Surface enzymes and receptors are sensitive to changes in surface potential, directly or by interaction with cations attracted to the surface, and it is useful to be able to probe changes in surface potential during biological processes.

A number of chromophores have been used to probe the surface potential of membranes. The principle of the experiment is straightforward: amphiphilic chromophores will bind to cell surfaces at 'hydrophobic' sites, the altered solvent environment directly affects fluorescence/absorbance so that it is possible optically to distinguish free and membrane-associated dye. If the dye is charged, the surface potential has a direct effect on the dye − membrane affinity (positively charged dyes are bound strongly, negatively charged dyes weakly). If the surface potential changes, then the affinity of the dye for the surface will change and this can be assessed. Thus 9-aminoacridine (a positively charged dye) has been used to probe changes in chloroplast membrane surface charge (19, 41), membrane-associated dye having negligible fluorescence; 1-Anilino-8-naphthalene sulphonate (ANS; a negatively charged dye), has been used with mitochondrial and other membranes (42−44), in this case free dye has negligible fluorescence. A decrease in surface potential increases 9-aminoacridine fluorescence (liberation of quenched dye into the medium) and ANS fluorescence (binding of dye to a site which increases fluorescence quantum yield).

The disadvantage of these methods is that it is known that many factors, including the transmembrane distribution of the dye, affect their fluorescence and it is difficult to ensure that only the contributions of surface charge are being detected. Unlike transmembrane potential, which can be modulated in a predictable way by ionophores, surface potential cannot easily be modulated in a straightforward fashion to allow accurate calibration of the signal.

4. FUTURE PROSPECTS

The elegance and simplicity of optical experiments ensures that considerable efforts will continue to be expended to develop photometric and fluorometric assays of cellular membrane function. Increasingly protocols are being devised for the study of individual cells in complex networks. Thus optical probes of membrane potential are useful for tracing synaptic networks in simple (45, 46) and more complex (47) systems. Likewise new indicators of pH and cell calcium will help to unravel the role of these cations in cell stimulation and differentiation (48), again with increasing emphasis on single cell studies. Finally existing procedures will be used, on account of the wide applicability of the technique, to systems inaccessible to more elaborate experiments permitting a proper investigation of the full range of biological diversity.

5. ACKNOWLEDGEMENTS

I would like to thank Mrs B.J.Bashford for her patient preparation of the diagrams and the Royal Society and the Cell Surface Research Fund for financial support.

6. REFERENCES

1. Keilin,D. (1925) *Proc. R. Soc. Lond. B*, **98**, 312.
2. Jackson,J.B. and Crofts,A.R. (1969) *FEBS Lett.*, **4**, 185.
3. Cohen,L.B., Salzberg,B.M., Davila,H.V., Ross,W.N., Landowne,D., Waggoner,A.S. and Wang,C.H. (1974) *J. Membr. Biol.*, **19**, 1.
4. Harris,D.A. and Bashford,C.L., eds (1987) *Spectrophotometry and Spectrofluorimetry − A Practical Approach*. IRL Press, Oxford and Washington, DC.
5. Johnson,L.V., Walsh,M.L., Bockus,B.J. and Chen,L.B. (1981) *J. Cell Biol.*, **88**, 526.
6. Reijngoud,D.J. and Tager,J.M. (1973) *Biochim. Biophys. Acta*, **297**, 174.
7. Rottenberg,H. (1979) *Methods Enzymol.*, **55**, 547.
8. Tsien,R.Y., Pozzan,T. and Rink,T.J. (1982) *J. Cell Biol.*, **94**, 325.
9. Rogers,J., Hesketh,T.R., Smith,G.A. and Metcalfe,J.C. (1983) *J. Biol. Chem.*, **258**, 5994.
10. Kauppinen,R.A. and Hassinen,I.E. (1984) *Am. J. Physiol.*, **247**, H508.
11. Montecucco,C., Pozzan,T. and Rink,T.J. (1979) *Biochim. Biophys. Acta*, **552**, 552.
12. Miller,J.N. (1981) *Standards in Fluorescence Spectrometry*. Chapman and Hall, London.
13. Bangham,A.D., De Gier,J. and Greville,G.D. (1967) *Chem. Phys. Lipids*, **1**, 225.
14. Chance,B. and Williams,G.R. (1955) *J. Biol. Chem.*, **217**, 395.
15. Chance,B., Schoener,B., Oshino,O., Itshak,F. and Nakase,Y. (1979) *J. Biol. Chem.*, **254**, 4764.

16. Harbig,K., Chance,B., Kovach,A.G.B. and Reivich,M. (1976) *J. Appl. Physiol.*, **41**, 480.
17. Waggoner,A.S. (1979) *Methods Enzymol.*, **55**, 689.
18. Geisow,M.J. (1984) *Exp. Cell Res.*, **150**, 29.
19. Barber,J. (1986) In *Ion Interactions in Energy Transfer Biomembranes*, Papageorgiou,G.C., Barber,J. and Papa,S. (eds), Plenum Press, London, p. 15.
20. Akerman,K.E.O. and Wikstrom,M.K.F. (1976) *FEBS Lett.*, **68**, 191.
21. Bashford,C.L., Alder,G.M., Gray,M.A., Micklem,K.J., Taylor,C.C., Turek,P.J. and Pasternak,C.A. (1985) *J. Cell. Physiol.*, **123**, 326.
22. Estabrook,R.W. (1956) *J. Biol. Chem.*, **223**, 781.
23. Jones,C.W. and Poole,R.K. (1985) In *Methods in Microbiology*. Gottschalk,G. (ed.), Academic Press, New York, Vol. 18, p. 285.
24. Poole,R.K. and Chance,B. (1981) *J. Gen. Microbiol.*, **126**, 277.
25. Bashford,C.L., Barlow,C.H., Chance,B., Haselgrove,J.C. and Sorge,J. (1982) *Am. J. Physiol.*, **242**, C265.
26. Jobsis,F.F., Keizer,J.H., LaManna,J.C. and Rosenthal,M. (1977) *J. Appl. Physiol.*, **43**, 858.
27. Wollenberger,A., Ristau,O. and Schoffa,G. (1960) *Pflugers Arch. ges. Physiol.*, **270**, 399.
28. Epstein,F.H., Balaban,R.S. and Ross,B.D. (1982) *Am. J. Physiol.*, **243**, F356.
29. Bashford,L. and Stubbs,M. (1986) *Biochem. Soc. Trans.*, **14**, 1213.
30. Bashford,C.L., Barlow,C.H., Chance,B. and Haselgrove,J. (1980) *FEBS Lett.*, **113**, 78.
31. Bashford,C.L. (1981) *Biosci. Rep.*, **1**, 183.
32. Bashford,C.L., Chance,B. and Prince,R.C. (1979) *Biochim. Biophys. Acta*, **545**, 46.
33. Smith,J.C., Russ,P., Cooperman,B.S. and Chance,B. (1978) *Biochemistry*, **15**, 5094.
34. Hoffman,J.C. and Laris,P.C. (1974) *J. Physiol. (Lond)*, **239**, 519.
35. Geisow,M.J. and Evans,W.H. (1984) *Exp. Cell Res.*, **150**, 36.
36. Junge,W.W., Auslander,W., McGeer,A.J. and Runge,T. (1979) *Biochim. Biophys. Acta*, **546**, 121.
37. Bashford,C.L., Alder,G.M., Micklem,K.J. and Pasternak,C.A. (1983) *Biosci. Rep.*, **3**, 631.
38. Bangham,A.D., Flemans,R., Heard,D.H. and Seaman,G.V.F. (1958) *Nature*, **182**, 642.
39. Bangham,A.D., Pethica,B.A. and Seaman,G.V.F. (1958) *Biochem. J.*, **69**, 12.
40. Heard,D.H. and Seaman,G.V.F. (1960) *J. Gen. Physiol.*, **43**, 635.
41. Chow,W.S. and Barber,J. (1980) *J. Biochem. Biophys. Methods*, **3**, 173.
42. Haynes,D.H. (1974) *J. Membr. Biol.*, **17**, 341.
43. Wojtczak,L. and Nalecz,M.J. (1979) *Eur J. Biochem.*, **94**, 99.
44. Robertson,D.E. and Rottenberg,H. (1983) *J. Biol. Chem.*, **258**, 11039.
45. Grinvald,A., Salzberg,B.M. and Cohen,L.B. (1977) *Nature*, **268**, 140.
46. Grinvald,A., Cohen,L.B., Lesher,S. and Boyle,M.B. (1981) *J. Neurophysiol.*, **45**, 829.
47. Blasdel,G.G. and Salama,G. (1986) *Nature*, **321**, 579.
48. Poenie,M., Alderton,J., Tsien,R.V. and Steinhardt,R.A. (1985) *Nature*, **315**, 147.
49. Clayton,R.K. (1980) *Photosynthesis: Physical Mechanisms and Chemical Patterns*. Cambridge University Press, Cambridge.
50. Telfer,A. (1986) In *Ion Interactions in Energy Transfer Biomembranes*. Papageorgiou,G.C., Barber,J. and Papa,S. (eds), Plenum Press, New York, p. 201.
51. Schuldiner,S., Rottenberg,H. and Avron,M. (1972) *Eur. J. Biochem.*, **25**, 64.
52. Kogure,K., Alonso,O.F. and Martinez,E. (1980) *Brain Res.*, **195**, 95.
53. LaManna,J.C. and McCracken,K.A. (1984) *Anal. Biochem.*, **142**, 117.
54. Ashley,C.C. and Campbell,A.K. (1979) *Detection and Measurement of Free Ca2*$^+$ *in Cells*. Elsevier/North Holland, Amsterdam.
55. Scarpa,A. (1972) *Methods Enzymol.*, **24**, 322.
56. Azzi,A., Chance,B., Radda,G.K. and Lee,C.P. (1969) *Proc. Natl. Acad. Sci. USA*, **62**, 612.
57. Brocklehurst,J.R., Freedman,R.B., Hancock,D.J. and Radda,G.K. (1970) *Biochem. J.*, **116**, 721.
58. Bashford,C.L., Radda,G.K. and Ritchie,G.A. (1975) *FEBS Lett.*, **50**, 21.
59. Bashford,C.L., Casey,R.P., Radda,G.K. and Ritchie,G.A. (1976) *Neuroscience*, **1**, 399.
60. Flanagan,M.T. and Hesketh,T.R. (1973) *Biochim. Biophys. Acta*, **298**, 535.
61. Shinitzky,M. and Barenholz,Y. (1974) *J. Biol. Chem.*, **249**, 2652.
62. Shinitzky,M. and Inbar,M. (1974) *J. Mol. Biol.*, **85**, 603.
63. Bashford,C.L., Harrison,S.J., Radda,G.K. and Mehdi,Q. (1975) *Biochem. J.*, **146**, 473.
64. Bashford,C.L., Morgan,C.G. and Radda,G.K. (1976) *Biochim. Biophys. Acta*, **426**, 157.
65. Thulborn,K.R. and Sawyer,W.H. (1978) *Biochim. Biophys. Acta*, **511**, 125.
66. Bashford,C.L., Foster,K.A., Micklem,K.J. and Pasternak,C.A. (1981) *Biochem. Soc. Trans.*, **9**, 80.
67. Kerr,S.E. (1935) *J. Biol. Chem.*, **110**, 625.
68. Dutton,P.L., Petty,K.M., Bonner,H.S. and Morse,S.D. (1975) *Biochim. Biophys. Acta*, **387**, 536.
69. Chance,B. and Graham,N. (1971) *Rev. Sci. Instrum.*, **42**, 941.

CHAPTER 8

Biophysical approaches to the study of biological membranes

DIETER SCHUBERT

1. INTRODUCTION

Biophysical methods have contributed much to the progress of membrane research, and undoubtedly they will continue to do so. However, the complexity of the theoretical background, the instrumentation and the evaluation procedures of most biophysical techniques preclude a detailed treatment of the commonly-used methods in a text such as this. Since most of the important types of biophysical experiment require the thorough involvement of specialists, this chapter will concentrate not on the details of such experiments, but on their potential and limitations. Preference will be given to the description of those techniques which are of general applicability to membrane proteins (as compared with those which, for example require the presence of special intrinsic chromophores — see Chapter 7). The chapter is not intended to serve the expert in biophysics. Rather, it is devoted to those who have little experience in this field and who are unaware of the information that can be obtained.

2. SIZE, SHAPE AND CONFORMATION OF MEMBRANE PROTEINS AND THEIR COMPLEXES

Most of the techniques described in this section cannot be applied to the study of membrane proteins in their native environment but require solubilization and, in many instances, purification. It is clear, however, that meaningful structural information can only be obtained if the solubilization and purification procedures do not lead to major changes in protein conformation. Fortunately, this is not a serious problem with peripheral membrane proteins, but in the case of integral membrane proteins there are considerable hazards to be overcome. Isolation of the latter proteins in solutions of suitable non-ionic detergents seems to be the best way to achieve this since, in these solutions, both the folding and the protein—protein associations of the membrane polypeptides seem to be identical or at least very similar to the situation in the native membranes (see Chapter 5). If this approach is used, the results of biophysical studies on solubilized membrane proteins can be as meaningful as those on 'water-soluble' proteins (1).

2.1 Molecular weight of monomeric proteins or subunits

Polypeptide molecular weight is now almost exclusively determined by SDS—polyacrylamide gel electrophoresis. In contrast to the classical methods like analytical ultracentrifugation, light scattering, X-ray small-angle scattering or osmometry, SDS—gel electrophoresis does not need expensive equipment or sophisticated evaluation techniques. There is no requirement for highly purified proteins or large quantities since molecular weight determinations can be carried out on protein species which are minor components of a sample. Furthermore, SDS is the most effective reagent for the solubilization of membrane proteins and the de-polymerization of protein complexes. Methods are fully described in another volume of this series (2) (see also Chapter 6).

It should be noted that determining polypeptide molecular weight by SDS−gel electrophoresis, though undoubtedly a most useful and important method, is not only without any theoretical justification but has also failed in some cases (1). In particular, glycoproteins with a high carbohydrate content tend to yield inaccurate results. It also cannot be taken for granted that treatment with SDS will in all cases yield monomeric protein subunits only [(1), see Chapter 6]. In the case of glycoproteins, reliable molecular weight determinations can be obtained by analytical ultracentrifugation in SDS solutions and accompanying measurements of detergent binding (1). If the proteins should be monomeric in non-ionic detergents, the techniques described below for protein complexes can also be applied. The problem of whether complete depolymerization has been achieved by SDS is a more arduous one. When in doubt, a combination of SDS−gel electrophoresis, analytical ultracentrifugation under several denaturing conditions and quantitative end group analysis may be necessary.

2.2 Particle weight of protein complexes

2.2.1 *Chemical cross-linking*

The technique of chemical cross-linking of protein complexes in the intact membrane and subsequent molecular weight determinations of the cross-linked aggregates by SDS−gel electrophoresis yield information on the size of the complex in its native environment; it has contributed significantly to our present knowledge of the organization of membrane proteins. The procedure can also be applied to solubilized complexes of membrane proteins. A detailed description of the method is given in Chapter 6; an introduction to the procedures and problems can also be found in ref. 3. Two comments on the approach are worth mentioning. (i) Whereas the positive demonstration of pair formation between two protein molecules has to be considered as hard information, negative results do not always mean that no association exists. (ii) In the study of membrane systems, it is difficult or nearly impossible to distinguish between stable complexes and equilibrium situations. With solubilized complexes, the distinction can be made by studying the influence of protein concentration on the cross-linking reaction. On the other hand, cross-linking during the random collision of protein molecules, which in the past was thought to be a serious problem, does not seem to be so common an event (4,5).

2.2.2 *Analytical ultracentrifugation*

Analytical ultracentrifugation, the study of the movement and distribution of solubilized macromolecules in centrifugal fields, represents the classical method of determining macromolecule molecular weight. Though less fashionable than in earlier days, it still seems to be the best method for determining the particle weight of complexes of membrane proteins, especially if the amount of protein available is limited. A detailed description of the procedures can be found in another volume of this series (6). Here, the description will be restricted to a summary of those aspects which are of special importance in the study of membrane proteins.

(i)   If the centrifuge is equipped with a u.v. scanning system, solutions of protein of below 100 μg/ml can be studied (at a sample volume per cell of ~150 μl). Besides reducing the amount of purified material necessary, this allows us, in most cases, to treat the solutions as being ideal and thus greatly simplifies the evaluation of the experimental data.

(ii)   With peripheral membrane proteins, which can be studied in the absence of detergents, particle weights can be determined both by combining sedimentation velocity and diffusion measurements and by sedimentation equilibrium analysis. For intrinsic membrane proteins, which require the presence of non-ionic detergents, sedimentation equilibrium analysis is the method of choice, since it is much easier with this technique to correct for the contributions from protein-bound detergent.

(iii)   Sedimentation equilibrium experiments on proteins in non-ionic detergents allow the determination of the particle weight of both the protein/detergent complex and of the protein moiety alone. The procedure involves variation of solvent density by using appropriate H_2O/D_2O mixtures. Determination of the particle weight of the protein alone requires a solvent density which matches that of the detergent, or extrapolation of the data to that density. In addition, knowledge of the partial specific volume $\bar{v}$ of the protein is required. For density matching, the density of the detergent has to be between 1.0 and 1.1 g/ml. Fortunately, this is true for a large number of commonly used non-ionic detergents, including those of the Brij series and Triton X-100 (1).

(iv)   Mixtures of oligomeric aggregates of a single polypeptide chain or complex can be studied by analysing the concentration-versus-radius profiles from sedimentation equilibrium runs by least-squares techniques. This allows the determination of the size and the relative contribution of the different oligomers present. In addition, experiments performed at different initial protein concentrations or different rotor speeds can distinguish whether the oligomers are stable or in an association equilibrium (7). An example is shown in *Figure 1*.

2.2.3 *Solution scattering techniques*

Measurements of the small-angle scattering of X-rays or of thermal neutrons by solutions of homogeneous macromolecules or their complexes, can give the molecular weight of the particles. The information is obtained from the lowest angle portion of the scattering curves, the central maximum of the scattering intensity, by extrapolating the data to zero angle. Light scattering experiments (at any angle) will yield the same information. In all cases, knowledge of the partial specific volume of the particles (or of its light scattering equivalent) will be necessary. As in the case of analytical ultra-centrifugation, the contribution of protein-bound detergent causes problems. It is very difficult to correct for this detergent effect in X-ray small-angle scattering and light scattering experiments. In neutron scattering, however, the contribution of the detergent can be cancelled by using appropriate H_2O/D_2O mixtures, the scattering density of which matches that of the detergent. For detailed descriptions of the scattering techniques see ref. 9.

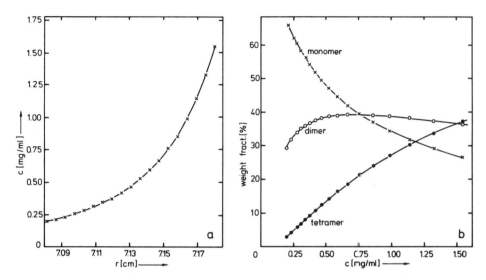

Figure 1. Analysis of a sedimentation equilibrium experiment with the anion transport protein (band 3) from human erythrocyte membranes in Triton X-100. Solvent density was adjusted to match that of the detergent. (**a**) Experimental concentration-versus-radius data (×), and least-squares fit to the data based on a monomer/dimer/tetramer model of self-association (—). (**b**) Relative contributions of the different oligomers to local protein concentration. Supplementary data demonstrate that the oligomers are in an association equilibrium (from ref. 8 by permission).

With respect to particle weight determinations, the solution scattering techniques possess two distinct disadvantages as compared with analytical ultracentrifugation.

(i)   The protein concentration required is higher by a factor of approximately 100.

(ii)   There is no simple way to detect heterogeneity of the material.

The existence of an association equilibrium can be detected, however, by measurements at different solute concentrations. Normally, the major aim of small-angle scattering experiments is to learn about the shape of the molecules (see later).

In a related type of light scattering measurement, changes in light scattering from a membrane vesicle suspension are utilized to follow association and dissociation reactions among proteins at the membrane surface (10).

2.2.4 *Sucrose gradient centrifugation and gel filtration*

A popular method of determining particle weight of detergent-solubilized membrane proteins is to combine, in a rearranged Svedberg equation, sedimentation coefficients measured by zonal centrifugation in a sucrose gradient (relative to standard water-soluble proteins) and values of the apparent Stokes radius measured by gel filtration (1). This method is simple and does not require sophisticated apparatus or highly purified material. However, it must be stressed that the method is of dubious validity not only because it lacks a theoretical foundation but also because it will, in an uncontrolled way, be subject to the influence of protein-bound detergent on both the sedimentation coefficient and the Stokes radius and to preferential hydration of the protein (1). Application of the method to integral membrane proteins is therefore not recommended.

2.2.5 *Radiation inactivation analysis* (see also Chapter 6)

When enzymes or receptors are subjected to ionizing radiation, their functional activity decreases. Target theory relates the dependency of the activity loss on radiation dose to the size of the target and hence to its molecular weight. Since what is measured is biological activity, the molecular weight determined should, in principle, be that of the functional unit rather than of the physical complex (11).

Target theory assumes that the damage to functional units is caused entirely by primary ionization. Since indirect effects can occur in aqueous solutions, radiation inactivation analysis is usually applied to dried or, preferably, frozen specimens. As purification of the molecules is not required, the method is applicable to membrane-bound enzymes or receptors in their native environment.

This approach can yield useful results but is also accompanied by a range of serious problems, for example energy transfer (references 11, 12 and Chapter 6). It is best used in combination with a second independent form of analysis.

2.3 Molecular dimensions

There are two methods which can provide detailed and unambiguous information on the shape of proteins and their complexes: X-ray diffraction from three-dimensional crystals, and three-dimensional image reconstruction from tilt series of electron micrographs of two-dimensional crystals of the protein (see Section 2.4). Depending on the resolution achieved, these techniques can also yield detailed information on protein conformation. However, for the vast majority of membrane proteins, which have not yet been obtained in crystalline form, less satisfactory methods have to be employed. Among these, the small-angle scattering techniques, especially neutron scattering, and conventional electron microscopy are the most useful.

2.3.1 *Small-angle scattering*

The lowest-angle portion of the X-ray or neutron scattering curves of homogeneous macromolecules or their complexes yields not only their molecular weight (Section 2.2.3) but also a useful measure of the overall molecular dimensions, namely the radius of gyration R. R is the root-mean-square of the distances of the scattering centres from the centre of gravity (weighted with respect to scattering density). Its value is close to that of its mechanical analogue (where mass density replaces scattering density). R increases with increasing anisotropy of the particles and with increasing hollow space inside and is thus a measure of their compactness. If suitable additional information is available, for example on the gross shape of the particles, knowledge of R may even allow conclusions on particle dimensions (13).

More information on the shape of the particles can be obtained from the outer section of the scattering curve, a segment which is particularly sensitive to shape. Unfortunately, there is no ready relationship between particle shape and the form of the scattering curve, nor is any relationship unique. What is normally done is to compare the experimental scattering curves with theoretical curves for, first, simple bodies (with random orientation). Where some measure of agreement is apparent, the model may be refined by testing more sophisticated variants. Since quite different models may yield rather similar scattering curves, the search for the most appropriate model should be guided by other data, for example from electron microscopy. Despite these problems, there are examples of remarkable success in this model building approach (13).

In general, protein-bound detergent will affect the scattering curves (see Section 2.2.3). However, in neutron scattering experiments its influence can be removed by using appropriate H_2O/D_2O mixtures. In the absence of detergents, X-ray small-angle scattering is as useful a technique as neutron scattering.

Small-angle scattering methods can also be applied to intact and reconstituted membranes, preferably 'stacked' (14,15). They yield the cylindrically averaged (and time averaged) protein and lipid arrangement perpendicular to the plane of the membrane. Again, neutron diffraction is the more useful technique, especially when combined with isomorphous labelling, by deuteration, of defined membrane components. The maximal spatial resolution will be approximately 10 Å; the positions of certain labelled atomic sites within the cylindrically averaged structure may be determined to within 1 Å. An example using sarcoplasmic reticulum is shown in *Figure 2*. In addition, time-resolved structure determinations can be obtained by utilizing intense synchrotron or laser-plasma X-ray sources (15). These techniques may in the future provide considerable insight into structure − function relationships in biological membranes.

2.3.2 *Electron microscopy*

The discrimination limit in the electron microscope lies in the size range 20−40 Å, which corresponds to the average dimensions of proteins of molecular weight below 20 000. The rough size and shape of most membrane proteins, therefore, should be revealed by this technique. Unfortunately, the procedures for sample preparation may seriously distort the particles. This holds both for negative staining and for shadow-casting, the standard techniques used for contrasting individual particles (17). Nevertheless, electron microscopy is a major source of information on the shape and dimensions of solubilized membrane proteins, especially in the case of elongated particles. Convincing examples are the studies on spectrin, the main component of the erythrocyte membrane cytoskeleton, and on its interactions with ankyrin (18). These studies, which used low-angle unidirectional and rotary shadowing of the isolated proteins, revealed clearly the dimensions, the flexible character and the configuration of the spectrin heterodimer, the arrangement of the dimers within the spectrin tetramer and the location of the ankyrin binding sites (*Figure 3*).

2.3.3 *Hydrodynamic methods*

The frictional coefficient of a solubilized protein particle, together with its molecular weight, can be used to obtain the protein's 'shape factor' or 'Perrin factor', F, which is the ratio of the frictional coefficient of the particle to that of a sphere with equal volume. Values of F can easily be calculated for ellipsoids of different axial ratio (9,19). A comparison between the actual value of F and those obtained from calculations will therefore yield information on the relative dimensions of the 'equivalent' ellipsoid and thus on the gross shape of the particle. The frictional coefficient may be determined either from the diffusion coefficient (which is measured in the analytical ultracentri-

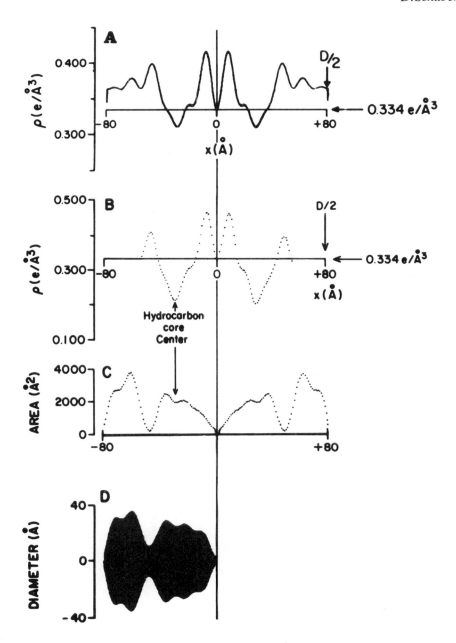

Figure 2. The sarcoplasmic reticulum (SR) membrane profile at 10 Å resolution, obtained from a combination of X-ray and neutron diffraction techniques. (**A**) Electron density profile (on an absolute scale) for the unit cell of the oriented SR membrane multilayer containing the two opposed single-membrane profiles of the flattened unilamellar SR vesicle; a single-membrane profile is contained within the interval $0 \text{ Å} \leq |\times| \leq 80$ Å. (**B**) Separate electron density profile for the asymmetric lipid bilayer. (**C**) Separate protein profile, expressed as the area occupied by the protein in the membrane plane. This protein area profile can also be expressed as a protein diameter profile (**D**) for the SR membrane protein structure cylindrically averaged about the normal to the membrane plane (from ref. 16 by permission).

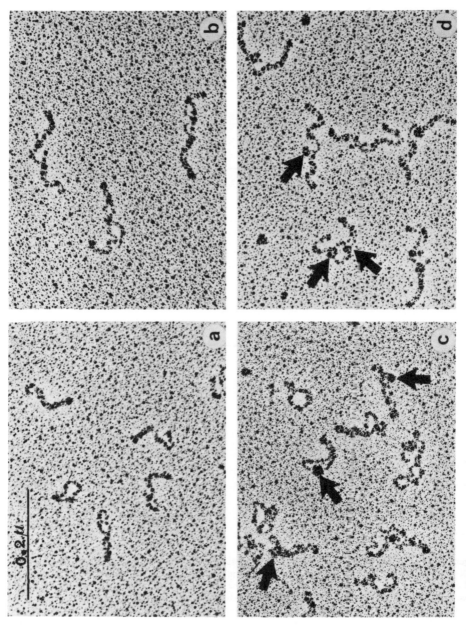

Figure 3. Electron micrographs of rotary-shadowed samples of spectrin and spectrin/ankyrin from human erythrocyte membranes. (**a**) Spectrin dimers; (**b**) spectrin tetramers; (**c**) spectrin dimers with bound ankyrin; (**d**) spectrin tetramers with bound ankyrin. The arrows indicate the location of the ankyrin (from ref. 18 by permission).

fuge or, preferably, by inelastic light scattering), or from the sedimentation coefficient of the particles (measured in the analytical ultracentrifuge) (9,19). Viscosity measurements yield a parameter of similar meaning to F ('Simha factor' v) (9,19). Since the experimental techniques mentioned are relatively easy to apply, determination of F or v may be worthwhile if little else is known on the particle shape. However, with integral membrane proteins no simple way exists to correct for the contribution of protein-bound detergent, which could be substantial.

2.4 **Polypeptide chain conformation**

2.4.1 *X-ray diffraction analysis of three-dimensional crystals*

Analysis of the X-ray diffraction from single crystals of a protein is the classical technique for the determination of the three-dimensional structure of proteins. A successful X-ray study leads to the establishment of the three-dimensional electron density distribution of the protein. The amount of information actually available from the electron density maps is governed by their resolution, which in turn depends mainly on the quality of the protein crystal and the extent of data collected and used in the calculations. At $6-7$ Å resolution, the detailed shape of a protein particle can be determined, the subunit organization and the presence of helices becomes visible, and striking structural features like clefts or holes can be discerned. At 4 Å resolution, the elements of secondary structure can be identified with respect to their nature, length and orientation. Refinement of the electron density maps to below 2 Å permits the identification and placement of individual amino acid side chains. A detailed description of the methods can be found in references 9 and 20.

Application of the technique to intrinsic membrane proteins has for long been prevented by the lack of methods for the formation of three-dimensional crystals of the proteins (see ref. 21 and Chapter 5). However, great success has now been achieved with the matrix porin from *Escherichia coli* (22), and the photosynthetic reaction centre from a purple bacterium (21). For the latter protein complex, the complete three-dimensional structure at 3 Å resolution has been established (23).

The procedures involved in the determination of the detailed structure of a membrane protein by X-ray crystallography are very arduous. The number of proteins which will be studied by this technique will therefore probably remain small. However, it is the only approach which will permit a detailed understanding of the structural features of membrane proteins and of their structure—function relationships.

Recent progress in the field of neutron crystallography now makes it feasible to perform, in addition to X-ray diffraction analysis, high-resolution neutron diffraction studies on crystals of membrane proteins. Neutron diffraction is similar to X-ray diffraction both in the experimental methodology and in the resulting structural information. However, the neutron density maps are dominated by the contributions from hydrogen (or deuterium) atoms, whereas the electron density maps do not show hydrogen positions. The two techniques could thus complement one another (24).

2.4.2 *Electron microscopical analysis of tilted two-dimensional crystals*

Some membrane proteins form two-dimensional crystal arrays in their native environment; in other cases crystal array can be induced either in the membrane or after reconstitution in lipid bilayers. Electron microscopy of tilt series of these two-dimensional crystals, when combined with electron diffraction patterns, allows the three-dimensional structure of the proteins to be reconstructed (for methods see references 20, 25, 26). In general, staining the specimen, which is essential in the electron microscopy of individual particles, has to be avoided in order to access the internal structure of the protein. However, in some cases parallel studies on stained samples may reveal important structural features, for example membrane-spanning channels (25). At present, the limit of resolution of the technique seems to be around 6 Å (and thus distinctly lower than that of X-ray crystallography of three-dimensional crystals).

The outstanding example of this approach is seen in the pioneering work of Henderson and Unwin with bacteriorhodopsin, the light-driven proton pump from the purple membrane of *Halobacterium halobium* (27). The retinal-containing protein (mol. wt 26 000) forms extended crystalline patches composed of some 10 000 protein molecules which are arranged in a hexagonal lattice. The authors have taken 15 electron diffraction patterns and 18 electron micrographs of the purple membrane at different tilt angles and have calculated from them a three-dimensional contour map of atomic densities. The resolution achieved was 7 Å in the plane of the membrane and 14 Å perpendicular to it. The map shows seven rod-like polypeptide segments, which are presumed to represent α-helices extending completely through the membrane (*Figure 4*). It should be noted that the structure of the chain segments linking the helices, the helix connectivity, the arrangement of the amino acid side chains and the position of the retinal has not yet been resolved by this approach.

2.4.3 *Other methods*

The conformation of membrane proteins can also be studied by measurements of circular dichroism, optical rotary dispersion (o.r.d.), and by two other spectroscopic methods closely related to each other, infrared and Raman spectroscopy. In general, these methods yield global rather than detailed structural information. They have the great advantage of not requiring protein crystals but are usually applied to membrane suspensions or solubilized proteins. Spectral measurements are rapid and simple. The problems for which these methods are suitable are (i) determination of the content of the elements of secondary structure: α-helix, β-sheet and random coil, and (ii) the study of conformational change.

(i) *Circular dichorism.* The term circular dichroism (c.d.) describes the phenomenon whereby samples show differences in the absorption of left and right circular polarized light (28). O.r.d. describes, in principle, the corresponding differences in the refractive indices. Both effects are very sensitive to secondary structure. In most cases, application of c.d. is recommended over o.r.d. Its use in membrane research is detailed in ref. 29.

Determination of a protein's content of α-helix, β-sheet and random coil is based on c.d. measurements in the wavelength range 185−240 nm. To this experimental data is fitted, by least-squares procedures, a linear combination of three basic functions representing the different secondary structure features. This directly yields the percentage of the different structural elements present. Although the interpretations can be fairly accurate, problems can arise with proteins embedded in a bilayer and/or forming

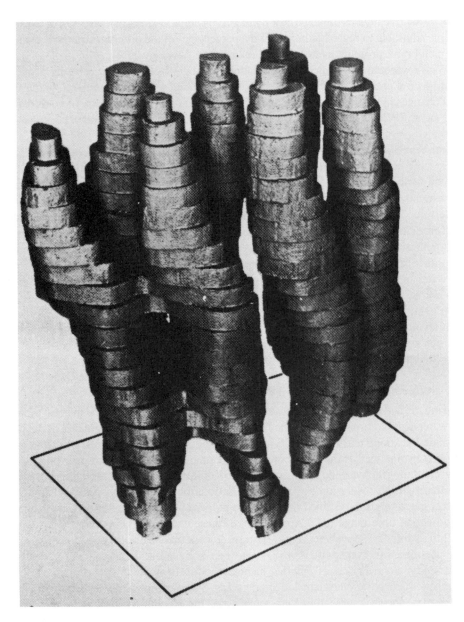

Figure 4. Three-dimensional model of bacteriorhodopsin, as deduced from electron microscopical analysis. The protein is viewed parallel to the plane of the membrane. The top and bottom of the model correspond to those parts of the molecule in contact with the aqueous phase; the rest is in contact with lipid. Seven rods, presumed to be α-helices, are visible, which make up approx. 75% of the polypeptide. Three of them (in the background) run nearly perpendicular to the plane of the membrane whereas the others are somewhat tilted. The overall dimensions of the rod cluster are $25 \times 35 \times 45$ Å (from ref. 27 by persmission).

aggregates (29), which increase the uncertainty of the calculated values. With bacterio-rhodopsin, for example, one group deduced that the seven rods are virtually all α-helix (29), whereas another group concluded that the α-helix content of the protein is sufficient for five helices only and that there is substantial β-sheet present (30). It therefore seems advisable to regard estimates of the percentages of the different structural elements in membrane proteins as being semiquantitative only.

With respect to the study of conformational changes by c.d. measurements, it is clear that spectral changes are indicative of structural changes. The interpretation of the spectral changes in terms of defined changes in secondary structure suffers, of course, from the same difficulties as the estimation of the α-helix or β-sheet content.

(ii) *Infrared and Raman spectroscopy.* Infrared absorption spectroscopy and Raman scattering both measure the vibrational spectra of the protein. The information is obtained from the positions and intensities of the so-called amide I and amide III bands (for reviews see references 31,32). It seems that the content of β-sheet can be reliably determined by these techniques. It is difficult to distinguish between α-helix and random coil from infrared spectra; with Raman spectra the situation seems to be more favourable.

Application of infrared spectroscopy to aqueous dispersions or solutions of membranes and membrane proteins has for long been severely impeded by the strong absorption of water in the wavelength region of interest. Recent technical progress, like the introduction of Fourier transform infrared spectroscopy, has considerably improved the situation (31). Raman scattering of water is weak and does not present serious problems.

It should be noted that the protein concentrations needed for measuring infrared or Raman spectra of aqueous samples are around 10 mg/ml and thus exceed those used in c.d. measurements by a factor of 100.

2.5 Segmental mobility of proteins

During the last few years, considerable insight has been obtained into the internal mobility of water-soluble proteins (33). The flexibility and mobility of the segments of the polypeptide backbone and the mobility of amino acid side chains in protein crystals were studied by analysis of the temperature effects on electron density, as derived from X-ray crystallography. Side chain rotations of proteins in aqueous solution have been studied by high resolution n.m.r. (33). The rotations of tryptophan side chains were investigated by time-resolved fluorescence polarization spectroscopy down to the subnanosecond time range (34). Application of these techniques to membrane proteins is still in its infancy. However, recent ^{2}H-n.m.r. work on bacteriorhodopsin demonstrates the usefulness of systematic investigations on the dynamics of membrane proteins. In this study, a variety of ^{2}H-labelled bacteriorhodopsin samples was prepared by biosynthetic incorporation of labelled amino acids. Sharp components in the corresponding ^{2}H-n.m.r. spectra, indicating high mobility of the labelled groups, were ascribed to surface residues, whereas broad components, indicating constrained motion, were assigned to residues located in the interior of the protein. The locations of the different labelled residues determined by this procedure showed good agreement with those determined by other methods (35).

3. LIPID STRUCTURE IN BIOLOGICAL MEMBRANES AND IN MODEL SYSTEMS

In biophysical studies on the lipid component of biological membranes, model systems consisting only of lipids play a very important role. This is due to three factors.

(i)   Pure and well-defined samples suitable for physical studies can be much more easily obtained with lipids (by isolation or synthesis) than with membrane proteins.
(ii)   Pure lipids spontaneously form extended structures closely resembling the lipid structures in biological membranes, which makes studies on these samples of high biological relevance.
(iii) Structural studies on the isolated lipids are easier to interpret than those on bio-membranes; they can thus greatly advance our understanding of the lipid structure of native biomembranes and its possible functional significance.

3.1 Detection of phase transitions

Lipids can exist in a variety of different phases (36,37). The first step in the study of such systems normally is to locate the different phases in the temperature- and concentration-dependent phase diagrams. The simplest way to accomplish this is to look for thermotropic phase transitions, using a number of methods (37).

3.1.1 *Calorimetry*

Most heat-induced phase changes can be detected by calorimetric measurements. The modern techniques of differential thermal analysis (d.t.a.) and, especially, differential scanning calorimetry (d.s.c.) have greatly increased the ease and, at the same time, the sensitivity of such measurements. They now represent the method of choice for the detection of phase transitions and for studies on the influence of internal and external parameters (e.g. fatty acid chain length and cholesterol content, ionic strength, Ca^{2+} concentration, etc.) on these transitions. A classical example, a d.s.c. study on the influence of fatty acid chain length on two phase transition temperatures, is shown in *Figure 5*. It should be noted that the transitions will be broad and thus difficult to detect by calorimetry with lipid mixtures of strongly heterogeneous fatty acid composition or high cholesterol content (37–39).

3.1.2 *X-ray and neutron scattering*

These techniques will reliably detect phase changes, since configurational changes both in the headgroup and in the fatty acid region will distinctly influence the scattering patterns. However, they are more suitable for detailed studies on lipid configuration than for the simple detection of phase transitions (see below).

3.1.3 *Other methods*

Most spectroscopic methods are able to detect lipid phase transitions. ^{31}P-n.m.r. spectroscopy probes configurational changes in the headgroup region of phospholipids. ^{2}H-n.m.r. of lipids with deuterated headgroups or fatty acids, probes structural changes in both the polar and apolar regions of the bilayer. The same holds for electron spin resonance (e.s.r.) of spin-labelled fatty acids and for fluorescence spectroscopy of fatty acids bearing fluorescent labels (free or covalently attached to phospholipids). With the latter method phase transitions are also sensed by small lipid-soluble fluorescent molecules such as 8-amino-1-naphthalene sulphonate (ANS). Infrared and Raman spectroscopy also allow phase changes to be monitored, by observing the stretching frequencies of the C-H (or C-^{2}H) bonds in fatty acids (37).

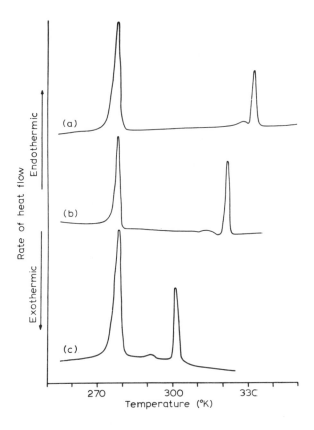

Figure 5. DSC heating curves for distearoyl phosphatidylcholine (**a**) and the corresponding dipalmitoyl (**b**) and dimyristoyl (**c**) derivatives in water, at a lipid concentration of 50% (w/w). The peaks represent the melting of ice (left) and the so-called 'pre-transition' ($L_{\beta'} \rightarrow P_{\beta'}$) (small peak) and the chain melting transition ($P_{\beta'} \rightarrow L_{\alpha}$) of the lipid (from ref. 38 by permission).

3.2 **Structure of the phases**

3.2.1 *X-ray and neutron diffraction*

Homogeneous, anhydrous or slightly hydrated lipids may form three-dimensional crystals, which sometimes are of sufficient quality to allow detailed analysis by X-ray or neutron crystallography. This will yield the three-dimensional structure of the lipids at atomic resolution. Often, features of the structure will also be relevant for less ordered phases which are usually physiologically more important. These predominantly lamellar phases can only be studied by X-ray or neutron scattering, as described for proteins in Section 2.3. Due to the relatively high short range order of the hydrocarbon chains, significant scattering is observed not only in the low-angle but also in the wide-angle range. Application of these methods to lipids yields much more detailed information than for proteins, especially if stacked lamellae are studied. Thus, the two scattering methods allow a quite detailed characterization of the following:

(i)   the lipid phases in terms of the gross arrangement of the lip molecules;
(ii)   the dimensions of the bilayer;

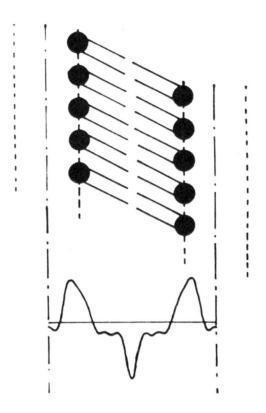

Figure 6. Electron density distribution (lower frame) and structural model (upper frame) of the phase L$_{\beta'}$ of dipalmitoyl phosphatidylcholine, as derived from X-ray scattering. The fatty acid chains are fully extended and tilted with respect to the normal to the lamellae (from ref. 36 by permission).

(iii)   the angles of tilt of the fatty acid chains with respect to the plane of the lamellae;
(iv)   the average order of the chains;
(v)   the area per lipid molecule;
(vi)   the localization of the water molecules (14,20,36).

The earlier results from X-ray scattering, *Figure 6*, suffered from uncertainties in the phases of the scattered radiation, which made the derived models ambiguous. However, this difficulty does not exist with the more modern technique of neutron scattering, where in principle substitution of a single ^{1}H-nucleus in the lipid molecules by ^{2}H is sufficient for correct phase determination (40). Instead of using deuterated lipids, it may even suffice to hydrate with D_2O instead of H_2O (40).

More detailed information on the configuration of the phospholipid molecules can be obtained from neutron diffraction experiments if, for a given lipid species, a number of derivatives deuterated at different positions in the molecule are available. In this case, the different label positions, cylindrically averaged about the perpendicular to the membrane plane, can be determined with an accuracy of approximately $\pm$ 1 Å. This leads to a determination of the mean configuration of the phospholipid molecules at segmental resolution and may even reveal subtle details of phospholipid structure,

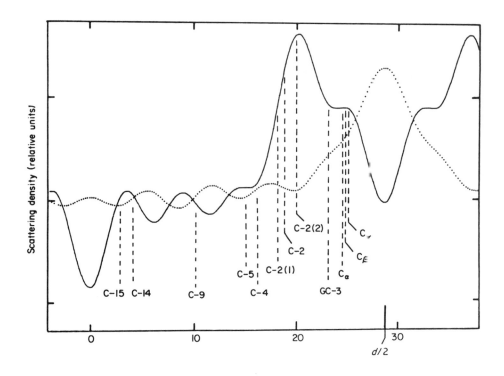

Figure 7. Neutron-scattering density profiles of undeuterated dipalmitoyl phosphatidylcholine in the $L_{\beta'}$ phase (—), mean label position determined with the corresponding deuterated lipids (vertical lines), and water distribution profile (·····). The distances are measured from the centre of the bilayer. The sample contained 6% (w/w) water and was studied at 20°C. Note that the (projected) positions of the different labels within the phosphocholine head group are close together, which shows that the average orientation of the group is almost parallel to the membrane surface. Water is shown to penetrate to the glycerol backbone (from ref. 40 with permission).

like the orientation of the polar headgroup with respect to the plane of the membrane. When phospholipid samples that were deuterated in both fatty acid chains separately are studied, even differences in the configuration of the two chains can be identified (40,41). An example is shown in *Figure 7*; it should be compared with the earlier results from X-ray small-angle scattering on the same lipid material and phase shown in *Figure 6*. The width of the label distribution is a measure of the orientation disorder and angular fluctuations of the chain segments (41) (see below).

The two scattering techniques can also be applied to mixed lamellar phospholipid—cholesterol systems, in order to determine the position of the steroid relative to the centre of the bilayer (14). However, the methods cannot yield high-resolution information on the arrangement of the two types of molecules in the plane of the membrane, which is still an open question.

Application of the methods involving selectively deuterated phospholipid molecules is not restricted to model membranes, since these lipid molecules can also be incorporated, via phospholipid exchange proteins, into biological membranes. It should be noted, however, that synthesis of deuterated phospholipids is quite time-consuming.

3.2.2 *Nuclear magnetic resonance*

N.m.r., that is absorption spectroscopy in the radio-frequency range which monitors the transitions between the various spin states of certain nuclei in high magnetic fields, is a most powerful method in membrane research. It derives its success from the ability to focus on nuclei of a single species and to probe their environment and motion (for a general introduction see ref. 42). Earlier studies on biomembranes and lipid systems used mainly ^{1}H-n.m.r. In recent years, however, studies applying the nuclear spins of ^{2}H, ^{13}C and ^{31}P have become more fruitful (for detailed treatments see references 43,44).

(i) *Static order parameters from ^{2}H-n.m.r.* With lamellar phospholipids in the so-called 'gel' state, the fatty acids are in a completely ordered configuration (all-*trans*). Above the chain-melting transition, rapid *trans−gauche* isomerizations occur which lead to a loss of order. The time-averaged conformation of an individual chain segment is commonly described by the 'local order parameter' S_1, defined by:

$$S = \frac{1}{2} \overline{(3\cos^2\theta - 1)} \qquad \text{Equation 1}$$

where θ is the instantaneous angle between the segment and the direction of the bilayer normal and where the bar denotes the time average. In ^{2}H-n.m.r. applied to specifically deuterated lipids in model systems or biomembranes, S_1 is proportional to the so-called deuterium quadrupole splitting, which can be taken directly from the ^{2}H-n.m.r. spectra (43).

(ii) *Detection of hexagonal lipid phases by ^{31}P-n.m.r.* The ^{31}P-n.m.r. spectra of phospholipids in non-bilayer phases are distinctly different from those arranged in bilayers. ^{31}P-n.m.r. can be used as a sensitive tool for studying the occurrence of non-bilayer phases and of bilayer−non-bilayer transitions (45). This is of special interest in the case of phosphatidylethanolamine, which, under physiological conditions, shows a hexagonal arrangement (H_{II}) but, nevertheless, is a major component of the lipid bilayer of many biological membranes. Cardiolipin in the presence of Ca^{2+} is another example for such a behaviour (45). The massive influence of the thermotropic lamellar-hexagonal phase transition of phosphatidylethanolamine on the peak position and shape of the ^{31}P-n.m.r. spectra is demonstrated in *Figure 8*. Application of the effect shown in the figure to problems of bilayer stability is described in ref. 47.

3.2.3 *Other methods*

Historically, the first determinations of order parameters were achieved by e.s.r. measurements using spin-labelled fatty acids (for review see reference 48). It should be noted, however, that the dependency of the order parameter on the position in the fatty acid chain as determined by e.s.r. differs from that obtained from ^{2}H-n.m.r. Probably a perturbation of order by the bulky spin labels is responsible for most of these differences (49), although part may also be due to difference in the time-scale of the two techniques (50).

Order parameters can also be determined by fluorescence depolarization measurements in which suitable fluorescent molecules, either 'dissolved' in the membrane lipid or covalently coupled to lipid molecules, are excited by a pulse of polarized light, and the time-

257

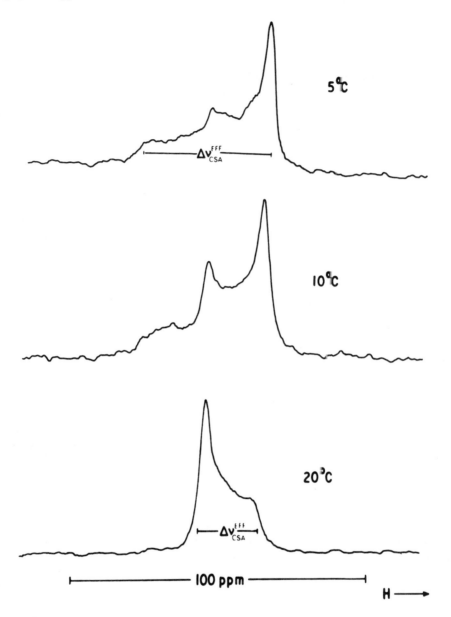

Figure 8. [^{31}P]n.m.r. spectra (proton-decoupled) obtained from aqueous dispersions of dioleoyl phosphatidyl-ethanolamine at different sample temperatures. The lipid is in a lamellar phase (L$_\alpha$) at 5°C and in a hexagonal one (H$_{II}$) at 20°C (from ref. 46 with permission).

of the so-called 'fluorescence anisotropy' is measured (51,52). It has been shown that there is a unique relationship between the limiting long-time value of the fluorescence anisotropy and the order parameter. A similar relationship holds if, instead of the data from time-resolved fluorescence spectroscopy, the anisotropy value from steady-state measurements of fluorescence (which is more easily obtainable) is used (53,54). As in the case of ^{2}H-n.m.r., the determination of a complete order profile will require the use of a number of different lipids which are specifically labelled at different positions along the fatty acid chains.

3.3 Phase separation and domain structure

During the construction of phase diagrams of lipid mixtures, the co-existence of regions of different structural organization may be revealed. Several methods allow a more direct demonstration of phase separations and of the resulting domains.

3.3.1 X-ray and neutron diffraction

The techniques of X-ray and neutron scattering are not only able to reveal lipid organization in a single phase (Section 3.2.1) but can also be applied for studying heterogeneities in lipid distributions and thus phase separations. As in related cases, neutron scattering using contrast matching is the more useful technique. For an instructive application on a phosphatidylcholine–cholesterol mixture see ref. 55.

3.3.2 E.s.r. of spin-labelled probes

The e.s.r. line shape of nitroxide spin labels (the commonly used labels in e.s.r.) is determined primarily by the spin exchange broadening, which at high levels of probe can be a measure of the degree of interaction between the labelled molecules. If, in a fluid membrane, the concentration of label is chosen such that when homogeneously distributed there should be negligible spin exchange, any occurrence of strong line broadening indicates that the lipids have undergone phase separation and that the spin-labelled molecules are now concentrated in distinct domains. The general approach is illustrated in *Figure 9* for model membranes containing high concentrations of spin-labelled lipid. It is also possible to estimate the size of these domains (56). However, extension of this method to biological membranes is hampered by the high label concentrations normally required which will almost inevitably lead to a perturbation of the membrane's structure.

3.3.3 Fluorescence spectroscopy

There are simple fluorescence techniques which indicate the formation of a solid or a fluid domain. Some fluorescent molecules, like *trans*-parinaric acid, partition preferentially into solid phases of phospholipids, where they exhibit a strongly enhanced quantum yield (57). Others, like pyrene decanoic acid, show a strong preference for fluid phases, where their fluorescence is increased by 'excimer' formation (56). In both cases, the formation of the respective domains by phase separation will thus lead to an increase in fluorescence intensity.

In lipid monolayers at the air–water interface which are doped with fluorescent dyes, temperature- and pressure-induced phase separation and the formation of extended fluorescent and non-fluorescent domains can be studied by fluorescence microscopy (58).

3.3.4 Freeze-fracture electron microscopy

Electron microscopy of freeze-fractured lipid bilayers reveals, after shadowing with heavy metal atoms, the morphology of the inner surfaces of the two monolayers (59). One of the many applications of the technique concerns the study of model systems

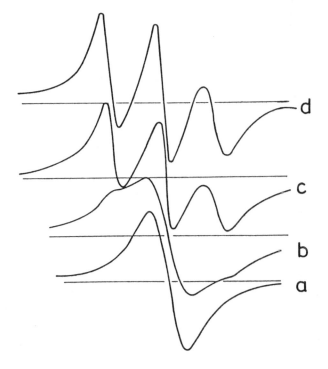

Figure 9. E.s.r. spectra of aqueous dispersions of spin-labelled dipalmitoyl phosphatidylcholine, (**a**) without added unlabelled lipid and with unlabelled (**b**) dipalmitoyl phosphatidic acid, (**c**) dipalmitoyl phosphatidyl-ethanolamine, and (**d**) dipalmitoyl phosphatidylcholine, at a molar ratio of 2:8. pH 9; T = 59°C. The shape of spectra **a** and **b** is dominated by spin exchange broadening (from ref. 56 with permission).

consisting, in part, of the phase $P_{\beta'}$, which shows characteristic 'rippled' pattern (36). In mixed systems involving this phase, smooth and rippled regions may occur together, and along with other morphological features (60). Changes which affect the observed phase patterns can be investigated with freeze-fracture electron microscopy. This technique can also be used to detect, in lipid model systems or biomembranes, domains containing a high content of sterol. For this purpose, the domains are cytochemically labelled with filipin or saponins. By binding specifically to 3β-hydroxysterols, these agents induce distinct deformations and lesions in the fracture faces (61). Similarly, domains with a high content of anionic lipids can be visualized by labelling with poly-myxin (61).

Phase separations in biological membranes may be accompanied by segregation and, possibly, subsequent aggregation of intrinsic membrane proteins. This will be clearly visible in freeze-fracture electron micrographs as accumulations of the intramembranous particles.

3.4 Segmental motion of lipids

The segmental mobility of phospholipids is responsible for the 'fluid' state of the lipid bilayer. Its molecular basis is that the C-C bonds of fatty acids can exist either in the *trans* or in the *gauche* configuration, the transformation being equivalent to a rotation

(62). The resulting disorder can be characterized by an order parameter (Section 3.2.2). The dynamics of the *trans–gauche* isomerization can, in principle, be described by a rotational correlation time τ_c, which is the time needed by a group or molecule to rotate through an angle of 1 rad. Several physical techniques are sensitive to phospholipid segmental motion and can thus be applied to their study. However, a quantitative description is complicated by several factors.

(i)   Other types of lipid motion can occur simultaneously with *trans–gauche* isomerizations and may contribute to the experimental data (e.g. motions of the whole molecule or of the entire bilayer unit).
(ii)   The segmental motions occurring are anisotropic and restricted in space and thus often cannot be characterized adequately by a single τ_c value.
(iii)   The interpretation of the experimental data in terms of physical motion is not straightforward but involves complex theories of disputed validity.

References 43, 44, 50 and 63 contain detailed discussions of these points.

3.4.1 *Nuclear magnetic resonance*

N.m.r. can be used to study phospholipid segmental motion by measuring either the spin-lattice relaxation time, T_1, or the spin-spin relaxation time, T_2, of suitable nuclei in the fatty acid chains, or the linewidth of the corresponding absorption bands (44,64). T_1 measurements are the least problematic.

The spin-lattice relaxation time T_1, also called the longitudinal relaxation time, describes (for an ensemble of identical nuclei) the return to equilibrium of the magnetization directed parallel to the applied static field (after disturbance of the equilibrium). It can be accurately measured by special pulse techniques (64). Both ^{2}H- and ^{13}C-n.m.r. of specifically labelled lipids can be used for T_1 determinations. Although in simple cases T_1 is inversely proportional to τ_c, in general the relationship between T_1 and τ_c is much more complex (44,64). The range of T_1 experimentally accessible corresponds to approximate τ_c values between 10^{-4} and 10^{-11} sec (65).

3.4.2 *E.s.r. of spin-labelled phospholipids*

The e.s.r. spectra of nitroxide spin labels are dependent on both the orientation and the motional state of the label. The spectra of a label which is rigidly fixed to a segment of a phospholipid molecule in a lipid model system or a biomembrane will thus contain information on the motion of the segment. The time scale of the motions which, in principle, can be studied by e.s.r. is $\tau_c = 10^{-7}$ to 10^{-11} sec for conventional e.s.r. and 10^{-3} to 10^{-7} sec for saturation transfer e.s.r. (48,66). Lipid segments showing slower motion are usually classified as 'immobile'. Unfortunately, interpretation of e.s.r. data requires complex calculations (48,66).

3.4.3 *Fluorescence depolarization spectroscopy*

Measurements of the fluorescence depolarization of small, lipid-soluble fluorescent probes like perylene dissolved in a lipid bilayer can be used to determine bilayer viscosity η in the immediate surroundings of the probe (52,67). From the value of η, the segmental rotational correlation time τ_c of the lipids (at the average location of the probe) can be calculated from straightforward equations. The relationships can also be applied to calculate η from τ_c values which were determined by e.s.r. or n.m.r. according to Sections 3.4.1 and 3.4.2. The approach of describing phospholipid motion by membrane viscosity values derived from fluorescence depolarization experiments is quite popular. There are, however, a number of complicating factors (43,52,67), which in part explains the fact that the membrane viscosity values obtained by different experimental techniques differ by more than an order of magnitude (43).

4. ROTATIONAL, LATERAL AND TRANSVERSE DIFFUSION OF PROTEINS AND LIPIDS

In a 'fluid' membrane, the components can undergo considerable rotation and lateral diffusion. Depending on the system there is also the possibility of some components moving from one monolayer to the other.

4.1 Rotational diffusion of proteins

The rotational correlation times of integral membrane proteins in the lipid bilayer will be longer than a few microseconds, as can be estimated from the membrane viscosity values obtained by fluorescence or magnetic resonance techniques (Section 3.4.3). They may even exceed one second if the protein is 'fixed' by protein – protein interactions. Both microsecond and millisecond correlation times can be obtained by optical spectroscopy using triplet probes (68 – 72) or by saturation transfer e.s.r. (66,68).

4.1.1 *Optical spectroscopy with triplet probes*

Measurement of protein rotation in the micro- to millisecond time range by optical spectroscopy requires a spectroscopic state of correspondingly long lifetime. Such conditions are met by the 'triplet' state of certain luminescent organic molecules. These molecules can be exogenous probes (e.g. eosin derivatives) covalently coupled to a protein; or they can be endogenous ligands (e.g. retinal in rhodopsin, bacteriorhodopsin). In each case, an initially oriented population of excited probe molecules is produced by a linearly polarized pulse of light. This oriented population of molecules shows anisotropy:

(i)   of the absorption of the (depleted) ground state,
(ii)   of the absorption of the triplet state,
(iii)   of the emission of the triplet state (phosphorescence), and
(iv)   of the (depleted) ground state prompt fluorescence.

Rotation of the protein to which the probe is attached will lead to a decrease in anisotropy of both absorbed and emitted radiation. Measurement of the time course of the anisotropy will yield information on the rate of rotation.

In principle and in practice, all four effects (i) to (iv) can be used for measurements of the anisotropy decay. Among the corresponding physical techniques, those measuring absorption changes have the lowest sensitivity (68). The minimum probe concentration required for them is approximately 1 μM (68). The sensitivity of the phosphorescence measurement is higher by a factor of 100 (71). The recently described method of 'fluorescence depletion', based on the effect (iv), shows by far the highest sensitivity: it is claimed to be more sensitive than the phosphorescence method by a factor of approximately 10^6 and is able to study rotational motion of membrane proteins on the surface of a single cell (71,73). Using this method, the time-decay of anisotropy from as little as 27 000 molecules of the anion transport protein (band 3) in part of a single erythrocyte membrane could be measured, the results being similar to those obtained by the other methods (73).

As with segmental rotation of lipids (Section 3.4), the main difficulty in studying the rotational diffusion of membrane proteins seems to be the interpretation of the measured time-decay of anisotropy. The anisotropy decay of a sphere in an isotropic medium is monoexponential; however, its correct description for an ellipsoid of revolution requires three, and of an irregularly shaped body, five exponentials. Even more complex equations will be needed to describe relaxation in an anisotropic medium (as a lipid bilayer) or in the presence of different oligomeric states of the protein. Due to the limited accuracy of the experimental data, a reliable determination of the many parameters involved in this description is impossible. The anisotropy decay is therefore analysed by empirical equations, in most cases involving two exponentials plus an additive constant (68,69,72). As a result, the 'rotational relaxation times' or similar parameters determined by this procedure may describe only very imperfectly the actual rotational behaviour of the protein (see, for example, reference 74). Nevertheless, triplet probe spectroscopy has greatly contributed and will further contribute to an understanding of the qualitative and semiquantitative aspects of protein rotation in biomembranes.

4.1.2 Saturation transfer e.s.r.

In conventional e.s.r., care is taken to use radiation power levels far below saturation of the transitions under study. The opposite is the case with saturation transfer e.s.r. This method makes use of the fact that, at any time during a spectral scan, only probes corresponding to a narrow angular range interact with the saturating radiation and that rotational diffusion will transfer these probes out of this saturated angular range (and transfer other probes into it) (66,68,75). In this way, rotational diffusion may strongly influence the shape of the e.s.r. spectrum. The effect can be enlarged by special modulation and detection methods (66,75).

A demonstration of the ability of saturation transfer e.s.r. in discriminating between different rates of rotational motion is given in *Figure 10*. In this case, haemoglobin, containing a rigidly attached spin-label, was studied at different solvent viscosities and thus at different rotational correlation times (τ_c) for the protein. The amplitudes at several positions in the spectra show drastic and, at some of the positions, monotonic changes with changes in viscosity. Since, in this simple case, the τ_c values can be accurately calculated by means of the Debye equation, a calibration curve can be established relating the spectral parameters to the rotational correlation time for isotropic rotation (76). No analogous procedure is, however, available for anisotropic rotation, so that evaluation of the spectra in terms of correlation times has to rely on computer simulations. Unfortunately, the relationships between the spectral parameters and the type of motion which have emerged so far are much less clear and unique than for isotropic motion (66,75). It therefore seems, at present, that saturation transfer e.s.r. studies on protein rotation in membranes will yield qualitative or, at best, semiquantitative results only.

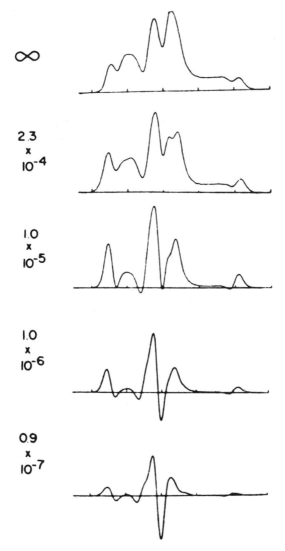

Figure 10. Saturation transfer e.s.r. spectra (second harmonic) from haemoglobin spin-labelled with a maleimide nitroxide, at different rotational mobilities of the protein. From top: precipitated Hb; Hb in 90% glycerol at $-12°C$; Hb in 80%, 60% and 40% glycerol, in each case at 5°C. The figures at the left represent the rotational correlation times (in sec) calculated from the average diameter of the protein and the solvent viscosity (from ref. 76 with permission).

4.2 Rotational diffusion of lipid molecules

Besides segmental motion, phospholipids may also show rotation of the whole molecule (rigid body rotation) around their long axis. In addition, the direction of the long axis may reorient about an average director (44,50). For lipid systems in the liquid-crystalline state, these motions have been treated in different ways. Some authors prefer to describe them within the framework of the rotational isomeric model, that is rigid body motion is described as the superposition of segmental motions (43). In other models, both rigid body motions and chain isomerizations are considered simultaneously, using the same set of experimental data. Thus, the experimental techniques used in the studies are those described in Sections 3.4.1 and 3.4.2.

Below the main phase transition of phospholipid membranes, where *trans—gauche* isomerizations are rare events, rigid body motions of the lipid molecules are expected to dominate. These slow motions are best studied by saturation transfer e.s.r. of spin-labelled lipids (77,78). The techniques and problems are the same as in the corresponding studies on protein rotation (Section 4.1.2).

The motion of cholesterol in biological membranes is almost exclusively rigid body motion [the exception being the terminal part of the aliphatic tail, which shows segmental motion (79)]. In lipid bilayers above the main phase transition, rigid body motion of cholesterol is in the same time range as segmental motion of the phospholipids. It can be studied by two of the techniques applied in the study of the latter, (^{2}H- and ^{13}C-) n.m.r. and conventional e.s.r. (Sections 3.4.1 and 3.4.2). In gel state membranes, saturation transfer e.s.r. can be applied (78). Again, the interpretation of the experimental data is complicated by the anisotropy of the motion.

4.3 Lateral diffusion of proteins

The field of lateral diffusion measurements on proteins in biological membranes or reconstituted bilayers is dominated by a single technique, fluorescence recovery after photobleaching (FRAP). There exist a number of variations of the technique, called a variety of names [e.g. fluorescence microphotolysis (FM) or fluorescence photobleaching recovery (FPR)].

In this approach, the proteins are first covalently and, if possible, specifically labelled with a fluorescent probe, typically a fluorescein or rhodamine derivative. The sample, in most cases a single cell, is then observed in a fluorescence microscope. In the most widely used version of the method, part or all of the fluorophore in a small area of the membrane is photochemically destroyed ('bleached') by an intense laser pulse. A typical spot size is around 10 μm. Immediately afterwards, measurements are made of the time course of the recovery of fluorescence in the irradiated area, due to diffusion of the protein from outside the area. For this, the fluorescence in the originally bleached region is excited by highly attenuated light from the laser (weak enough not to induce significant further bleaching), and the microscope is adjusted such that the recovered fluorescence intensity can be measured by a photomultiplier (68,80,81). In the version called 'continuous fluorescence microphotolysis' (CFM), bleaching and diffusion measurement are performed simultaneously, keeping the laser light at a constant and intermediate level (80—82).

Evaluation of FRAP measurements is much simpler than of those from the CFM technique. In most cases, curve fitting procedures will have to be applied; in others, straightforward formulae may be used. Evaluation of the CFM data requires simultaneous consideration of diffusion (or flow) and of the bleaching kinetics and thus the introduction of additional parameters. The method, therefore, is best suited to systems with simple diffusion behaviour. The CFM method, however, is more sensitive than the standard FRAP technique by several orders of magnitude (82). It may thus allow diffu-

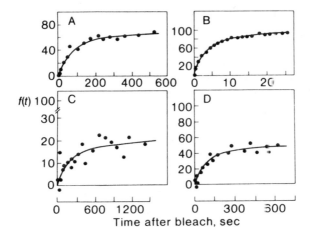

Figure 11. Experimental (●) and fitted (——) FRAP curves from eosin-labelled erythrocyte ghosts at pH 7.4 and at different temperatures and salt concentrations. The label was virtually exclusively located in the band 3 protein. The parameter f(t) represents the fractional recovery of fluorescence at time t. (A) 21°C, 5 mM sodium phosphate; (B) 37°C, 13 mM sodium phosphate; (C) 21°C, 46 mM sodium phosphate; (D) 37°C, 46 mM sodium phosphate. Note the large differences in the time scale (i.e. diffusion coefficient) and in the size of the mobile fraction. Variations in the experimental parameters predominantly effect the structure of the red cell cytoskeleton, which in turn controls the lateral diffusion of band 3 (from ref. 84 with permission).

sion measurements on proteins which represent only a minor component of the membrane protein, for example hormone or virus receptors.

The information which can be obtained from such studies includes the identification of the type of transport process (diffusion versus flow), the determination of the diffusion constant D and/or the flow velocity, and the fraction of total fluorophore which is mobile (83). With both methods, the range of accessible D values is 10^{-6} to 10^{-12} cm²/sec (80,81). Particles characterized by values of D less than 10^{-12} cm²/sec are usually regarded as 'immobile'. Mobile membrane proteins have D values up to 10^{-8} cm²/sec.

Besides giving numerical values for D, the FRAP technique has permitted insight into the processes which control the mobility of membrane proteins. *Figure 11*, for example, illustrates that both the time course of the fluorescence recovery and the size of the immobile fraction of band 3 in erythrocyte membranes are affected by conditions which alter the state of the red cell cytoskeleton. Other examples are given in reference 80.

4.4 Lateral diffusion of lipids

In contrast to the situation with proteins, a variety of different experimental techniques are being used for studying the lateral diffusion of lipids.

4.4.1 Fluorescence recovery after photobleaching

The FRAP technique described above is equally suited for similar studies on lipid molecules bearing a fluorophore and is, at present, probably the most widely used one. The range of diffusion coefficients observed for lipids, 10^{-7} to 10^{-11} cm²/sec (80), is better suited to the FRAP technique than is the case with proteins. Under most experimental conditions, a molar ratio of labelled to unlabelled lipid of around 0.01 yields a sufficiently high signal-to-noise ratio; much lower molar ratios will suffice when the CFM method is used. The D values obtained are those of the labelled lipid molecules in an environment of unlabelled lipid but they are expected to be closely related to those for unlabelled lipid if the structural differences between the two types of molecule are small. Nevertheless, it must be appreciated that a bulky fluorophore will produce a relatively large perturbation in its immediate environment. This perturbation may, at least in part, be reflected in the calculated D values, especially at the low mobilities characteristic of the gel state.

4.4.2 E.s.r. of spin-labelled lipids

When spin-labelled lipids diffuse randomly in the plane of the membrane, their collisions will lead to exchange broadening of the e.s.r. lines (Section 3.3.2). Since the number of collisions per second is directly related not only to the concentration of the label but also to the value of its diffusion coefficient D, measurements of the e.s.r. line width can be used for a determination of D. Typical molar ratios of spin-labelled and unlabelled lipid applied in the measurements are $0.1-0.4$. Relatively simple equations are obtained for the relationship between D and line width, so that sophisticated evaluation procedures are not required (66,85,86). As in the case of FRAP (Section 4.4.1), it should be noted that the bulky spin label will perturb the organization of the lipid and may thus influence lipid diffusion.

4.4.3 N.m.r. techniques

As compared with the other methods discussed in Section 4.4, n.m.r. has the advantage that the dynamic properties of the lipid to be studied will not be perturbed by the presence of any probe. However, other problems exist.

Pulsed ^{1}H-n.m.r., applying the 'magnetic field gradient spin-echo method', is able to yield precise D values in a straightforward manner but the technique is applicable to lipid bilayers only if the bilayers have been macroscopically oriented. In addition, lipids in the gel state cannot be studied by this approach (44,87). Thus, the applicability of the method is severely limited.

Lateral diffusion coefficients of lipids have also been determined from the dependency of the ^{31}P-n.m.r. line width on solvent viscosity (88) but rather gross assumptions had to be made in the calculations.

4.4.4 Other methods

In a new technique for measuring lipid lateral diffusion in planar multilayers, a spin-labelled lipid in regions of the sample not protected by a 'mask' is photochemically inactivated at two different times. In between the light pulses, the spin-labelled lipid diffuses out of the protected area. The D value of the lipid is determined from the decrease in e.s.r. signal intensity caused by the second light pulse (89). The method, which resembles FRAP, is claimed to be versatile and applicable over a wide range of motional rates.

The lateral diffusion of lipids bearing a fluorophore can also be studied by fluorescence correlation spectroscopy (FCS). The experimental system is essentially identical to that for FRAP experiments except that the FCS technique uses not bleaching of the probe but laser-induced fluorescence to detect the spontaneous concentration fluctuations in microscopic surface areas. The diffusion coefficient is determined from the time-correlation function of the fluctuations (90). The technique is not practical for slow diffusion (as that of membrane proteins) and is limited by systematic fluctuations. Nevertheless, it has been successfully applied to black lipid membranes and to lipid multilayers in the liquid-crystalline and gel state (91).

4.5 Transverse diffusion of lipids

Transverse diffusion or 'flip-flop' is the process by which membrane constituents cross from one half of the lipid bilayer to the other. It has been observed with fatty acids, phospholipids and cholesterol and also with peptide 'carriers' like valinomycin but not with membrane proteins. The following treatment, therefore, is relevant only to membrane lipids.

The study of lipid flip-flop is a thorny problem since one has to disentangle 'native' flip-flop from flip-flop which may be induced during experimental manipulation of the membranes. Indeed, induced flip-flop may be the dominating process, and apparently it is extremely difficult to prevent (92). In general, it is assumed that those results showing slower flip-flop will be more reliable than those obtained with the same system which show a faster rate. Flip-flop can be studied by biochemical methods (see Chapter 4 and references 92–94). Biophysical methods include e.s.r. and n.m.r. spectroscopy.

4.5.1 *E.s.r. of spin-labelled lipids*

In a typical experiment, spin-labelled lipids are inserted symmetrically into the membranes of sealed vesicles or cells. An asymmetric distribution of label is then created either by destroying the outward-facing label with ascorbate (thereby removing the external label's contribution to the total e.s.r. signal) (95), or by allowing the destruction of the inward-facing label by reducing agents in the cell (96). The half-time of lipid flip-flop is determined from the subsequent change with time of the remaining signal (95,96). Unfortunately, the results obtained by this procedure can conflict, for example the reported half-times of the flip-flop of spin-labelled phosphatidylcholine analogues in human erythrocyte membranes vary, in the same temperature range, between a few minutes (68) and greater than 4 h (96). Perturbation of the lipid configuration by some spin-labels and by the added ascorbate seem to contribute to this variability.

4.5.2 *N.m.r. measurements*

Investigations on the flip-flop of lipids by n.m.r. are based on measurements of signals arising from their polar headgroups and on the use of membrane-impermeable 'shift reagents'. These reagents, in most cases paramagnetic cations like Dy^{3+} or Nd^{3+}, have the property of shifting to another magnetic field strength and, at the same time, of broadening the resonance, of those nuclei with which they are in contact. Thus, addition of a shift reagent to sealed lipid vesicles will influence the position and shape of the resonances resulting from the lipid head groups in the outer monolayer of the vesicles, whereas the corresponding resonances from the inner monolayer will not be affected. This effect has been widely applied in structural studies on small lipid vesicles using

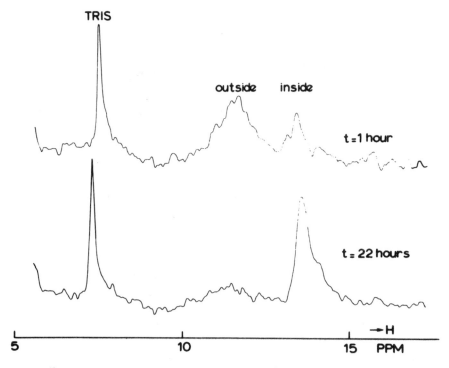

Figure 12. [^{13}C]n.m.r. spectra of dioleoyl phosphatidylcholine, ^{13}C-labelled in its headgroup, in vesicles of dimyristoyl phosphatidylcholine. The spectra were recorded 1 h and 22 h after introducing the labelled lipid into the outer monolayer of the vesicles by means of a phospholipid exchange protein. Prior to data accumulation, DyCl$_3$ solution was added to a final concentration of 2 mM (from ref. 99 with permission).

^{31}P- and ^{13}C-n.m.r. (97,98). In the latter technique, it can also be applied for studying phospholipid flip-flop. For this purpose, an asymmetry in the ^{13}C-distribution in the lipid headgroups has first to be created by inserting the appropriate ^{13}C-labelled lipids into the outer monolayer by means of phospholipid exchange proteins (see Chapter 4). Transverse diffusion of the ^{13}C-labelled lipids to the inner monolayer is then monitored by measuring the intensity of the ^{13}C-signal in the absence and presence of a shift reagent (99) (see *Figure 12*). It should be noted that the relatively fast flip-flop evident from the figure is due to a non-equilibrium distribution of a small amount of dioleoyl phosphatidylcholine in dimyristoyl phosphatidylcholine vesicles and that it is not observed in vesicles composed only of the former or the latter lipid (99).

The method is not applicable to whole cells, due to insufficient resolution of the spectra.

5. PROTEIN−LIPID INTERACTIONS

The lipid bilayer surrounding an intrinsic membrane protein might appear, at first glance, as a fairly inert solvent for a protein with a hydrophobic surface. However, in many cases distinct protein−lipid interactions seem to confer special properties to those lipid molecules which are in immediate contact with the protein (e.g. reduced mobility, changed order, selective association). The protein may even influence the configuration and motion of those lipid molecules which are not in direct contact with it. As a corollary, the protein−lipid contacts may also influence the conformation of the protein. In a few atypical cases, the protein may even possess a small number of specific and functionally-important lipid binding sites of high affinity. This holds for both integral and peripheral membrane proteins.

A more detailed description of the methods outlined below can be found in a recent review (100).

5.1 Characterization of protein−lipid associations

This section deals with the detection of protein-associated lipid and the characterization of the complexes with respect to lifetime, stoichiometry and specificity.

5.1.1 *Electron spin resonance*

E.s.r. of spin-labelled lipids was the first technique by which the influence of an intrinsic membrane protein on the properties of the surrounding lipid could be demonstrated (101). It is now well established that most e.s.r. spectra of native or reconstituted membranes containing a small percentage of spin-labelled lipid can be decomposed into contributions from 'immobile' and 'mobile' lipid molecules, the latter showing nearly the same spectral properties as in protein-free bilayers. An example is shown in *Figure 13*. Most authors now assume that the 'immobile' lipid fraction represents the innermost layers of lipid molecules surrounding the protein ('lipid annulus') (66,100,101) rather than lipid 'trapped' between aggregated protein (102−105).

It is important to appreciate that e.s.r. measurements will classify lipid molecules or segments thereof as 'immobile' when the rotational correlation times are larger than approximately 10^{-7} sec [(48,66), see also Section 3.4.2]. 'Immobile' lipid, therefore, may still have considerable mobility; in particular, it may exchange with the bulk lipid at a rate up to 10^7 sec^{-1} (43,48,66). N.m.r. measurements have shown that this is indeed the case (43,100). If lipid exchange becomes faster than 10^7 sec^{-1} (e.g. due to an increase in temperature), the e.s.r. signals arising from annular lipid will be partially averaged with those of the bulk lipid. This will lead to an apparent decrease in the relative contribution of the motionally restricted component (100).

Assuming that the motionally restricted component of the e.s.r. signal represents the 'annular' component, the number of lipid molecules in immediate contact with the protein can be determined. In reconstituted protein−lipid systems this can be done at different protein to lipid ratios, which makes the interpretation of the measured spectra more reliable. In addition, varying the spin-labelled lipids allows one to determine the relative affinities of the protein for the various labelled lipid species and thereby assess the specificity of the protein− lipid interactions (100,104). It should be noted that the relative affinities determined in this way will be average values. Since the number of annular lipids per protein molecule in most systems is 20−60 (66,100,101,105), the presence of a single or a small number of lipid binding sites of much higher than average affinity (as described in references 106,107) may remain undetected (100).

The determination of the size of the motionally restricted fraction of the membrane lipid encounters two major problems.

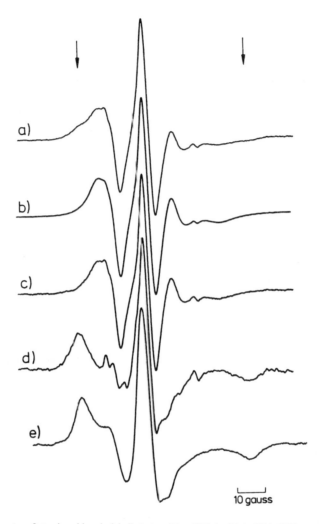

Figure 13. E.s.r. spectra of stearic acid, spin-labelled at position C14, in (Na^+, K^+)-ATPase-containing membranes from shark rectal gland and in isolated membrane lipids. (**a**) Membranes; (**b**) aqueous dispersion of extracted membrane lipids; (**c**) fluid component difference spectrum obtained by subtracting the immobilized spectrum given in (**e**) from the membrane spectrum, (**d**) immobilized component difference spectrum obtained by subtracting the lipid spectrum from the membrane spectrum. T = 0°C. The 'immobilized' spin-label component is seen in the outer wings of spectra (**a**), (**d**) and (**e**) (arrows) (from ref. 104 by permission).

(i)   The use of different methods for dividing the e.s.r. spectra into the contributions from a motionally unrestricted and a motionally restricted lipid fraction may, according to some authors, yield distinctly different results (100).

(ii)   With any evaluation method, the *apparent* size of the motionally restricted fraction depends not only on temperature (see above) but also on the position of the spin-label on the lipid molecule. In addition, it may depend on the flexibility of the labelled molecule (100).

5.1.2 *Nuclear magnetic resonance*

N.m.r. is characterized by a much longer time-scale than e.s.r.; the motional averaging takes place over a time interval of about 10^{-4} sec, in contrast to 10^{-7} sec in e.s.r. measurements. Since exchange between a molecule of the lipid annulus and the bulk lipid in both native and reconstituted systems apparently occurs within less than 10^{-4} sec, n.m.r. spectra do not reveal the existence of two lipid subpopulations of different motional behaviour but show only one lipid population of average motional state. On the other hand, the longer time-scale should make n.m.r. suited for detecting and studying lipid subpopulations which are more strongly and thus more specifically bound to an intrinsic membrane protein than the average 'annular' lipid. This potential of the n.m.r. measurements has not yet been widely utilized, perhaps because of difficulties in synthesizing sufficient quantities of suitable ^{2}H- or ^{13}C-labelled lipid analogues or because of problems associated with the low sensitivity of the method.

5.1.3 *Fluorescence measurements*

The naturally occurring fluorescence of tryptophan residues as well as of fluorescent labels introduced into the protein can be quenched by spin-labelled or brominated lipids (9,108,109). Since quenching will only occur if the distance between the fluorescent group and the quencher does not exceed about 7 Å, only those lipid molecules which are in direct contact with the protein (i.e. among the 'annular' lipids) will act as quenchers. The lifetime of the excited states both of tryptophan and of typical extrinsic fluorophores is less than 10^{-7} sec, and so quenching will not be significantly influenced by the rate of exchange between annular and bulk lipids. It will, however, depend on the number and position of the fluorescent groups in the protein and on the location of the quenching group(s) in the lipid molecules. The best way to apply the effect to the study of protein–lipid interactions seems to be to compare, in mixtures of labelled and unlabelled lipids, the quenching efficiency of phospholipids with different headgroups but identical fatty acid chains which contain the quenching group(s). This will yield the relative binding affinities of the different phospholipids to those regions on the protein in which the fluorescent groups are located, and the number of lipid molecules surrounding the fluorophores. The main advantage of the method is its high sensitivity, especially if extrinsic chromophores with excited states of long lifetime can be used (100,108,109).

Measuring the enhancement (or, in some cases, the decrease) of protein fluorescence which arises from association of the protein with lipids represents another possibility for investigating protein–lipid interactions. This method is suited mainly for investigations on peripheral membrane proteins. Binding of the protein to both single lipid molecules and lipid mono- or bilayers (vesicles) can be studied. Measurements of the fluorescence intensity as a function of lipid concentration, at fixed protein concentration, will allow the determination of binding constants and stoichiometries (100,110). An example concerning the interactions of spectrin, the main cytoskeletal protein of the erythrocyte membrane, with phosphatidylserine vesicles is shown in *Figure 14*.

Measurements of fluorescence energy transfer (9) can also be utilized in studies on the specificity of protein–lipid interactions. In this technique, the effect monitored is the enhancement of lipid fluorescence by energy flow from a tryptophan residue of a protein to a lipid bearing a suitable fluorophore. Alternatively, the quenching of the

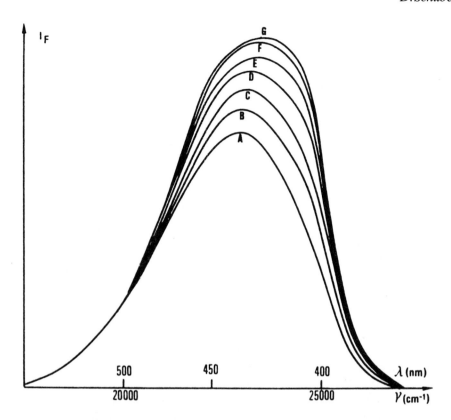

Figure 14. Fluorescence emission spectra (intensity I_F versus wavenumber ν) of anilinonaphthyl-spectrin in the absence of lipid (**A**) and in the presence of phosphatidylserine vesicles, at a lipid concentration of 0.095 mM (**B**), 0.185 mM (**C**), 0.36 mM (**D**), 0.53 mM (**E**), 0.69 mM (**F**) and $0.85-1.0$ mM (at saturation) (**G**). Spectrin concentration was 0.2 μM. T = 30°C. From the data, a stoichiometry (at saturation) of one spectrin dimer per 1500 phosphatidylserine molecules and a binding constant of about 3×10^7 M^{-1} was calculated. Addition of phosphatidylcholine vesicles to the protein did not cause any changes in the protein's fluorescence emission spectrum (from ref. 110 by permission).

fluorescence of a labelled lipid by energy flow to a protein bearing a suitable chromophore (the absorption spectrum of which overlaps the emission spectrum of the fluorophore) can be measured. The important feature of the method is that energy transfer can effectively occur over a distance of up to more than 70 Å. The method is ideally suited for studies on the binding of solubilized peripheral proteins to lipid bilayers: if the fluorophores are located in the inner regions of the bilayer and thus far from the site of protein binding, possible perturbations of the protein−lipid interactions by the bulky probe will be minimized. For applications see references 100, 111 and 112.

5.1.4 *Other methods*

In reconstituted membranes, information on the specificity of protein−lipid interactions can be obtained from Raman and infrared spectroscopy. Thus, for proteins in binary lipid mixtures in which the fatty acid chains of one of the components are deuterated, determination of the width of the corresponding C- ^{1}H and C- ^{2}H vibrational bands allows one to determine the relative affinities of the protein for the two lipids (113).

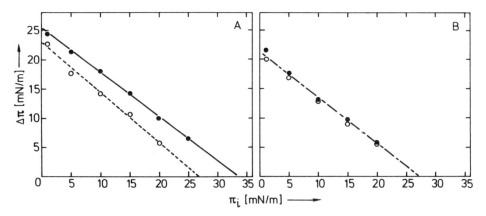

Figure 15. Interactions of the spectrin polypeptides (bands 1 and 2) from human erythrocyte membranes with monolayers of bovine brain phosphatidylserine (**A**) and egg yolk phosphatidylcholine (**B**) at the air-water interface: dependency of the protein-induced increase $\Delta\pi$ in surface pressure on the initial surface pressure π_i. (○) Band 1; (●) band 2. The initial protein concentration in the subphase was 7.5 μg/ml. The interactions can only be of physiological relevance if π_{io}, the limiting value of π_i for which $\Delta\pi$ is zero, will exceed 31 mN/m. Only the combination of band 2 with phosphatidylserine fulfils this condition (from ref. 114 by permission).

Valuable information on specificities in the interactions between peripheral membrane proteins and lipids can be obtained from studies on lipid monolayers at the air−water interface. In such studies, the solubilized protein is added to the aqueous subphase of the monolayer, and the changes in monolayer surface pressure ($\Delta\pi$) which follow the binding to or the partial penetration into the lipid are measured. $\Delta\pi$ decreases with increasing initial surface pressure π_i of the monolayer. The limiting value of π_i above which $\Delta\pi$ is zero, π_{io}, seems to be especially useful for characterizing the strength of the interactions; its value may also allow one to draw conclusions as to their biological relevance (114). An example is shown in *Figure 15*. The method requires protein which is completely free of detergents. It is applicable to intrinsic membrane proteins only if they can be isolated in a detergent-free form in which the native capacity for lipid binding is essentially preserved (115).

High-affinity lipid binding sites on a solubilized, detergent-free peripheral or intrinsic membrane protein may also be detected by binding studies employing equilibrium dialysis. In such studies, it may be necessary to apply lipid derivatives which show a higher solubility in aqueous solutions than the lipid of interest (e.g. deoxycholic acid instead of epicoprostanol) (116).

In fatty acid auxotroph bacterial strains grown in the presence of a single fatty acid, or in reconstituted membrane systems, a fraction of the lipid may not take part in thermotropic phase transitions. From the size of this fraction, the number of 'annular' lipid molecules per intrinsic protein molecule can be determined. The figures obtained are higher than those derived from e.s.r. measurements (117, 118) (see also Section 5.3).

5.2 Influence of the lipid on protein structure

There is little doubt that the specific folding of all intrinsic and some peripheral membrane proteins is in part determined and stabilized by the surrounding lipid. For the study of this effect, reconstituted membranes offer better chances than native ones, since they allow one to vary the composition of the lipid. The dependency of a protein's functional properties on lipid composition is the most direct and most sensitive measure for lipid-induced changes in protein conformation and their specificity. However, the functional changes do not reveal the underlying structural events. For the study of structural changes, physical methods need to be applied. In principle, any technique is applicable which responds to changes in protein structure or dynamics (Sections 2.4 and 2.5), notably c.d. and fluorescence. The application of these methods to serum apolipoproteins is described in references 119 and 120.

Studies on lipid-induced changes in protein structure and function using reconstituted membranes are subject to two serious difficulties. (i) Any lipid specificity observed in functional studies may be due not to lipid requirements for protein function but to requirements related to the process of incorporating the isolated protein into the lipid bilayer (121). Similarly, any change in protein conformation observed during reconstitution may represent a lipid-induced recovery from a partial denaturation which occurred during protein isolation, instead of a property of the native protein – lipid interaction. (ii) Spectral changes following a protein's incorporation into or binding to a lipid bilayer, especially in fluorescence and near u.v. c.d. measurements, are not necessarily due to conformational changes. Instead, they may simply be caused by the change in polarity of the environment.

5.3 Influence of the protein on lipid structure and mobility

According to the available experimental data, the following effects can be exerted by membrane proteins on membrane lipid:

(i)   a change of the molecular order and a restriction of the mobility of the 'annular' lipid,
(ii)   similar changes in putative 'secondary' lipid domains adjacent to the annulus,
(iii)   the induction of changes in order and of low-frequency motions in the bulk lipid,
(iv)   the induction of lipid phase separations, and
(v)   the promotion or stabilization of lipid asymmetries.

5.3.1 *Mobility of the annular lipid*

The motional restriction of the annular lipid can be best studied by e.s.r. measurements on spin-labelled lipids, as described in Section 5.1.1.

5.3.2 *Lipid regions adjacent to the annular lipid*

Insight into the properties of the lipid regions adjacent to the annular lipid can be obtained from calorimetric measurements. For example, in high-sensitivity differential scanning calorimetry of the reconstituted Ca^{2+}-ATPase of sarcoplasmic reticulum in phosphatidylcholine vesicles, three classes of excess heat capacity peaks were observed one of which was assigned to 'secondary domains' of disrupted lipid packing. Analogous findings can be obtained from fluorescence anisotropy measurements on diphenylhexatriene or similar fluorescent probes immersed in the lipid (122). It seems that earlier differential scanning calorimetry measurements may have failed to resolve these secondary domains from the annular lipids, thus yielding higher values for the number of annular lipids than obtained from e.s.r. measurements (see Section 5.1.1).

5.3.3 *Order and low-frequency motions in bulk lipids*

The influence of the protein on the lipid outside the annulus and the 'secondary' lipid domains can be inferred from e.s.r. measurements on spin-labelled lipids or from ^{2}H-n.m.r. of deuterated ones. In e.s.r. studies the information is obtained by comparing the 'mobile' spectral component of the protein−lipid system (Section 5.1.1) and the spectrum of the pure lipid (66,101,123). Similarly, in n.m.r. measurements the (broadened) signal of the protein−lipid system is compared with that of the pure lipid (43,44). In the case of e.s.r., the 'mobile' component may contain contributions from the 'secondary' domains, and in n.m.r. both the annular lipid and the 'secondary' domains will contribute to the signal. Nevertheless, conclusions on the motions and order in the bulk lipid have been drawn from both types of studies (43,66,101,123−125).

Recently it was demonstrated that information on long-range protein−lipid interactions can also be obtained from fluorescence depolarization measurements (126). The method makes use of the effect that the fluorescence of a probe immersed in the lipid will be quenched by energy transfer in the surroundings of a protein molecule if the latter bears a suitable energy acceptor. Determinations of the lipid order parameter from the end value of the fluorescence anisotropy of the probe (Section 3.2.3) will therefore not be influenced by contributions from the immediate surroundings of the protein, but refer only to the lipid outside the annulus and the 'secondary' domains (126).

5.3.4 *Lipid phase separations*

For studies on protein-induced lipid phase separations, the methods described in Section 3.3 for pure lipid systems can be applied. In addition, ^{31}P-n.m.r. measurements may be useful: membrane proteins may induce the formation of lipid bilayer domains in an otherwise hexagonal lipid phase (127) or the inverse effect (128), and these transformations can be easily detected by the latter technique (see Section 3.2.1).

5.3.5 *Lipid asymmetries*

Protein-induced or protein-stabilized asymmetric distributions of lipid molecules between the two monolayers of a biological or reconstituted membrane are mostly being studied by biochemical methods, as are the corresponding asymmetries in artificial lipid vesicles (129). With respect to physical methods, the ^{31}P- and ^{13}C-n.m.r. techniques described in Section 4.5.2 and references 97 and 98 can be applied to reconstituted systems (but not to whole cells). In addition, the quenching of the fluorescence of suitable sterols can be used for studying sterol transbilayer distribution (100,130). This problem can also be approached by analysing, using absorbance measurements, the time-course of the binding of the polyene antibiotic filipin to the membrane sterol (131) — although there are pitfalls (100). While these techniques may detect asymmetric lipid distributions, they should be supplemented by other measurements (including biochemical ones) for further characterization (see references 132 and 133).

6. FUTURE DEVELOPMENTS

In the near future, both the number and the sensitivity of the biophysical techniques available for membrane research probably will not increase substantially. Some improvement may, however, be achieved with respect to the evaluation methods (e.g. in saturation transfer e.s.r.), which could make application of these techniques more fruitful. Major progress in the future could result, however, from advances in the preparative biochemical aspects such as the increased availability of three-dimensional crystals of membrane proteins suitable for high-resolution X-ray diffraction analysis, of specifically deuterated proteins and lipids for neutron scattering and n.m.r. work, and of reconstituted membrane transport systems which are structurally more homogeneous than those available at present. Such advances could make biophysical approaches even more important to the progress in biomembrane research than in the past.

7. ACKNOWLEDGEMENTS

I gratefully acknowledge the numerous helpful suggestions on this chapter from Drs J.B.C.Findlay, H.Passow, H.Ruf and P.Wood, and my wife's patience in typing the manuscript.

8. REFERENCES

1. Tanford,C. and Reynolds,J.A. (1976) *Biochim. Biophys. Acta*, **457**, 133.
2. Hames,B.D. (1981) In *Gel Electrophoresis of Proteins — A Practical Approach*. Hames,B.D. (ed.), IRL Press, Oxford and Washington, p. 1.
3. Peters,K. and Richards,F.M. (1977) *Annu. Rev. Biochem.*, **46**, 523.
4. Dorst,H.-J. and Schubert,D. (1979) *Hoppe-Seyler's Z. Physiol. Chem.*, **360**, 1605.
5. Ji,T.H. and Middaugh,C.R. (1980) *Biochim. Biophys. Acta*, **603**, 371.
6. Eason,R. (1984) In *Centrifugation — A Practical Approach*. Rickwood,D. (ed.), IRL Press, Oxford and Washington, p. 251.
7. Fujita,H. (1975) *Foundations of Ultracentrifugal Analysis*. Wiley, New York.
8. Schubert,D., Boss,K., Dorst,H.-J., Flossdorf,J. and Pappert,G. (1983) *FEBS Lett.*, **163**, 81.
9. Cantor,C.R. and Schimmel,P.R. (1980) *Biophysical Chemistry*. Part II, Freeman and Company, San Francisco.
10. Kühn,H., Bennett,N., Michel-Villaz,M. and Chabre,M. (1981) *Proc. Natl. Acad. Sci. USA*, **78**, 6873.
11. Kempner,E.S. and Schlegel,W. (1979) *Anal. Biochem.*, **92**, 2.
12. Verkman,A.S., Skorecki,K. and Ausiello,D.A. (1984) *Proc. Natl. Acad. Sci. USA*, **81**, 150.
13. Pilz,I. (1973) In *Physical Principles and Techniques of Protein Chemistry*. Part C, Leach,S.J. (ed.), Academic Press, New York and London, p. 141.
14. Franks,N.P. and Levine,Y.K. (1981) In *Membrane Spectroscopy*. Grell,E. (ed.), Springer-Verlag, Berlin, Heidelberg and New York, p. 437.
15. Blasie,J.K., Herbette,L. and Pachence,J. (1985) *J. Membr. Biol.*, **86**, 1.
16. Blasie,J.K., Herbette,L., Pascolini,D., Skita,V., Pierce,D.H. and Scarpa,A. (1985) *Biophys. J.*, **48**, 9.
17. Slayter,E.M. (1969) In *Physical Principles and Techniques of Protein Chemistry*. Part A, Leach,S.J. (ed.), Academic Press, New York and London, p. 1.
18. Branton,D., Cohen,C.M. and Tyler,J. (1981) *Cell*, **24**, 24.
19. Tanford,C. (1961) *Physical Chemistry of Macromolecules*. Wiley, New York.
20. Makowski,L. and Li,J. (1984) In *Biomembrane Structure and Function*. Chapman,D. (ed.), Verlag Chemie, Weinheim, p. 43.
21. Michel,H. (1983) *Trends Biochem. Sci.*, **8**, 56.
22. Garavito,R.M., Jenkins,J., Jansonius,J.N., Karlsson,R. and Rosenbusch,J.P. (1983) *J. Mol. Biol.*, **164**, 313.
23. Deisenhofer,J., Epp,O., Miki,K., Huber,R. and Michel,H. (1985) *Nature*, **318**, 618.
24. Kossiakoff,A.A. (1985) *Annu. Rev. Biochem.*, **54**, 1195.
25. Engel,A., Massalski,A., Schindler,H., Dorset,D.L. and Rosenbusch,J.P. (1985) *Nature*, **317**, 643.
26. Glaeser,R.M. (1982) In *Methods of Experimental Physics, Vol. 20 (Biophysics)*. Ehrenstein,G. and Lecar,H. (eds), Academic Press, New York and London, p. 391.
27. Henderson,R. and Unwin,P.N.T. (1975) *Nature*, **257**, 28.
28. Sears,D.W. and Beychok,S. (1973) In *Physical Principles and Techniques of Protein Chemistry*. Part C, Leach,S.J. (ed.), Academic Press, New York and London, p. 446.
29. Long,M.M. and Urry,D.W. (1981) In *Membrane Spectroscopy*. Grell,E. (ed.), Springer-Verlag, Berlin, Heidelberg and New York, p. 143.

30. Jap,B.K., Maestre,M.F., Hayward,S.B. and Glaeser,R.M. (1983) *Biophys. J.*, **43**, 81.
31. Amey,R.L. and Chapman,D. (1984) In *Biomembrane Structure and Function*. Chapman,D. (ed.), Verlag Chemie, Weinheim, p. 199.
32. Lord,R.C. and Mendelsohn,R. (1981) In *Membrane Spectroscopy*. Grell,E. (ed.), Springer-Verlag, Berlin, Heidelberg and New York, p. 377.
33. Gurd,F.R.N. and Rothgeb,T.M. (1979) *Adv. Protein Chem.*, **33**, 73.
34. Munro,I., Pecht,I. and Stryer,L. (1979) *Proc. Natl. Acad. Sci. USA*, **76**, 56.
35. Keniry,M.A., Gutowsky,H.S. and Oldfield,E. (1984) *Nature*, **307**, 383.
36. Tardieu,A., Luzatti,V. and Reman,F.C. (1973) *J. Mol. Biol.*, **75**, 711.
37. Chapman,D. (1975) *Q. Rev. Biophys.*, **8**, 185.
38. Chapman,D., Williams,R.M. and Ladbrooke,B.D. (1967) *Chem. Phys. Lipids*, **1**, 445.
39. Bach,D. (1984) In *Biomembrane Structure and Function*. Chapman,D. (ed.), Verlag Chemie, Weinheim, p. 1.
40. Bueldt,G., Gally,H.U., Seelig,J. and Zaccai,G. (1979) *J. Mol. Biol.*, **134**, 673.
41. Zaccai,G., Bueldt,G., Seelig,A. and Seelig,J. (1979) *J. Mol. Biol.*, **134**, 693.
42. Schuh,J.R. and Chan,S.I. (1982) In *Methods of Experimental Physics. Vol. 20 (Biophysics)*, Ehrenstein,G. and Lecar,H. (eds), Academic Press, New York and London, p. 1.
43. Seelig,J. and Seelig,A. (1980) *Q. Rev. Biophys.*, **13**, 19.
44. Chan,S.I., Bocian,D.F. and Petersen,N.O. (1981) In *Membrane Spectroscopy*. Grell,E. (ed.), Springer-Verlag, Berlin, Heidelberg and New York, p. 1.
45. Cullis,P.R. and de Kruijff,B. (1979) *Biochim. Biophys. Acta*, **559**, 399.
46. Cullis,P.R. and de Kruijff,B. (1976) *Biochim. Biophys. Acta*, **436**, 523.
47. Cullis,P.R. and de Kruijff,B. (1978) *Biochim. Biophys. Acta*, **507**, 207.
48. Schreier,S., Polnaszek,C.F. and Smith,I.C.P. (1978) *Biochim. Biophys. Acta*, **515**, 375.
49. Seelig,J. and Niederberger,W. (1974) *Biochemistry*, **13**, 1585.
50. Petersen,N.O. and Chan,S.I. (1977) *Biochemistry*, **16**, 2657.
51. Chen,R.F., Edelhoch,H. and Steiner,R.F. (1969) In *Physical Principles and Techniques of Protein Chemistry*. Part A, Leach,S.J. (ed.), Academic Press, New York and London, p. 171.
52. Yguerabide,J. and Foster,M.C. (1981) In *Membrane Spectroscopy*. Grell,E. (ed.), Springer-Verlag, Berlin, Heidelberg and New York, p. 199.
53. Jaehnig,F. (1979) *Proc. Natl. Acad. Sci. USA*, **76**, 6361.
54. Heyn,M.P. (1979) *FEBS Lett.*, **108**, 359.
55. Knoll,W., Schmidt,G., Ibel,K. and Sackmann,E. (1985) *Biochemistry*, **24**, 5240.
56. Gebhardt,C., Gruler,H. and Sackmann,E. (1977) *Z. Naturforsch.*, **32c**, 581.
57. Sklar,L.A., Hudston,B.S. and Simoni,R.D. (1977) *Biochemistry*, **16** 819.
58. Peters,R. and Beck,K. (1983) *Proc. Natl. Acad. Sci. USA*, **80**, 7183.
59. Zingsheim,H.P. and Plattner,H. (1976) In *Methods in Membrane Biology*. Korn,E.D. (ed.), Plenum Press, New York and London, Vol. 7, p. 1.
60. Lentz,B.R., Barrow,D.A. and Hoechli,M. (1980) *Biochemistry*, **19**, 1943.
61. Severs,N.J. and Robenek,H. (1983) *Biochim. Biophys. Acta*, **737**, 373.
62. Traeuble,H. and Haynes,D.H. (1971) *Chem. Phys. Lipids*, **7**, 324.
63. Brown,M.F. and Williams,G.D. (1985) *J. Biochem. Biophys. Methods*, **11**, 71.
64. Lee,A.G., Birdsall,N.J.M. and Metcalfe,J.C. (1974) In *Methods in Membrane Biology*. Korn,E.D. (ed.), Plenum Press, New York and London, Vol. 2, p. 1.
65. Smith,R.L. and Oldfield,E. (1984) *Science*, **225**, 280.
66. Marsh,D. (1981) In *Membrane Spectroscopy*. Grell,E. (ed.), Springer-Verlag, Berlin, Heidelberg and New York, p. 51.
67. Shinitzky,M. and Barenholz,Y. (1978) *Biochim. Biophys. Acta*, **515**, 367.
68. Cherry,R.J. (1979) *Biochim. Biophys. Acta*, **559**, 289.
69. Corin,A.F., Matayoshi,E.D. and Jovin,T.M. (1985) In *Spectroscopy and the Dynamics of Molecular Biological Systems*. Bayley,P.M. and Dale,R.E. (eds), Academic Press, New York, p. 53.
70. Cherry,R.J. (1985) In *Spectroscopy and the Dynamics of Molecular Biological Systems*. Bayley,P.M. and Dale,R.E. (eds), Academic Press, New York, p. 79.
71. Garland,P.B. and Johnson,P. (1985) In *Spectroscopy and the Dynamics of Molecular Biological Systems*. Bayley,P.M. and Dale,R.E. (eds), Academic Press, New York, p. 95.
72. Hoffmann,W. and Restall,C.J. (1984) In *Biomembrane Structure and Function*. Chapman,D. (ed.), Verlag Chemie, Weinheim, p. 257.
73. Johnson,P. and Garland,P.G. (1981) *FEBS Lett.*, **132**, 252.
74. Cherry,R.J. and Godfrey,R.E. (1981) *Biophys. J.*, **36**, 257.
75. Thomas,D.D., Eads,T.M., Barnett,V.A., Lindahl,K.M., Momont,D.A. and Squier,T.C. (1985) In

Spectroscopy and the Dynamics of Molecular Biological Systems. Bayley,P.M. and Dale,R.E. (eds), Academic Press, New York, p. 239.

76. Thomas,D.D., Dalton,L.R. and Hyde,J.S. (1976) *J. Chem. Phys.*, **65**, 3006.
77. Delmelle,M., Butler,K.W. and Smith,I.C.P. (1980) *Biochemistry*, **19**, 698.
78. Marsh,D. (1980) *Biochemistry*, **19**, 1632.
79. Opella,S.J., Yesinowski,J.P. and Waugh,J.S. (1976) *Proc. Natl. Acad. Sci. USA*, **73**, 3812.
80. Peters,R. (1981) *Cell Biol. Int. Rep.*, **5**, 733.
81. Axelrod,D. (1985) In *Spectroscopy and the Dynamics of Molecular Biological Systems*. Bayley,P.M. and Dale,R.E. (eds), Academic Press, New York, p. 163.
82. Peters,R., Bruenger,A. and Schulten,K. (1981) *Proc. Natl. Acad. Sci. USA*, **78**, 962.
83. Axelrod,D., Koppel,D.E., Schlessinger,J., Elson,E.L. and Webb,W.W. (1976) *Biophys. J.*, **16**, 1055.
84. Golan,D.E. and Veatch,W. (1980) *Proc. Natl. Acad. Sci. USA*, **77**, 2537.
85. Traeuble,H. and Sackmann,E. (1972) *J. Am. Chem. Soc.*, **94**, 4499.
86. Sackmann,E., Traeuble,H., Galla,H.-J. and Overath,P. (1973) *Biochemistry*, **12**, 5360.
87. Kuo,A.-L. and Wade,C.G. (1979) *Biochemistry*, **18**, 2300.
88. Cullis,P.R. (1976) *FEBS Lett.*, **70**, 223.
89. Sheats,J.R. and McConnell,H.M. (1978) *Proc. Natl. Acad. Sci. USA*, **75**, 4661.
90. Koppel,D.E., Axelrod,D., Schlessinger,J., Elson,E.L. and Webb,W.W. (1976) *Biophys. J.*, **16**, 1315.
91. Fahey,P.F. and Webb,W.W. (1978) *Biochemistry*, **17**, 3046.
92. Rothman,J.E., Tsai,D.K., Dawidowicz,E.A. and Lenard,J. (1976) *Biochemistry*, **15**, 2361.
93. Dicorleto,P.E. and Zilversmit,D.B. (1979) *Biochim. Biophys. Acta*, **552**, 114.
94. Roseman,M., Litman,B.J. and Thompson,T.E. (1975) *Biochemistry*, **14**, 4826.
95. Kornberg,R.D. and McConnell,H.M. (1971) *Biochemistry*, **10**, 1111.
96. Rousselet,A., Guthmann,C., Matricon,J., Bienvenue,A. and Devaux,P.F. (1976) *Biochim. Biophys. Acta*, **426**, 357.
97. Bystrov,V.F., Shapiro,Y.E., Viktorov,A.V., Barsukov,L.I. and Bergelson,L.D. (1972) *FEBS Lett.*, **25**, 337.
98. Shapiro,Y.E., Viktorov,A.V., Volkova,V.I., Barsukov,L.I., Bystrov,V.F. and Bergelson,L.D. (1975) *Chem. Phys. Lipids*, **14**, 227.
99. De Kruijff,B. and Wirtz,K.W.A. (1977) *Biochim. Biophys. Acta*, **468**, 318.
100. Devaux,P.F. and Seigneuret,M. (1985) *Biochim. Biophys. Acta*, **822**, 63.
101. Jost,P.C., Capaldi,R.A., Vanderkooi,G. and Griffith,O.H. (1973) *J. Supramol. Struct.*, **1**, 269.
102. Hoffmann,W., Pink,D.A., Restall,J. and Chapman,D. (1981) *Eur. J. Biochem.*, **114**, 585.
103. Andersen,J.P., Fellmann,P., Moller,J.V. and Devaux,P.F. (1981) *Biochemistry*, **20**, 4928.
104. Marsh,D., Watts,A., Pates,R.D., Uhl,R., Knowles,P.F. and Esmann,M. (1982) *Biophys. J.*, **37**, 265.
105. East,J.M., Melville,D. and Lee,A.G. (1985) *Biochemistry*, **24**, 2615.
106. Awasthi,Y.C., Chuang,T.F., Keenan,T.W. and Crane,F.L. (1971) *Biochim. Biophys. Acta*, **226**, 42.
107. Anderson,R.A. and Marchesi,V.T. (1985) *Nature*, **318**, 295.
108. London,E. and Feigenson,G.W. (1981) *Biochemistry*, **20**, 1939.
109. East,J.M. and Lee,A.G. (1982) *Biochemistry*, **21**, 4144.
110. Bonnet,D. and Begard,E. (1984) *Biochem. Biophys. Res. Commun.*, **120**, 344.
111. Teissie,J. (1981) *Biochemistry*, **20**, 1554.
112. Sato,S.B. and Ohnishi,S.I. (1983) *Eur. J. Biochem.*, **130**, 19.
113. Mendelsohn,R., Dluhy,R.A., Crawford,T. and Mantsch,H.H. (1984) *Biochemistry*, **23**, 1498.
114. Schubert,D., Herbst,F., Marie,H. and Rudloff,V. (1982) In *Protides of the Biological Fluids*. Proc. 29th Coll. 1981, Peeters,H. (ed.), Pergamon Press, Oxford and New York, p. 121.
115. Klappauf,E. and Schubert,D. (1979) *Hoppe-Seyler's Z. Physiol. Chem.*, **360**, 1225.
116. Passing,R. and Schubert,D. (1983) *Hoppe-Seyler's Z. Physiol. Chem.*, **364**, 219.
117. Traeuble,H. and Overath,P. (1973) *Biochim. Biophys. Acta*, **307**, 491.
118. Rigell,C.W., de Saussure,C. and Freire,E. (1985) *Biochemistry*, **24**, 5638.
119. Lux,S.E., Hirz,R., Shrager,R.I. and Gotto,A.M. (1972) *J. Biol. Chem.*, **247**, 2598.
120. Jonas,A. and Krajnovich,D.J. (1977) *J. Biol. Chem.*, **252**, 2194.
121. Eytan,G.D. (1982) *Biochim. Biophys. Acta*, **694**, 185.
122. Lentz,B.R., Clubb,K.W., Alford,D.R., Hoechli,M. and Meissner,G. (1985) *Biochemistry*, **24**, 433.
123. Marsh,D. (1983) *Trends Biochem. Sci.*, **8**, 330.
124. Rice,D.M., Meadows,M.D., Scheinman,A.O., Goni,F.M., Gómez-Fernández,J.C., Moscarello,M.A., Chapman,D. and Oldfield,E. (1979) *Biochemistry*, **18**, 5893.
125. Borle,F. and Seelig,J. (1983) *Biochemistry*, **22**, 5536.
126. Rehorek,M., Dencher,N.A. and Heyn,M.P. (1985) *Biochemistry*, **24**, 5980.
127. Tarashi,T.F., De Kruijff,B., Verkleij,A.J. and Van Echteld,C.J.A. (1982) *Biochim. Biophys. Acta*, **685**, 153.

128. Van Echteld,C.J.A., Van Stigt,R., De Kruijff,B., Leunissen-Bijvelt,J., Verkleij,A.J. and De Gier,J. (1981) *Biochim. Biophys. Acta,* **648**, 287.
129. Etemadi,A.-H. (1980) *Biochim. Biophys. Acta,* **604**, 423.
130. Hale,J.E. and Schroeder,F. (1982) *Eur. J. Biochem.,* **122**, 649.
131. Bitman,R. (1978) *Lipids,* **13**, 686.
132. Dressler,V., Haest,C.W.M., Plasa,G., Deuticke,B. and Erusalimsky,J.D. (1984) *Biochim. Biophys. Acta,* **775**, 189.
133. Seigneuret,M. and Devaux,P.F. (1984) *Proc. Natl. Acad. Sci. USA,* **81**, 3751.

APPENDIX I

Properties of materials used in density gradient separations

1. Constants for sucrose solutions

Concentration		Molarity	Density (g/ml)		Ref. index		Viscosity (centipoises)	
w/w	w/v		5°	20°	5°	20°	5°	20°
0	0	0	1.0004	0.9882	1.3345	1.3330	1.515	1.000
2	2.01	0.059	1.0078	1.0060	1.3375	1.3359	1.585	1.053
4	4.06	0.118	1.0157	1.0139	1.3407	1.3388	1.674	1.112
6	6.13	0.179	1.0237	1.0218	1.3437	1.3418	1.780	1.177
8	8.24	0.241*	1.0318	1.0299	1.3467	1.3448	1.905	1.251
10	10.38	0.303	1.0400	1.0381	1.3497	1.3478	2.057	1.333
12	12.56	0.367	1.0483	1.0465	1.3530	1.3509	2.220	1.426
14	14.77	0.431	1.0568	1.0549	1.3562	1.3541	2.410	1.531
16	17.02	0.497	1.0653	1.0635	1.3593	1.3573	2.635	1.650
18	19.30	0.564	1.0740	1.0722	1.3625	1.3605	2.875	1.786
20	21.62	0.632	1.0829	1.0810	1.3646	1.3638	3.137	1.945
22	23.98	0.700	1.0918	1.0899	1.3692	1.3672	3.460	2.124
24	26.38	0.771	1.1009	1.0990	1.3724	1.3706	3.838	2.331
26	28.81	0.842	1.1101	1.1081	1.3757	1.3740	4.282	2.573
28	31.29	0.914	1.1195	1.1175	1.3792	1.3775	4.807	2.855
30	33.81	0.988	1.1290	1.1270	1.3826	1.3811	5.435	3.187
32	36.37	1.063	1.1386	1.1366	1.3864	1.3847	6.187	3.581
34	38.98	1.139	1.1484	1.1464	1.3900	1.3884	7.106	4.052
36	41.62	1.216	1.1583	1.1562	1.3937	1.3921	8.234	4.621
38	44.32	1.295	1.1683	1.1663	1.3975	1.3959	9.651	5.315
40	47.06	1.375	1.1785	1.1765	1.4018	1.3998	11.44	6.167
42	49.84	1.456	1.1889	1.1867	1.4055	1.4037	13.76	7.234
44	52.68	1.539	1.1994	1.1973	1.4096	1.4076	16.77	8.579
46	55.56	1.623	1.2100	1.2078	1.4139	1.4117	20.72	10.30
48	58.49	1.709	1.2208	1.2186	1.4181	1.4158	25.99	12.51
50	61.48	1.796	1.2327	1.2296	1.4225	1.4199	33.18	15.43
52	64.51	1.885	1.2428	1.2406	1.4268	1.4242	43.18	19.34
54	67.60	1.975	1.2541	1.2519	1.4313	1.4284	57.42	24.68
56	70.74	2.067	1.2655	1.2632	1.4356	1.4328	78.27	32.12
58	73.94	2.160	1.2770	1.2748	1.4400	1.4372	109.5	42.78
60	77.19	2.255	1.2887	1.2865	1.4445	1.4417	159.1	58.49

2. Comparison of physico-chemical properties of various gradient materials

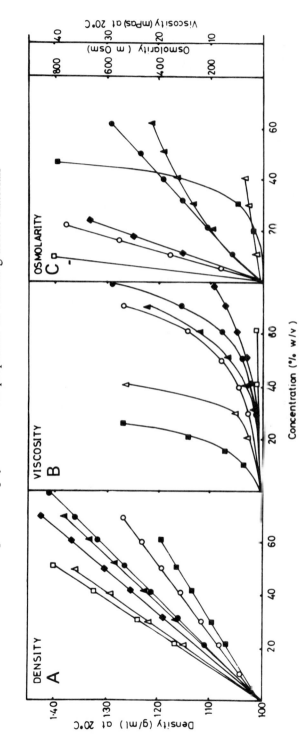

The relationships between the concentration (% w/v) of gradient solutes and their density, viscosity and osmolarity are shown for Nycodenz (● ——— ●), metrizamide (▲ ——— ▲), sodium metrizoate (◆ ——— ◆), CsCl (□ ——— □), sucrose (○ ——— ○), Ficoll (■ ——— ■) and Percoll (△ ——— △). Reproduced from (1).

3. Properties of solutions of Nycodenz

Concentration % (w/v)	Concentration (mol/l)	Refractive Index (η) 20°C	Density (g/ml) 20°C	Osmolaity (mOsm)	Viscosity (mPas)
0	0	1.3330	0.999	0	1.0
10	0.122	1.3494	1.052	112	1.3
20	0.244	1.3659	1.105	211	1.4
30	0.365	1.3824	1.159	299	1.8
40	0.487	1.3988	1.212	388	3.2
50	0.609	1.4153	1.265	485	5.3
60	0.731	1.4318	1.319	595	9.5
70	0.853	1.4482	1.372	1045	17.2
80	0.974	1.4647	1.426	–	30.0

4. Properties of aqueous Ficoll solutions at 4°C (2).

Concentration

Percent w/w	Percent w/v	Refractive index	Density (gm/ml)	Viscosity (cP)
0.00	0.000	1.3346	1.0004	1.564
3.00	3.032	1.3382	1.0106	2.788
4.00	4.058	1.3392	1.0145	3.397
5.00	5.090	1.3408	1.0180	4.102
6.00	6.129	1.3420	1.0215	5.014
7.00	7.177	1.3437	1.0253	6.017
8.00	8.232	1.3454	1.0290	7.372
9.00	9.295	1.3469	1.0328	8.579
10.00	10.37	1.3484	1.0365	10.35
12.00	12.53	1.3514	1.0441	14.27
14.00	14.73	1.3538	1.0518	20.21
16.00	16.96	1.3579	1.0597	27.42
18.00	19.21	1.3608	1.0673	38.33
20.00	21.50	1.3645	1.0752	52.31
22.00	23.84	1.3680	1.0837	69.28
24.00	26.21	1.3714	1.0922	95.03
26.00	28.61	1.3748	1.1004	125.7
28.00	31.06	1.3786	1.1093	172.0
30.00	33.53	1.3201	1.1176	225.6
32.00	36.04	1.3856	1.1263	308.8
34.00	38.57	1.3890	1.1345	407.2
36.00	41.19	1.3930	1.1442	565.4
38.00	43.83	1.3970	1.1534	762.2
40.00	46.52	1.4030	1.1629	1020.0

References

1. Rickwood,D. (1984) in *Centrifugation — A Practical Approach,* (2nd edition), Rickwood,D. (ed.), IRL Press, Oxford, Washington DC, p. 29.
2. Pretlow,T.G., Boone,C.W., Shrager,R.I. and Weiss,G.H. (1969) *Anal. Biochem.,* **29**, 230.

Enzymic subcellular markers

Fraction		Marker	References	EC number
Plasma membrane	baso-lateral	adenylate cyclase	1	4.6.1.1
		Na⁺ K⁺ ATPase	2	3.6.1.37
		receptors e.g. asialoglycoprotein	3	
	apical	5′-nucleotidase	2,4,5	3.1.3.5
		leucineaminopeptidase	6	3.4.11.1
		γ-glutamyltranspeptidase	7	2.3.2.12
Endoplasmic reticulum		glucose-6-phosphatase	8,22	3.1.3.9
		NADPH-cyt.c reductase	9	1.6.2.4
		epoxide hydrolase	10	3.3.2.3
Golgi apparatus	*trans* and middle region	galactosyltransferase	11,12	2.4.1.38
		sialyltransferase	11,12	2.4.99.1
		NADP-phosphatase	13	3.6.1.22
Mitochondria	inner membrane	succinate dehydrogenase	14	1.3.99.1
		cytochrome oxidase	15	1.9.3.1
		rotinone-insenstive NADH-cyt.c reductase	16	1.6.99.1
	outer membrane	monoamine oxidase	17	1.4.3.4
		kynurenine-3-hydroxylase	18	1.14.13.9
Lysosomes		acid phosphatase	19	3.1.3.2
		β-glucuronidase	19	3.2.1.31
		aryl sulphatase	20	3.1.6.1
Endosomes		monensin-activated Mg²⁺-ATPase plus intact ligands	21	3.6.1.3
Peroxisomes		catalase	23	1.11.1.6
		carnitine palmitoyl transferase	24	2.3.1.21
Cytosol		lactate dehydrogenase	25	1.1.1.22

Detailed descriptions of the assay procedures are to be found in the references quoted, and many of the procedures are described together in reference 9. For a discussion of the distribution of baso-lateral and apical plasma membranes see reference 4 in Chapter 1. No specific enzyme markers exist for nuclear membranes, and the *cis* region of the Golgi apparatus. For further details of the reliability of these markers, see Chapter 1 and 2.

References

1. Hardman,J.G. and O'Malley,B.W. (1974) *Methods Enzymol.* **38C**, 49.
2. Avruch,J. and Wallach,D.F.H. (1971) *Biochim. Biophys. Acta,* **233**, 334.
3. Hubbard,A.L., Wall,D.A. and Ma,A. (1983) *J. Cell Biol.,* **96**, 217.
4. Chatterjee,S.K., Battacharya,M. and Barlow,J.J. (1979) *Anal. Biochem.,* **95**, 497.
5. Ipata,P.L. (1967) *Anal. Biochem.,* **20**, 30.
6. Goldbarg,J.A. and Rutenberg,A.M. (1958), *Cancer,* **11**, 283.
7. Inoue,M.S., Horiuchi,S. and Morino,Y. (1977) *Eur. J. Biochem.,* **73**, 335.
8. Aronson,N.N. and Touster,O. (1974) *Methods Enzymol.,* **31**, 90.

9. Sottocasa,G.L., Kuylenstierna,B., Ernster,L. and Bergstrand,A. (1967) *J. Cell Biol.*, **32**, 415.
10. Galteau,M.M., Antoine,B. and Reggio,H. (1985) *EMBO J.*, **4**, 2793.
11. Vischer,P. and Reutter,W. (1978) *Eur. J. Biochem.*, **84**, 363.
12. Bergeron,J.J.M., Ehrenreich,J.H., Siekevitz,P. and Palade,G.E. (1973) *J. Cell Biol.*, **59**, 73.
13. Navas,P., Minnifield,N., Sun,I. and Morré,D.J. (1986) *Biochim. Biophys. Acta,* **881**, 1.
14. Earl,D.C.N. and Korner,A. (1965) *Biochem. J.*, **94**, 721.
15. Sun,F.F. and Crane,F.L. (1969) *Biochim. Biophys. Acta,* **172**, 417.
16. Fleischer,S. and Fleischer,B. (1967) *Methods Enzymol.* **10**, 427.
17. Schnaitman,C., Erwin,V.G. and Greenawalt,J.W. (1967) *J. Cell Biol.*, **32**, 719.
18. Hayashi,O. (1962) *Methods Enzymol.*, **5**, 807.
19. Gianetto,R. and de Duve,C. (1955) *Biochem. J.*, **59**, 433.
20. Chang,P.L., Rosa,N.E. and Davidson,R.G. (1981) *Anal. Biochem.*, **117**, 382.
21. Saermark,T., Flint,N. and Evans,W.H. (1985) *Biochem. J.*, **225**, 51.
22. Gierow,P. and Jergil,B. (1980) *Anal. Biochem.*, **101**, 305.
23. Thompson,J.F., Nance,S.L. and Tollaksen,S.L. (1978) *Proc. Soc. Exp. Biol. Med.*, **157**, 33.
24. Markwell,M.A.K., McGroarty,E.J., Bieber,L.L. and Tolbert,N.E. (1973) *J. Biol. Chem.*, **248**, 3426.
25. Neilands,J. (1955) *Methods Enzymol.*, **1**, 449.

APPENDIX III

Assay of receptors

A. Sivaprasadarao and J.B.C. Findlay

1. MEMBRANE-BOUND RECEPTORS

The assay of membrane-bound receptors is usually carried out by incubating cells or membranes with radiolabelled ligand to equilibrium and then separating the free from the bound radioligand by a variety of techniques which include filtration through suitable filters (glass, cellulose acetate, nylon, teflon etc.) or centrifugation. The cells/membranes retained by the filters or pelleted by centrifugation are then washed extensively to remove excess ligand (provided the dissociation constants are below 10^{-7} M). Parallel incubations in the presence of a 200- to 600-fold molar excess of unlabelled ligand are included to determine the non-saturable (non-specific) binding. Filtration assays are rapid and give reliable results provided that the binding of radioligand to the filters (background) is small compared to the membranes and that the washing steps do not cause significant dissociation of the ligand-receptor complex. Centrifugation of the cells or membranes can also be successful, particularly when a water-immiscible layer of oil is employed to separate the free from bound radioligand. This procedure has several advantages. Low backgrounds are obtained without the requirement for washing steps and the rapidity of the process enables detection of fast dissociating receptor-ligand systsems.

The choice of the oil density is important to ensure quantitative recovery of cells or membranes. This density should be intermediate between that of the cells and the aqueous media. Different densities can be obtained by mixing phthalate esters of short (2−4) and long (8−9) chain alcohols in varying proportions. Although the common practice is to transfer an aliquot of the incubation mixture on to the top of the oil and then centrifuge, it is often possible to carry out the initial incubation on the top of the oil layer without deleterious effect. A protocol for measuring the membrane-bound receptor for the plasma retinol-binding protein (RBP) is described below and can be easily modified for most receptor systems.

1.1 Reagents

[125[I]-RBP.   Iodinate 0.5 nmol of RBP with 1 mCi Na 125[I] using enzymobeads (see Section 6.1.1, Chapter 6). Separate 125[I]-RBP from excess 125I using Sephadex G-75 (60 × 1 cm column) equilibrated and developed with 0.2% ovalbumin in 20 mM sodium phosphate, 150 mM NaCl, pH 7.4 (PBS). Dilute the 125[I]-RBP with assay buffer such that 50 μl contain approximately 50 000 c.p.m. (0.2−0.5 nmol protein).

Assay buffer: 0.2% ovalbumin in 20 mM sodium phosphate, 150 mM NaCl, pH 7.4.

Appendix III

RBP:	20 μM solution in ovalbumin-free assay buffer.
Placental membranes:	Suspend human placental brush border membranes (1) in assay buffer to a final concentration of 5.0 mg of protein/ml.
Oil mixture:	Mix 60 ml of dibutyl phthalate with 40 ml of dinonyl phthalate (final density = 1.012; Oils from Aldrich Chemical Co.).

1.2 Procedure

(i) Half-fill two sets of triplicate Microeppendorf tubes (Sarstedt 72/702) with the oil mixture.
(ii) Layer 50 μl of ^{125}[I]-RBP onto the oil layer.
(iii) Add 20 μl of PBS to set 1 (total binding).
(iv) Add 20 μl of RBP to set 2 (non-saturable binding).
(v) Spin for 15 sec at 12 000 g (optional − to clarify layers)
(vi) Add 30 μl of the membrane suspension straight into the assay medium (not down the walls of the tube).
(vii) Incubate at 22°C for 15 min.
(viii) Chill on ice for 2 min.
(ix) Spin for 2 min at 12 000 g in a Microcentaur centrifuge.
(x) Aspirate off the supernatant (retain an aliquot for counting if required).
(xi) Cut off the bottom of the Eppendorfs containing the pellets and count in a suitable γ-counter.
(xii) Calculate the specific binding by subtracting the non-saturable component from the total binding.

2. SOLUBLE RECEPTORS

The assay of solubilized receptors also requires separation of bound from free radioligand following incubation to equilibrium. Filtration or centrifugation usually necessitates precipitating the complex and polyethylene glycol (PEG) is often used for this purpose (2). The precipitated receptor−ligand complex is then washed with solutions containing PEG using filtration or centrifugation. Other methods rely on binding the receptor−ligand complex in some way e.g. ion-exchange or immunoabsorbents (see Section 5.2, Chapter 3 and Section 4, Chapter 5). The washing procedure of the original method has recently been modified (unpublished) to enable the detection of low affintiy and fast dissociating receptor−ligand systems, even those of low capacity. The method has been successfully used for the detection and quantitiation of the solubilized RBP-receptor, which could not be detected using conventional techniques (e.g. gel filtration, PEI filtration, adsorpion to hydroxyapatite etc.).

The method described here for the solubilized RBP is, in principle, applicable to any receptor system, almost regardless of the affinity constant, provided a suitable concentration of PEG can be found to precipitate the receptor−ligand complex. The assay tolerates precipitation of up to 10% of free ligand (background). The method involves initial centrifugation of the PEG-induced precipitate, then overlayering the mixture with dibutyl phthalate and spinning again. This allows the aqueous layer containing the free ligand to rise through the oil layer, thus achieving physical separation of almost all the free ligand from the preciptated complex without perturbing the equilibrium.

288

2.1 **Reagents**

[125][I]-RBP:	As described above.
RBP:	50 μM in PBS.
Assay buffer:	20 mM sodium phosphate, 150 mM NaCl, pH 7.4.
Solubilisation buffer:	2% Nonidet P40 in assay buffer containing 4 mM $MgCl_2$, 0.2 mM PMSF.
Crube soluble extract:	To 1 ml of a 10 mg protein/ml placental microvilli suspension in PBS, add 1 ml of solubilization buffer. Vortex briefly. Incubate at 25°C for 20 min with occasional mixing. Centrifuge at 105 000 g and 4°C for 60 min. Collect the supernatant and store at −20°C or on ice if to be used immediately.
γ-Globulins:	Goat γ-globulins (Sigma), 5 mg/ml in 0.1 M sodium phosphate, pH 7.4.
25% PEG solution:	25 g PEG 8000 (Sigma) in a final volume of 100 ml of distilled water.

2.2. **Procedure**

(i)   Add 50 μl [125][I]-RBP to three sets of triplicate 0.5 ml Eppendorfs (Sarstedt 72/699).

(ii)   Add 30 μl PBS to two sets of tubes [total (1) and backbround (2)].

(iii)   Add 30 μl of RBP to the third set [non-specific (3)].

(iv)   Add 20 μl of soluble extract to sets (1) and (3).

(v)   Add 20 μl of solubilization buffer to set (2).

(vi)   Incubate all tubes at 22°C for 15 min.

(vii)   Chill on ice then add 50 μl of ice-cold γ-globulin solution to all tubes.

(viii)   Add 100 μl of PEG solution (an Eppendorf multipette is convenient).

(ix)   Vortex. Keep on ice for about 20 min, then spin in a Microcentaur at 12 000 g for 8 min.

(x)   Overlay the incubation mixture in the tubes with ice cold dibutyl phthalate nearly to the top.

(xi)   Spin for 2 min as above then gently aspirate the aqueous supernatant and the lower oil layer.

(xii)   Cut off the bottom of the tubes, containing the pellets and count in a suitable γ-counter.

(xiii)   Calculate the specific binding as before.

3. REFERENCES

1.   Booth,A.G., Olaniyan,R. and Vanderpuye,O.A. (1980) *Placenta*, **1**, 327.
2.   Cuatrecasas,P. (1972) *Proc. Natl. Acad. Sci. USA*, **69**, 318.

Protein modifying agents

Listed below are a number of protein modifying agents which could have membrane applications but which so far have not been used for such studies (please see Chapter 6, Section 6 and Table 4).

Amino acid	Potentially permeable	Impermeable	Conditions/comments	References
Arginine	2,3-Butanedione		pH 7–8, 100-fold molar excess, 25°C, 1 h. Can react with amino groups	1
		Camphorquinone-10-sulphonic acid	pH 9, 37°C, up to 24 h, reversible	2
	1,2-cyclohexanedione		pH 9, 35°C, 2 h.	3
	p-Hydroxy and p-nitro phenylglyoxals		pH 9, 30°C, 30 min	4
Asparagine		Bis (1,1-trifluoroacet-oxy) iodobenzene	Used so far only for compositional analysis	5
Cysteine		5,5′-dithiobis (2-nitro-benzoic acid) [Ellmans Reagent]	pH 7.3, 25°C, μM, 1 h	6
		N-(iodophenyl) tri-fluoroacetamide	pH 8.0, 25–50°C, up to 50-fold molar excess. Donates free amino group	7
		Ammonium 4-chloro-7-sulpho benzofurazan (SBF-chloride)	pH 8.0, 37°C, 4 h, μM; fluorescent, reversible with 2-mercaptoethanol	8
Glutamine		Bis(1,1-trifluoroacet-oxy) iodobenzene	As for asparagine	5
Lysine	Methyl-3,5-diiodo and methyl p-hydroxy benzimidates (Woods reagent)		pH 8.0–9.0, 25–37°C, μM; incor-poration increases with time	9
	2-Iminothiolane (Traut's reagent)		Donates free thiol — can also be used as a cross-linker. Table ?. pH 8.0, mM, 4–25°C for 1 h.	10

o-Phthalaldehyde			pH 9.0−10.5, up to 10 mM, 10−50°C, up to 10 min 2-mercaptoethanol required for fluorophore.	12
Serine/ threonine	Acetic/succinic/ anhydrides		Not selective or efficient	13
Tryptophan		Ethoxyformic acid anhydride	Will also react with His, Lys, Ser and Tyr	14
Tyrosine		*p*-Nitrobenzene sulphonyl chloride	pH 7−8.0, 25°C; 30 min. 1 mM	15

REFERENCES

1. Yankeelov,J.A.Jr (1972) *Methods Enzymol.*, **25**, 566.
2. Pande,C.S., Pelzig,M. and Glass,J.D. (1980) *Proc. Natl. Acad. Sci. USA*, **77**, 896.
3. Patthy,L. and Smith,E.L. (1975) *J. Biol. Chem.*, **250**, 557.
4. Yamasaki,R.B., Shimer,D.A. and Feeney,R.E. (1981) *Analyt. Biochem.*, **111**, 220.
5. Soby,L.M. and Johnson,P. (1981) *Analyt. Biochem.*, **113**, 149.
6. Riddles,P.W., Blakeley,R.L. and Zerner,B. (1979) *Analyt. Biochem.*, **94**, 75.
7. Schwartz,W.E., Smith,P.K. and Royer,G.P. (1980) *Analyt. Biochem.* **106**, 43.
8. Andrews,J.L., Ternal,B. and Whitehouse,M.W. (1982) *Arch. Biochem. Biophys.*, **214**, 386.
9. Bright,G.R. and Spooner,B.S. (1983) *Analyt. Biochem.*, **131**, 301.
10. Birnbaumer,M.E., Sehrader,W.T. and O'Malley,B.W. (1979) *Biochem. J.*, **181**, 201.
11. Lambert,J.M., Baileau,G., Cover,J.A. and Traut,R.R. (1983) *Biochemistry*, **22**, 3913.
12. Benson,J.R. and Hare,P.E. (1975) *Proc. Natl. Acad. Sci. USA*, **72**, 619.
13. Allen,G. and Harris,J.I. (1976) *Eur. J. Biochem.*, **62**, 601.
14. Tsurushiin,S., Hirawatsu,A., Inamatsu,M. and Yasunolsu,K.T. (1975) *Biochim. Biophys. Acta*, **400**, 451.
15. Liao,T.-H., Ting,R.S. and Yeung,J.E. (1982) *J. Biol. Chem.*, **257**, 3637.

Charge-shift electrophoresis

Integral membrane proteins are presumed to have extended hydrophobic regions which facilitate their integration into the bilayer. On solubilization under non-denaturing conditions, these regions bind detergent, a feature which is exploited by the technique of charge-shift electrophoresis. Since detergents can be neutral, positively or negatively charged, they can, on binding to a protein, change its charge status and hence alter its electrophoretic mobility. The ability of detergents to induce this behaviour can be used as an indicator of whether the native protein has substantial exposed hydrophobic regions and hence as to whether it falls into the class known as integral membrane proteins. The procedure described below is relatively straightforward and can be combined with immunoelectrophoresis. The biggest single obstacle is the behaviour of the protein in various detergents, viz. whether it remains as a monodisperse entity or is prone to aggregation.

Method

(i)   Prepare gels on glass plates (10 or 20 × 20 cm) using 1% agarose dissolved in 37.5 mM Tris, 100 mM glycine, pH 8.7 containing 0.5−1% detergent. Neutral detergents include Triton X-100, octyl glucoside, Lubrol, emulphogene BC720 (most successful) and digitonin; positively charged detergents include CTAB (most successful) and DTAB and the most successful negatively charged detergent is deoxycholate. Charged detergents are usually employed as mixtures (1:1−1:10) with the neutral detergent.

(ii)   Place gels in a flat bed electrophoresis chamber (water-cooled if possible) containing the corresponding buffer. Connect gel and buffer by paper wicks.

(iii)   Pre-electrophorese for 15−20 min at 4−5 V/cm.

(iv)   Apply samples (10−15 μl containing about 10−20 μg protein in up to 5% single or mixed detergent) into slits (1 × 10 mm) cut into the agarose. The origin should be approximately equidistant from the wicks unless the protein has pronounced charge characteristics.

(v)   Electrophorese for 2−3 h at 4−5 V/cm.

(vi)   Remove gel, dry with a warm air fan (may need to fix and press), stain for protein with 0.1% Coomassie Blue, 40% methanol, 7% acetic acid for 1−3 h then destain in the same solution minus the dye.

(vii)   Proteins which bind DOC or CTAB should exhibit anodic and cathodic shifts respectively compared with their positions in a neutral detergent such as Triton.

This method is particularly powerful when combined with crossed immuno-electrophoresis to give crossed immuno charge-shift electrophoresis. In this technique, charge-shift electrophoresis is performed along one edge of a square glass plate as described above. The bulk of the agarose above the sample is then removed and replaced with antibody-containing agarose which also contains the neutral detergent alone. Second dimension electrophoresis in the manner conventional for crossed immuno-electrophoresis then gives precipitation patterns in which charge-shifts are particularly easy to detect.

References

Helenius,A. and Simons, K. (1977) *Proc. Natl. Acad. Sci. USA*, **74**, 529.
Norrild,B., Bjerrum,D.J. and Vestergaard,P.F. (1977) *Anal. Biochem.*, **81**, 432.
Bjerrum,O.J. (ed.) (1983) *Immunoelectrophoretic Analysis of Membrane Proteins*. Elsevier, Amsterdam.
Booth,A.G., Hubbard,L.M.L. and Kenny,A.J. (1979) *Biochem. J.*, **179**, 397

APPENDIX VI

Analysis of inositol phosphates by h.p.l.c.

Considerable progress has been made recently in understanding the role of inositol phosphates and in their analysis. Inositol-1,4,5-trisphosphate produced from phosphatidylinositol-4,-5-phosphate (see Figure 6, Chapter 4) is a second messenger, the major role of which is to mobilize calcium ions from the endoplasmic reticulum. In addition, inositol-1,4,5-trisphosphate is phosphorylated by a kinase to produce inositol-1,3,4,5- tetrakisphosphate, which apparently functions as a further second messenger acting at the plasma membrane and modulating calcium movements (1,2).

Analysis of the inositol phosphates has been facilitated by development of h.p.l.c. methods (3−6). The inositol-containing phospholipids are prelabelled with [^{32}P]- or [^{3}H]inositol and the production of labelled inositol phosphates determined by linking the h.p.l.c. system to a scintillation counter and/or a u.v. detector. When only small amounts of inositol phosphates are produced it is not possible to determine these except by appearance of radioactive peaks which correspond to those of known standards eluted from the h.p.l.c. column. It is essential therefore that appropriate control experiments are performed to ensure that labelled products are not impurities in the system under investigation.

After experimental stimulation of cells or tissue the reaction is stopped by addition of 10% trichloracetic acid. The samples may be analysed as described in section 3.3 Chapter 4, or by h.p.l.c. as follows. (i) Neutralize the sample with Tris base. (ii) Add 100 μl of 50 mM mannitol to aid recovery of the inositol phosphates. (iii) Freeze dry the sample, and then dissolve the sample in 2.0 ml of 1M EDTA, pH 7.0, or distilled water. (iv) Inject the samples onto h.p.l.c. columns of Whatman Partisil 10 SAX anion exchange resin.

The inositol phosphates are eluted using ammonium formate adjusted to pH 3.7 with phosphoric acid. Phosphate is necessary for sharp peak separation (3). Inositol phosphate, inositol-1,4-bisphosphate, inositol-1,3,4-trisphosphate and inositol-1,4,5-trisphosphate are eluted sequentially by a gradient of 100% water to 100% 1 M or 0.85 M ammonium formate/phosphate buffer (3,4,5). To separate higher inositol phosphates including inositol-1,3,4,5-tetrakisphosphate the elution with 0−1 M ammonium formate/phosphate is followed by a second steeper gradient of ammonium formate/phosphate from 1 M to 1.7 M (6) or 0.85 M to 3.4 M (5). Authentic samples of inositol phosphates should be used to standardize the columns and AMP, ADP or ATP markers are also useful to check that individual column runs are dependable.

REFERENCES

1. Irvine,R.F. and Moore,R.M. (1986) *Biochem. J.*, **240**, 917.
2. Irvine,R.F., Letcher,A.J., Heslop,J.P. and Berridge,M.J. (1986) *Nature*, **320**, 631.
3. Irvine,R.F., Anggard,E.A., Letcher,A.J. and Downes,C.P. (1985) *Biochem. J.*, **229**, 505.
4. Heslop,J.P., Irvine,R.F., Tashjiana,A. and Berridge,M.J. (1985) *J. Exp. Biol.*, **119**, 395.
5. Burgess,G.M., McKinney,J.S., Irvine,R.F. and Putney,J.W. (1985) *Biochem. J.*, **232**, 237.
6. Tennes,K.A., McKinney,J.S. and Putney,J.W. (1987) *Biochem. J.*, **242**, 797.